NONLINEAR OPTICAL AND ELECTROACTIVE POLYMERS

NONLINEAR OPTICAL AND ELECTROACTIVE POLYMERS

EDITED BY
PARAS N. PRASAD
State University of New York at Buffalo
Buffalo, New York

AND
DONALD R. ULRICH
Air Force Office of Scientific Research
Washington, D.C.

PLENUM PRESS • NEW YORK AND LONDON

Library of Congress Cataloging in Publication Data

American Chemical Society Symposium on Electroactive Polymers (1987: Denver, Colo.)
  Nonlinear optical and electroactive polymers.

  "Proceedings of an American Chemical Society Symposium on Electroactive Polymers, held April 6-10, 1987, in Denver, Colorado" — T.p. verso.
  Includes bibliographies and index.
  1. Polymers and polymerization — Optical properties — Congresses. 2. Polymers and polymerization — Electric properties — Congresses. I. Prasad, Paras N. II. Ulrich, Donald R. III. American Chemical Society. IV. Title.
QD381.9.066A47   1987   547.7   87-29151
ISBN 0-306-42768-0

Proceedings of an American Chemical Society symposium on Electroactive Polymers, held April 6-10, 1987, in Denver, Colorado

© 1988 Plenum Press, New York
A Division of Plenum Publishing Corporation
233 Spring Street, New York, N.Y. 10013

PREFACE

This treatise is a compendium of papers based on invited talks
presented at the American Chemical Society Symposium on Electroactive
Polymers which covered nonlinear optical polymers and conducting polymers,
the common denominator being the correlated pi-electron structures. The
improved understanding of the consequences of pi-electron delocalization
upon nonlinear optical properties and charge carrier dynamics has laid the
foundation for the rapid development and application of the electroresponse
of conjugated polymers. As a result, the area of electroactive and
nonlinear optical polymers is emerging as a frontier of science and
technology. It is a multidisciplinary field that is bringing together
scientists and engineers of varied background to interface their expertise.

The recent explosion of interest in this area stems from the prospect
of utilizing nonlinear optical effects for optical switching and logic
operations in optical computing, optical signal processing, optical sensing
and optical fiber communications. Polymers and organic are rapidly becoming
one of the major material classes for nonlinear optical applications along
with multiple quantum wells, ferroelectrics and other oxides, and direct
band-gap semiconductors. The reasons for this lie in the unique molecular
structures of polymers and organics and the ability to molecularly engineer
the architecture of these structures through chemical synthesis.

The uniqueness of polymers and organics lies in their large non-
resonant second and third order optical susceptibilities whose origin
resides in ultrafast, lossless virtual excitations of correlated pi-electron
states. Furthermore, polymers have outstanding mechanical, thermal and
environmental properties which can be designed into self-supporting films or
sheets, monolayer structures, fibers and conformable shapes. In addition to
molecular manipulation, structure can also be tailored on the
ultrastructural and microstructural levels to induce order, orientation and
anisotropy through processing. Examples are electric-field poled guest-host
systems, liquid crystalline polymers, and polymer blends and composites.
Furthermore, the scaling of structure and control of intra- and
intermolecular interactions has resulted in new and improved polymers and
organics which has already been demonstrated in spatial light modulators and
waveguides as viable candidates for optical systems.

Advances are reported in the processing and conduction mechanism
control of conducting and semiconducting polymers which for the first time
appear to be making these viable materials for use in electronics.

Considering the recent surge in research activity and its
multidisciplinary nature, there is a need for a comprehensive treatise which
would cover current important developments in this field. This book is
intended to fill such a void by providing a broad coverage of recent work by
leading international scientists in their respective areas. The book in its

scope is very comprehensive, covering the current status and providing for the first time the state-of-the-art in the following topics: (i) microscopic theoretical understanding of pi-electron conjugation effects on the electroresponse and non-linear optical properties; (ii) the relationship between microscopic properties and bulk properties; (iii) chemical synthesis of novel materials; (iv) molecular engineering of unique chemical structures and molecular assemblies, including self-ordering mechanisms; (v) multifunctional, macromolecular ultrastructures and composites for novel materials concepts and device structures; (vi) measurements and data base interpretation for second and third order nonlinear optical susceptibilities; (vii) charge carrier dynamics and the roles of novel excitations such as solitons, polarons, and bipolarons in conducting polymers; and (viii) device concepts for advanced microelectronics, modulators and optical switches.

We intend that the book will serve as a useful monograph for scientists and engineers entering this area as well as a valuable reference for those already working in this emerging field.

Paras N. Prasad
State University of
New York at Buffalo
Buffalo, New York

Donald R. Ulrich
Air Force Office of
Scientific Research
Bolling Air Force Base
Washington, D. C.

June, 1987

CONTENTS

ELECTROACTIVE POLYMERS

THEORY

SYNTHESIS

DEVICES

NONLINEAR OPTICAL AND ELECTROACTIVE POLYMERS: AN OVERVIEW

Donald R. Ulrich

Air Force Office of Scientific Research
Bolling Air Force Base
Washington, D.C.

Polymers have come a long way in being recognized as an important class
of materials for electronics and optical applications. Also, they have
begun to attract considerable attention of chemists and physicists as
challenging systems for basic understanding of ultrastructure and dynamics
of charge carriers and optical interactions. In the past year the field has
witnessed explosive growth at industries and universities alike with a
considerably widen research scope.

Until the early 1980's both the electronics and chemical industries had
a closed mindset toward the use of polymers in electronics. The integrated
circuit and semiconductor device industries considered polymers to be low
ambient temperature, structurally uncontrollable, environmentally unstable
materials whose outgassing and organic fragments would contribute to circuit
and device reliability problems. The chemical industry, on the other hand,
focused all resources and intellectual property on structural and
elastomeric applications rather than electronics.

In 1969-1970, while with the Integrated Circuit Products Department of
General Electric Company, I did some work on making ground planes for the
Perimenter Acquisition Radar of the ABM system that was under development.
I was using fluorinated ethylene propylene as the ground plane, and also as
a batch process for separation of and interconnection of integrated
circuits. I became very interested at the time in using various polymers,
in particular polyimides, as gate dielectrics in place of silicon dioxide in
Metal-Oxide-Semiconductor Field Effect Transistors. I constructed some
devices using polyimides instead of SiO_2, and built some C-MOS circuits,

1

Complementary MOS, and found that the turn-on voltages, flatband voltages and surface state densities were respectable in comparison to silicon dioxide gates. It is only now nearly 20 years later that major device and system manufacturers are looking to use polymers, in particular polyimides, in integrated circuits as passive dielectric layers. Their role for active functions has not apparently been developed at this time.

The development of polymers as structural materials has shown that they offer the flexibility of modifying molecular structure, conformation, order and morphology. Therefore, they can be tailored to suit a specific application. To start with, polymers can be formed in a rich variety of classes. Structural scale ordering has been the common denominator in the development of polymers with high degrees of orientation, order and

Table I

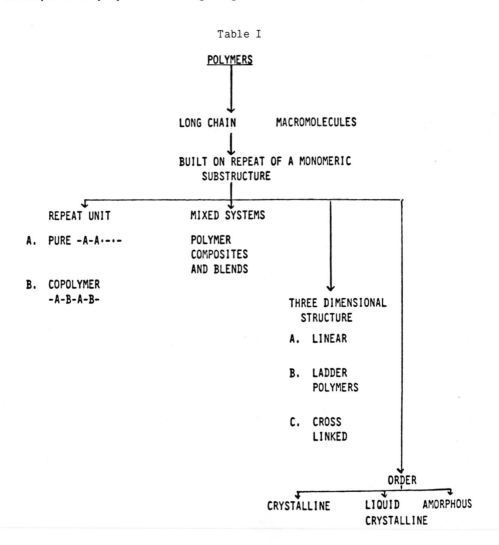

anisotropy. The hierarchial ordering manifested in the molecular, supramolecular and morphological aspects of structure and their implementation in structure-property-processing relationships have been fundamental to the development of structural polymers whose elastic modulus and tensile strength start to approach the upper limits predicted by theory. A classification of the dominant systems is suggested in Table 1. For example, three important polymer concepts have been rigid rod heterocyclic aromatic polymers, the liquid crystalline polymers, and copolymer blends and alloys.

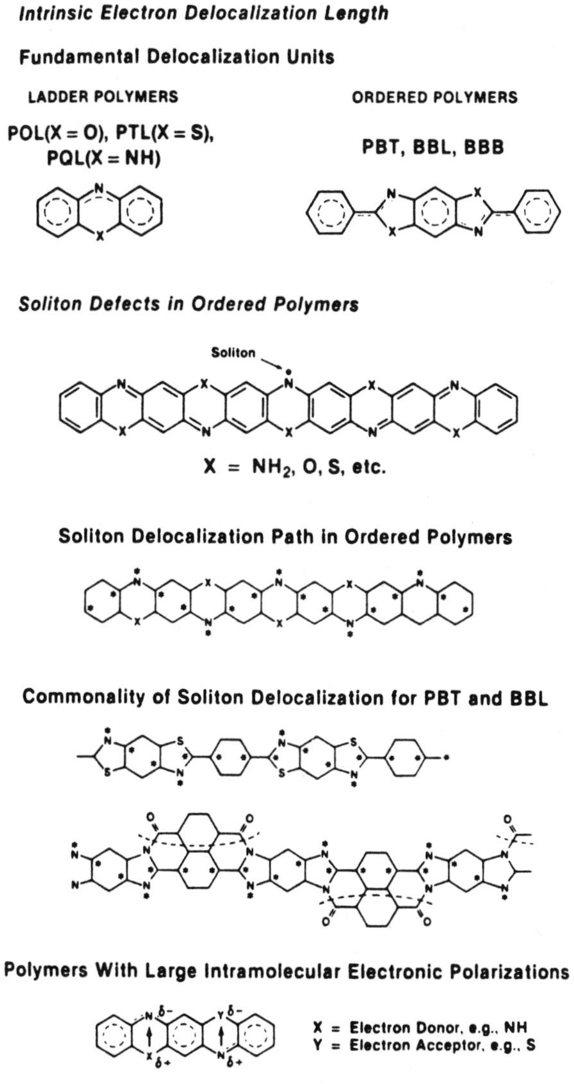

Intrinsic Electron Delocalization Length

Fundamental Delocalization Units

LADDER POLYMERS   ORDERED POLYMERS

POL(X = O), PTL(X = S),
PQL(X = NH)   PBT, BBL, BBB

        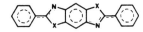

Soliton Defects in Ordered Polymers

Soliton

X = NH₂, O, S, etc.

Soliton Delocalization Path in Ordered Polymers

Commonality of Soliton Delocalization for PBT and BBL

Polymers With Large Intramolecular Electronic Polarizations

X = Electron Donor, e.g., NH
Y = Electron Acceptor, e.g., S

Figure 1. Electron dynamics defining the design of
controlled semiconductivity and conductivity
in ordered and ladder polymers

The rigid rod aromatic heterocyclic polymers such as PBT and PBO were developed in the Air Force Ordered Polymers Program with properties of outstanding high temperature thermal oxidative resistance and excellent environmental and damage threshold stabilities. Whereas other polymers are characterized by flexible and semiflexible molecular chains, much like the model of cooked spaghetti, the unique chain structures of PBT and PBO are shown in Figure 1. These can be described by the simple model of uncooked spaghetti. Because of their conjugated structures, the lyotropic liquid crystalline rigid chain poly-p-phenylenebenzobisthiazole (PBT), poly-p-phenylenebenzobisoxazole (PBO), and poly-2,2'(m-phenylene)-5,5' bibenzimidazole (PBI) have also shown promise as nonlinear optical (NLO) materials while maintaining their structural advantages.

In designing liquid crystalline polymers for nonlinear optical applications, a significant advancement has been the synthesis of polymers containing side chain liquid crystalline moities and nonlinear optical activity. Side-chain mesogens can contribute to NLO enhancement and particularly to the alignment of structures by the application of electric fields, or poling.

Another multifunctional approach is being designed around polymer alloys and blends. It is well known that high molar mass homopolymer pairs are generally not miscible. However, the opportunities for preparing miscible blending pairs is greatly enhanced where at least one of the components is a random copolymer and that further enhancement occurs when both components are copolymers rather than homopolymers. In research and scale-up already completed by the University of Massachusetts with a subcontract to the Celanese Research Company, a new family of high performance, miscible blends based on aromatic PBI and aromatic polyimides have been obtained. These results suggest that polymers of these generic types may be miscible over a wide range of compositions and structural variations and provide the format for new approaches to both structural and NLO polymers. Another advantage of the blends approach is the flexibility of processing which permits the ease of fabrication into optical waveguides and fibers. There is the very high potential for mass production because of the compatibility with in-place polymer manufacturing processes and future prospects of intelligent processing of materials (IPM) for quality control of order, anisotropy and sequential ultrastructural features. Waveguides with reasonable operating parameters have recently been demonstrated by Lockheed and Celanese based on blends.

Appropriate polymeric structures also promote the tailoring of Langmuir-Blodgett (LB) films and heterostructures. For polymers that demonstrate very high microscopic second order polarizabilities but are centrosymmetric, alternate layered structures deposited by the LB film technique can mimic noncentrosymmetry at the ultrastructure level and therefore translate the high β to macroscopic susceptibilities. Work at SUNY Buffalo has shown that the polymer conformation in monolayer films can be controlled by adjustment of the film forming conditions. This feature can play an important role in establishing the relation between microscopic nonlinearity and bulk susceptibility. For example, third harmonic generation from monolayer films of a conjugated centrosymmetric polymer, poly-4-BCMU, establishes a relation between the microscopic second hyperpolarizability, γ, and the third order nonlinear bulk susceptibility, $\chi^{(3)}$, using an oriented gas model.

The LB technique has been used by Celanese and the University of Pennsylvania to synthesize a polymer structure with second harmonic generation at 800 nanometers, the wavelength required for GaAs solid state lasers. Based on a quinoid polymer with large second order microscopic polarizability, i.e., $\beta = 200 \times 10^{-30}$ esu, noncentrosymmetric polymer structures with large SHG efficiency have been designed through asymmetric multilayer construction. In another experiment reported in Japan LB layers have been deposited based on PBO aromatic heterocyclics.

A hurdle in the application of polymeric materials for electronics and nonlinear optics has been the lack of processibility of conjugated polymeric structures by conventional film casting techniques. Many of these conjugated polymers are insoluble in common organic solvents. Understanding of ultrastructure and the role of solvent interaction has played an important role in building on the parent structure to improve the solubility. Work done at USC has established the use of selective derivatization along with the use of soluble precursor polymers and charge transfer solvents to successfully achieve processing of polymers. The strong advantage of derivatization is that lattice contamination by elimination products from the soluble precursor polymer is avoided and the electroactive polymer can be characterized and processed by standard techniques. The merit of derivatization of the parent polymer structure in improving solubility has also been demonstrated by other work such as that at the University of Massachusetts and the University of California at Santa Barbara.

5

Ultrastructure as a concept is defined as a recognizable solid state structure with characteristic dimensions of a few hundred angstroms. This was first demonstrated in ceramics and then extended later to polymers in the form of ordered block copolymer electroactive systems at the University of Massachusetts. The ultrastructure relates to the interaction topology at the molecular and domain levels. This understanding is important in relation to improving the processibility as well as in controlling the quality for meaningful conductivity measurements and nonlinear optical studies.

Application of polymers as electroactive materials dates back to the 1960's when photoconduction in the π electron complex polyvinylcarbazole - trinitrofluorenone was shown to have applications in photocopying. It was also recognized that the conventional band model used for inorganic semiconductors could not be used for explaining the photogeneration and motion of charge carriers. Consequently the Onsager model was proposed. Furthermore, the role of multiple trapping in these polymeric systems was recognized.

Conducting polymers first came into visibility as a research topic in the late 1970's. In 1978 at the National Science Foundation's Microstructures Workshop papers on electrically conducting polyacetylene polymers were highlighted which described the conduction mechanisms in terms of silicon semiconductor technology. The Air Force Office of Scientific Research questioned the approach of presenting conducting polymers in terms of silicon semiconductor integrated circuit concepts. It was pointed out that important differences exist between the behavior of a conducting polymers and a silicon based semi-conductor. Problems of processing were also pointed out.

A team effort was put in place by the Air Force in 1979-1981 to establish the principles for the design, synthesis, and processing of semiconductive and conductive polymers. It was determined that the two primary efforts should be to establish the electron dynamics required for conductivity in the ordered polymers and to theoretically predict dopant level effects in these materials.

Since the observation of large conductivity in doped polyacetylenes, many other conducting polymers such as polypyrrole, polythiophene and aromatic heterocyclic polymers such as poly-2H, 11H-bis-1,4-oxazino-2,3-b: 3',2'-m-triphenodioxazine-3,12-dinyl-2,11-diylene-11,12-bis methylidine

(POL), poly-1-6-dihydropyrazino-2,3-g-quinoxaline-2,3,8-triyl-7(2H)-ylidene-7,8 dimethylidene (PQL), poly-2H, 11H-bis-1,4-triazino-3,2-b:3',2'-m-triphenodithiazino-3,12-diyl-2,11-diylidene-11,12-bis methylidine (PTL), and poly-7-oxo-7,10-benzole-imidazo-4',5':5,6-benzimidazo-2,1-a-isoquinoline-3,4:10,11-tetrayl-10-carbonyl (BBL) have been reported. The work on BBL and PBT was as a result of the team put in place in 1979 and showed that doping, electron delocalization symmetry and solubility are dominated by the same mechanisms. Theory and experiment showed that approaches used in silicon are indeed possible using highly ordered polymeric substrates.

Novel conduction mechanisms have been proposed for these conducting polymers; they involve coupling of electronic excitations to nonlinear conformations of these polymers. In the presence of a degenerate ground state such as in polyacetylene and POL, this coupling leads to solitonic defects. The work at USC has shown that in the case of POL, the solitonic defect delocalization path is greater than 200 angstroms in the polymer backbone because of high perfection in polymer symmetry. On the other hand, polyacetylene often has a delocalization path of less than 100 angstroms because of symmetry breaking lattice perturbants such as cis-polyacetylene segments. In the case where the ground state is nondegenerate, the coupling of electronic excitations with conformations can lead to the formation of polarons and bipolarons which can be expected to be the dominant charge species. These structural defects can be present in the pristine polymers or they can be introduced by doping. It has been suggested that they can also be generated by excitation at energies above the band gap.

The work at USC, Cincinatti, Virginia and elsewhere have shown that PBT, PBO and BBL are characterized by extensively delocalized pi-electrons and that this π-electron delocalization profoundly influences electrical conductivity. Electron nuclear double resonance and relaxation measurements suggest electron-electron interactions are important for long range electron delocalization and thus for electrical conductivity. An extensively delocalized π-electron defect of metallic character is observed. Preliminary results indicate that intramolecular charge transfer plays an important role in defining intrinsic electrical properties. ENDOR and electron spin echo envelope modulation studies have established that existence and permitted the characterization of stable, reversible charge transfer complexes between polymers and dopants; such complexes are shown to change π-electron delocalization and alter polymer solubility. Spectroscopic measurements of delocalization and of both intra and intermolecular charge transfer have been correlated with electrical conductivity. Significant increases in electrical conductivity with exposure to electron donor or acceptor dopants is observed for the ordered polymers. Of exception and high interest is that the lack of such increase in conductivity for PBT with doping is likely due to a closely packed lattice which prevents dopant intercalation.

A number of applications have been proposed for these conducting polymers. They include field-effect transistors, light weight rechargeable batteries, electrochromic polymer displays, an erasable compact disc and photovoltaic devices. However, no potentially marketable device utilizing conductive polymers has emerged yet. The problem has been in finding an ideal conductive polymer which is processable, environmentally stable and which exhibits conductivity per unit weight superior to that of conventional metals. Recent reports on polyacetylene, prepared by BASF, in the doped state appears to have conductivity higher than that of copper when compared on the basis of per unit weight as well as volume. This form of polyacetylene also has enhanced environmental stability. However, the polymer is not processable and the long term stability is still lacking. The progress acheived on polyacetylene and the growing reported cases of soluble conducting polymers reveal that, through molecular engineering and material processing, commercial applications of conducting polymers should be realized in the near future.

The history of nonlinear optical effects in organics goes back to about two decades. Urea was one of the early organic systems to demonstrate the potential of achieving large optical nonlinearities in organic crystals and polymers. Since then most of the work focussed on the second order nonlinear optical effects in various organic crystals. An intrinsic requirement for second-order nonlinearity is a noncentrosymmetric structure. The material 2-methyl-4-nitroaniline has been found to exhibit second order nonlinearities more than two orders of magnitude larger than that of the KDP crystal. Various theoretical approaches to calculate microscopic second order nonlinearities (molecular hyperpolarizabilities, β) have been proposed. They include Pariser-Parr-Pople (PPP) calculations, CNDO(S) studies with configuration interactions, SCF-LCAO methods and a perturbation approach using a two-level model. It appears that existing theories can reliably predict molecular engineering of an organic structure with large second-order response. A conjugated structure with an electron-rich (donor) group on one end and a dificient (acceptor) group on the other end contains the asymmetric charge distribution in the π-system required for large β. Within the concept of perturbation expansion, a two-level system model predicts a correlation between β and the change in dipole moment between the ground state and the excited state. The low lying charge-transfer states in the donor-acceptor substituted structures such as given below

provide such a large change in the dipole moment upon excitation and hence a large β. In obtaining a large bulk second order nonlinear optical susceptibility ($\chi^{(2)}$) from the packing of molecules with a large β one often encounters the problem that highly dipolar molecules tend to align in an antiparallel centrosymmetric structure. Consequently $\chi^{(2)}$ vanishes. To overcome this problem, the molecular engineering approaches used have been chemical modification, electric field poled guest-host systems, and alternate layer deposited or asymmetric Langmuir-Blodgett films. For second harmonic generation, an additional problem is that of phase matching because of the dispersion of the linear refractive index. To deal with phase-matching, studies have focussed on experimenting with chemical modification, guest/host systems and using a guided wave geometry.

The recent surge of interest in the nonlinear optical effects is derived from their prospects in optical computing and optical signal processing. Proposed schemes of optical switching and logic operations utilize third-order optical nonlinearity. Compared to second order nonlinear effect, the history of third order optical nonlinearity in organic systems is relatively recent. Microscopic understanding of the third-order nonlinearity, γ, of organic systems is still in its infancy. Theoretical treatments have ranged from a simple one dimensional pseudopotential model to a sophisticated SCF ab initio calculation. The articles in this book contain approaches such as PPA (de Melo and Silbey) for large polymeric structures and ab initio SCF (Prasad) for small chain conjugated systems. One important structural feature emerging both from experimental and theoretical studies is the importance of π-electron conjugation. Therefore, the same π-electron conjugated polymers which exhibit large conductivity in doped states or upon photoexcitation can be suitable candidates for large γ and consequently large $\chi^{(3)}$. It should be noted that conductivity is a bulk property which is heavily enfluenced by intrachain as well as interchain charge transport. In constrast, the current status of understanding of third-order optical nonlinearity in conjugated polymers indicates the NLO behavior to be primarily a microscopic property (determined by γ).

Table II

WHERE WILL NLO POLYMERS FIT IN?

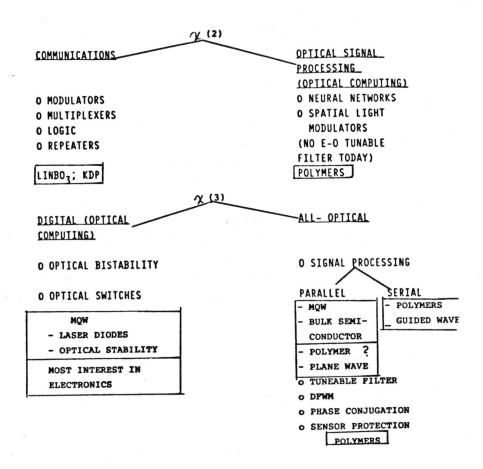

Polydiacetylenes are a class of polymers which exhibit very large non-resonant $\chi^{(3)}$ but are very poor conductors (semi-conductors). Another example is the heterocyclic polymer PBT which is an insulator but has a respectable $\chi^{(3)}$. Table II lists the schematics of possible application of $\chi^{(2)}$ and $\chi^{(3)}$ materials. Polymeric materials are expected to play an important role in spatial light modulators and serial processing.

Although conjugated organic polymers exhibit the largest non-resonant $\chi^{(3)}$ with response times in femtoseconds, the nonlinearity is still not sufficient to be utilized with low switching energies of diode lasers. The limiting value of $\chi^{(3)}$ appears to be 10^{-9} esu. The other problem from device application point-of-view appears to be the availability of processable conjugated polymeric materials for the fabrication of good optical quality integrated guided wave devices.

In my opinion the field of organic nonlinear optics will see a rapid growth in which scientists from various disciplines can be expected to contribute. Theorists can play an important role in refining the microscopic understanding of optical nonlinearities, especially the third order effect, in organic systems. A better microscopic understanding would help design molecular structures with enhanced nonlinearity. I also envision an interactive feedback where synthetic chemists can assist in our microscopic understanding by providing novel structures for measurements. Molecular engineering of ultrathin films can make a significant contribution in establishing the relation between microscopic nonlinearity and bulk nonlinearity. Physics of these processes such as the effect of resonance and presonances, the nature the local field effects can also be expected to be areas of future development.

The application of organic systems in devices, based on $\chi^{(3)}$, would require two areas to be developed: (1) enhancement of non-resonant $\chi^{(3)}$ so that very low switching energy can be used. Hopefully, a better microscopic understanding coupled with the availability of novel organic structures would help achieve this goal. (ii) Fabrication of optical quality organic films or fibers. Optical switching devices would utilize either a planar waveguide or an optical fiber geometry where long propagation distances can be used. Organic systems tend to be highly lossy. Materials scientists and chemical engineers can play an important role in developing processing and fabrication techniques for organic systems for making high optical quality planar wave guides and fibers. Finally, a strong input from device engineers is essential in implementing the concepts of optical modulators, optical switches, and optical logics using organic materials.

MICROSCOPIC ORIGIN OF SECOND ORDER

NONLINEAR OPTICAL PROPERTIES OF ORGANIC STRUCTURES

A.F. Garito, K.Y. Wong and O. Zamani-Khamiri

Department of Physics and Laboratory for
Research on the Structure of Matter
University of Pennsylvania
Philadelphia, PA 19104-6396

ABSTRACT

Microscopic descriptions and theoretical results for
second order nonlinear optical properties of organic and
polymer structures are presented. For cyclic polyenes and
quinoid structures, it is suggested that the frequency
dependent microscopic second order susceptibility β_{ijk} can
exceed 1000×10^{-30} esu over a broad frequency range.

INTRODUCTION

It is well established that organic and polymer
structures exhibit unusually large, ultrafast second order
$\chi_{ijk}^{(2)}(-\omega_3;\omega_1,\omega_2)$ and third order $\chi_{ijkl}^{(3)}(-\omega_4;\omega_1,\omega_2,\omega_3)$

nonlinear optical properties[1-8]. These outstanding properties
are important both to fundamental physics and chemistry and
to optical device technologies. Second and third order
properties have been investigated in a large number of
structures, phases and states[1-21]. Naturally, the results of
these studies taken together with developments in optical
bistability[22], phase conjugate wave generation[23] and

associative memory networks[24] have stimulated considerable activity in fundamental research and device applications.

The principal focus of fundamental studies of nonlinear optical organic and polymer structures has been the nature and characteristic features of the virtual excitations of the π-electron states that are responsible for the observed large nonresonant macroscopic $\chi_{ijk}^{(2)}(-\omega_3;\omega_1,\omega_2)$ and $\chi_{ijkl}^{(3)}(-\omega_4;\omega_1,\omega_2,\omega_3)$ and their relations to the material structure of the nonlinear optical medium[1-8]. Currently, unlike the traditional cases of inorganic dielectric insulators and semiconductors, the nonlinear optical excitations are viewed as occurring within the π-electron states of individual molecular, or polymer chain, sites in condensed assemblies and providing macroscopic sources of nonlinear optical response through the corresponding on-site microscopic nonlinear optical susceptibility. Thus, in a rigid lattice approximation, the macroscopic $\chi^{(2)}(-\omega_3;\omega_1,\omega_2)$ and $\chi^{(3)}(-\omega_4;\omega_1,\omega_2,\omega_3)$ responses are expressed by the equations

$$\chi_{ijk}^{(2)}(-\omega_3;\omega_1,\omega_2) = N\langle R_{im}\cdot R_{jn}\cdot R_{ko}\cdot f_{j'n'}^{\omega_1} f_{k'o'}^{\omega_2} f_{m'i'}^{\omega_3} \beta_{i'j'k'}(-\omega_3;\omega_1,\omega_2)\rangle$$

and

$$\chi_{ijkl}^{(3)}(-\omega_4;\omega_1,\omega_2,\omega_3) =$$

$$N\langle R_{im}\cdot R_{jn}\cdot R_{ko}\cdot R_{lp}\cdot f_{j'n'}^{\omega_1} f_{k'o'}^{\omega_2} f_{l'p'}^{\omega_3} f_{m'i'}^{\omega_4} \gamma_{i'j'k'l'}(-\omega_4;\omega_1,\omega_2,\omega_3)\rangle$$

where N is the number of molecular, or polymer, sites per unit volume, $\beta_{i'j'k'}$ and $\gamma_{i'j'k'l'}$ are the respective microscopic second and third order nonlinear optical susceptibilities, f^{ω_i} local field tensors, R the rotation matrix transforming the molecular frame to the laboratory frame, the brackets <...> an average over orientational distribution, and the unprimed (primed) coordinates designate the laboratory (molecular) fixed axes. Thus, the macroscopic nonlinear optical susceptibilities of a medium are related to their corresponding microscopic susceptibilities through orientational statistical averaging. The understanding of the nonlinear optical properties of the macroscopic medium is reduced to experimental and theoretical studies of the orientational distribution function of the molecular, or polymeric, units making up the nonlinear optical medium and their corresponding microscopic susceptibilities.

It has long been viewed that second and third order π-electron virtual excitations in conjugated organic and polymer structures are true many-body problems, and that, within a given structure, the correlated motion between π-electrons due to natural repulsive Coulomb interactions markedly determines microscopic nonlinear optical responses[3,25,26]. In calculations of the sign, magnitude and dispersion of microscopic higher order susceptibilities, suitable descriptions of the virtual excitations to the correlated π-electron excited states can be obtained by evaluating the susceptibility terms with self consistent field molecular orbital procedures (SCF-MO) that include single and double configuration interactions (SDCI) to account for electron correlations. Such a description differs considerably from independent particle models and simple phenomenological models of two-level systems and one-electron charge transfer; in addition, the many-electron description provides the necessary foundation for inclusion of possible additional interactions such as electron-phonon coupling if required by experimental results and demonstration.

In Section II of this article, a description of the microscopic second harmonic susceptibility is presented that is generally applicable to molecular and polymer structures. Section III illustrates the electronic origin of the second order susceptibility for the fundamentally important case of conjugated cyclic polyenes and quinoid structures.

THEORY FOR THE MICROSCOPIC SECOND ORDER NONLINEAR OPTICAL SUSCEPTIBILITY OF MOLECULAR STRUCTURES

The nonlinear response functions of a molecule interacting with a coherent electromagnetic radiation can be obtained by solving the Schrodinger equation. The electric polarization of a molecule results from the induced redistribution of electron density driven by an external electric field under the binding fields of the atomic Coulomb attraction. In this section, we will show that the nonlinear optical susceptibility can be obtained from time dependent perturbation solution to the Schrodinger equation by taking the external field as a small perturbing field in the Hamiltonian. In particular we will review the important example of second harmonic generation.

The Schrodinger equation for a molecule interacting with coherent electromagnetic radiation is

$$(H_O + H_I) \, |\psi(t)\rangle = i\hbar \, \frac{\partial}{\partial t} \, |\psi(t)\rangle \qquad (1)$$

where H_O is the unperturbed Hamiltonian. The molecule is characterized by a complete set of eigenstate $|n\rangle$ of the unperturbed Hamiltonian H_O, with energy eigenvalues $\hbar\omega_n$.

$$H_0 |n> = \hbar\omega_n |n>$$ (2)

The interaction Hamiltonian H_I is given, in the electric dipole approximation, by

$$H_I = -e\vec{E} \cdot \vec{r}$$ (3)

where $\vec{E}$ is the electric field operator. In the quantized field formalism, $\vec{E}$ is given by

$$\vec{E} = \frac{1}{V^{1/2}} \sum_{\vec{k}\,\alpha} (2\pi\hbar\omega)^{1/2} [a_{\vec{k},\alpha}\hat{\varepsilon}_\alpha e^{i(\vec{k}\cdot\vec{x}-\omega t)} - a^+_{\vec{k},\alpha}\hat{\varepsilon}_\alpha e^{-i(\vec{k}\cdot\vec{x}-\omega t)}]$$

(4)

where V is the normalization volume; $a^+_{k,\alpha}$ and $a_{k,\alpha}$ are the creation and annihilation operator for the photon field with mode $\vec{k}$ and α, where $\vec{k}$ is the propagation vector with $|\vec{k}| = \omega/c$ and $\hat{\varepsilon}_\alpha$ with $\alpha = 1$ or 2 are the unit vectors corresponding to the two transverse polarization. In the dipole approximation we can take $e^{i\vec{k}\cdot\vec{x}} \cong 1$. In the coherent representation the photon field is given by $|u_{k_1\alpha} u_{k_2\alpha}...>$ which is the eigenstate of the annihilation operator

$$a_{k_1\alpha} |u_{k_1\alpha} u_{k_2\alpha}...> = u_{k_1\alpha} |u_{k_1\alpha} u_{k_2\alpha}...>$$ (5)

The eigenvalue $u_{k\alpha}$ is related to the classical electric field amplitude by

$$u_{k\alpha} = -\frac{1}{2}\left(\frac{V}{2\pi\hbar\omega}\right)^{1/2} E^\omega_\alpha$$ (6)

With the input fundamental electromagnetic wave as a single mode with frequency ω and wave vector k, the process of second harmonic generation corresponds to the annihilation of two photons of frequency ω with the creation of a new photon of frequency 2ω. This process can be mathematically described by the action of the creation and annihilation operator. For a typical input laser power of 10 MW/cm^2, the photon flux is of the order of 10^{25}/cm^2. The annihilation of two photons of fundamental frequency thus produces negligible change in the input photon number. We can thus assume that the initial and final states of the fundamental photon field

are the same coherent state. Classically, this corresponds
to process where the phase of the electric field is
maintained.

We express the wave function $|\psi(t)\rangle$ by the complete set of
unperturbed wave functions $|m\rangle$

$$|\psi(t)\rangle = \sum_m C_m(t) \, |m\rangle \, e^{-i\omega_m t} \tag{7}$$

With Eq. (7) in the Schrodinger equation, we get a relation
for the expansion coefficient $C(t)$

$$\frac{d}{dt} C_l(t) = \frac{1}{i\hbar} \sum_m C_m(t) \, \langle l|H_l|m\rangle \, e^{-i\omega_{ml} t} \tag{8}$$

where $\omega_{ml} = \omega_m - \omega_l$.

Equation (8) can be solved by iteration. The zeroth
order solution of Eq. (8) corresponds to no perturbing field,
thus to zeroth order $C_1(t) = \delta_{1g}$, where g denotes the ground
state of the molecule. By putting this solution for $C_1(t)$ to
the right hand side of Eq. (8), the equation can be
integrated to obtain the first order correction for $C_1(t)$.
Higher order corrections are obtained by subsequence
iterations. Second harmonic generation corresponds to the
third order terms where the annihilation operator for the ω
photon operates twice, and the creation operator for the 2ω
photon operates once. There are three possible orderings of
the creation and annihilation operators, and they are
represented diagrammatically in Figure 1. The final state of
the molecule remains the ground state since no real
absorption or emission occurs.

The coefficient $C_{SHG}(t)$ for second harmonic generation,
which is the total transition amplitude for the scattering
processes depicted in Figure 1, is given by

$$C_{SHG}(t) = C_1(t) + C_2(t) + C_3(t) \tag{9}$$

with

$$C_1(t) = \left(\frac{e}{i\hbar}\right)^3 \sum_{n_1 n_2 \alpha_1 \alpha_2} \int_{-\infty}^{t} dt_3 \int_{-\infty}^{t_3} dt_2 \int_{-\infty}^{t_2} dt_1$$

$$\times \langle g|\hat{\varepsilon}_{\alpha_3} \cdot \vec{r} \, |n_2\rangle \, e^{i\omega' t_3} \, (2\pi\hbar\omega'/V)^{1/2} \, e^{i\omega_{gn_2} t_3}$$

$$x \langle n_2 | \hat{\varepsilon}_{\alpha_2} \cdot \vec{r} | n_1 \rangle \, e^{-i\omega t_2} (2\pi\hbar\omega/V)^{1/2} \, e^{i\omega_{n_2 n_1} t_2}$$

$$x \langle n_1 | \hat{\varepsilon}_{\alpha_1} \cdot \vec{r} | g \rangle \, e^{-i\omega t_1} (2\pi\hbar\omega/V)^{1/2} \, e^{i\omega_{n_1 g} t_1} \, u_{k\alpha_2} u_{k\alpha_1}$$

$$= -\left(\frac{e}{\hbar}\right)^3 \left(\frac{2\pi\hbar\omega'}{V}\right)^{1/2} \sum_{n_1 n_2 \alpha_1 \alpha_2} r_{gn_2}^{\alpha_3} \, r_{n_2 n_1}^{\alpha_2} \, r_{n_1 g}^{\alpha_1}$$

$$x \; \frac{E_{\alpha_1}^{\omega} E_{\alpha_2}^{\omega}}{(\omega_{n_1 g} - \omega)(\omega_{n_2 g} - 2\omega)} \int_{-\infty}^{t} e^{i(\omega' - 2\omega) t_3} \, dt_3$$

(10a)

where $r_{n_2 n_1}^{\alpha} = \langle n_2 | \hat{\varepsilon}_{\alpha} \cdot \vec{r} | n_1 \rangle$, etc. In evaluating the integral, the interaction Hamiltonian is assumed to be turned on adiabatically at $t = -\infty$. We have used Eq. (6) to express $u_{k\alpha}$ in terms of E_{α}^{ω}. Similarly,

$$C_2(t) = -\left(\frac{e}{\hbar}\right)^3 \left(\frac{2\pi\hbar\omega'}{V}\right)^{1/2} \sum_{n_1 n_2 \alpha_1 \alpha_2} r_{gn_2}^{\alpha_2} \, r_{n_2 n_1}^{\alpha_3} \, r_{n_1 g}^{\alpha_1}$$

$$x \frac{E_{\alpha_1}^{\omega} \, E_{\alpha_2}^{\omega}}{(\omega_{n_1 g} - \omega)(\omega_{n_2 g} + \omega)} \int_{-\infty}^{t} e^{i(\omega' - 2\omega) t_3} \, dt_3 \qquad (10b)$$

$$C_3(t) = -\left(\frac{e}{\hbar}\right)^3 \left(\frac{2\pi\hbar\omega'}{V}\right)^{1/2} \sum_{n_1 n_2 \alpha_1 \alpha_2} r_{gn_2}^{\alpha_2} \, r_{n_2 n_1}^{\alpha_1} \, r_{n_1 g}^{\alpha_3}$$

$$x \frac{E_{\alpha_1}^{\omega} \, E_{\alpha_2}^{\omega}}{(\omega_{n_1 g} + 2\omega)(\omega_{n_2 g} + \omega)} \int_{-\infty}^{t} e^{i(\omega' - 2\omega) t_3} \, dt_3 \qquad (10c)$$

The total transition rate for the scattering process into a solid angle $d\Omega$ is given by

$$\frac{dW}{d\Omega} = \int \frac{|C_{SHG}(t)|^2}{t} \rho_{\hbar\omega'} \, d(\hbar\omega') \tag{11}$$

where $\rho_{\hbar\omega'} = \dfrac{V\omega'^2}{\hbar(2\pi c)^3}$ is the density of state for photon with frequency ω'. The flux density of the second harmonic photon with wave vector k' and polarization α_3 is

$$\frac{<N_{\alpha_3}^{\omega'}>c}{V} = \frac{1}{r^2} \frac{dW}{d\Omega} \tag{12}$$

where $N_{\alpha}^{\omega'}$ is the number operator $a_{k'\alpha}^+ a_{k'\alpha}$. Thus

$$<N_{\alpha_3}^{\omega'}> = <u_{k'\alpha_3} | a_{k'\alpha_3}^+ a_{k'\alpha_3} | u_{k'\alpha_3}>$$

$$= u_{k\alpha_3}^* \, u_{k'\alpha_3}$$

$$= |- \frac{1}{2} \left(\frac{V}{2\pi\hbar\omega'} \right)^{1/2} E_{\alpha_3}^{\omega'} |^2 \tag{13}$$

The integral in Eq. (10) can be evaluated using the relation

$$\lim_{t\to\infty} | \int_{-\infty}^{t} e^{i(\omega'-3\omega)t_3} dt_3 |^2 = 2\pi t \, \delta(\omega'-3\omega) \tag{14}$$

We obtain, from Eq. (9)–(14)

$$E_{\alpha_3}^{2\omega} = \frac{1}{r} \left(\frac{2\omega}{c} \right)^2 \vec{p} \cdot \hat{\varepsilon}_{\alpha_3} \tag{15}$$

where

$$\vec{p} = \frac{-e^3}{2\hbar^2} \sum_{n_1 n_2 \alpha_1 \alpha_2} E_{\alpha_1}^{\omega} E_{\alpha_2}^{\omega}$$

$$\times \left\{ \frac{\vec{r}_{gn_2}^{\alpha_2}\, r_{n_2n_1}^{\alpha_1}\, r_{n_1g}}{(\omega_{n_1g}-\omega)(\omega_{n_2g}-2\omega)} + \frac{r_{gn_2}^{\alpha_2}\, \vec{r}_{n_2n_1}^{\alpha_1}\, r_{n_1g}}{(\omega_{n_1g}-\omega)(\omega_{n_2g}+\omega)} + \frac{r_{gn_2}^{\alpha_2}\, r_{n_2n_1}^{\alpha_1}\, \vec{r}_{n_1g}}{(\omega_{n_1g}+2\omega)(\omega_{n_2g}+\omega)} \right\}$$

$$(16)$$

In electromagnetic theory, the electric field of the radiation from a dipole $\vec{p}^{2\omega}$ oscillating at frequency 2ω is given by

$$\vec{E}^{2\omega} = \frac{1}{r}\left(\frac{2\omega}{c}\right)^2 (\hat{n}\times\vec{p}^{2\omega})\times\hat{n}$$

where $\hat{n}$ is the radial unit vector.
Thus

$$E_{\alpha_3}^{2\omega} = \varepsilon_{\alpha_3}\cdot\vec{E}^{2\omega}$$

$$= \frac{1}{r}\left(\frac{2\omega}{c}\right)^2 \vec{p}^{2\omega}\cdot\varepsilon_{\alpha_3} \qquad (17)$$

We can thus identify $\vec{p}$ in (15) as the microscopic second harmonic polarization.   In standard notation

$$p_i^{2\omega} = \beta_{ijk}(-2\omega;\omega,\omega)\, E_j^{\omega}\, E_k^{\omega} \qquad (18)$$

where $\beta_{ijk}(-2\omega;\omega,\omega)$ is the molecular second harmonic susceptibility which, by convention, is symmetrical under the permutation of the indices j and k.   We obtain, therefore,

$$\beta_{ijk} = -\frac{e^3}{4\hbar^2}\sum_{n_1 n_2} \left\{ \frac{r_{gn_2}^i r_{n_2n_1}^j r_{n_1g}^k + r_{gn_2}^i r_{n_2n_1}^k r_{n_1g}^j}{(\omega_{n_1g}-\omega)(\omega_{n_2g}-2\omega)} \right.$$

$$\left. + \frac{r_{gn_2}^j r_{n_2n_1}^i r_{n_1g}^k + r_{gn_2}^k r_{n_2n_1}^i r_{n_1g}^j}{(\omega_{n_1g}-\omega)(\omega_{n_2g}+\omega)} + \frac{r_{gn_2}^j r_{n_2n_1}^k r_{n_1g}^i + r_{gn_2}^k r_{n_2n_1}^j r_{n_1g}^i}{(\omega_{n_1g}+2\omega)(\omega_{n_2g}+\omega)} \right\}$$

$$(19)$$

This expression relates the second harmonic susceptibility

$\beta_{ijk}(-2\omega;\omega,\omega)$ to the transition matrix elements r_{nm} and excitation energies ω_{ng} and generally applies to any noncentrosymmetric molecular or polymer structure.

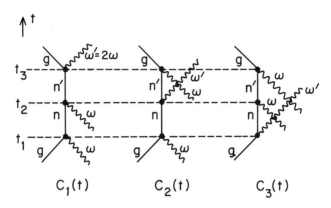

Figure 1: Space time diagram for second harmonic generation.

III. ELECTRONIC ORIGIN OF THE MICROSCOPIC SECOND ORDER OPTICAL SUSCEPTIBILITY OF CYCLIC POLYENES AND QUINOID STRUCTURES

In this section, recent progress is reviewed in understanding second order virtual excitation processes and the microscopic origin of large, nonresonant susceptibilities $\beta_{ijk}(-\omega_3;\omega_2,\omega_1)$ for noncentrosymmetric conjugated cyclic polyenes and quinoid structures. As in the cases of resonant ring structures such as PNA (p-nitroaniline) and MNA (2-methyl-4-nitroaniline)[3,25-27] and conjugated linear chains such as NMDVDA [trans-trans 1-(4-methylphenyl)-8-(4'-nitrophenyl)-1,7-diene-3,5-octadiyne][28-30], a principal approach is one of combined theoretical and experimental studies of the frequency dependence of the nonresonant second harmonic susceptibility $\beta_{ijk}(-2\omega;\omega,\omega)$ at frequencies below one photon electronic resonances but well above molecular vibrations. Results from recent studies of $\beta_{ijk}(-2\omega;\omega,\omega)$ of several important systems are reviewed, and, in these cases, the basic symmetry controlled mechanisms of second order virtual excitations and the important role of electron correlations are found essentially the same as originally presented[3,25-27].

The frequency dependent susceptibility $\beta_{ijk}(-2\omega;\omega,\omega)$ and singlet-singlet excitation spectra have been theoretically calculated for two experimental systems DCNQI [2-(4-dicyanomethylene cyclohexa-2,5-dienylidine)-imidazolidine][31-32] (Figure 2), MPC [N-methyl[4(1H)-pyridinylidene ethylidene]-2,5-cyclohexadien-1-one][33] (Figure 3), and a predicted structure DPHQ [2-(4'-dicyanomethylene bis cyclohexa-2,4,5,2',5'-pentaenylidine)-imidazolidine][32] (Figure 4). For calculating and analyzing the terms in $\beta_{ijk}(-2\omega;\omega,\omega)$, the theoretical method of direct summation is basically the same as originally described[3,25]. The computations of dipole moments and excitation energies are based on an all valence, self consistent field molecular orbital calculation (SCF-MO) in the rigid lattice CNDO/s approximation and configuration interaction theory with single and double excitations (SDCI) to account for electron correlations. The results of the theoretical calculations for DCNQI and MPC agree well with experimental values obtained by DC induced second harmonic generation studies of liquid solutions.

The measured quantity is the vector part $\beta_x(-2\omega;\omega,\omega)$ of susceptibility tensor $\beta_{ijk}(-2\omega;\omega,\omega)$ and without Kleinman symmetry is given by the relation

$$\beta_x = \beta_{xxx} + \frac{1}{3}\sum_{i \neq x}(\beta_{xii} + 2\beta_{iix})$$

where the x axis remains aligned along the dipolar axis of the molecule. Calculation of all components of $\beta_{ijk}(-2\omega;\omega,\omega)$ confirmed the basic symmetry for each of the three cases wherein DCNQI and DPHQ belong to the C_{2v} point group and MPC the C_s point group.

In the calculation of β_x for each case, the β_{xxx} component is the major contribution. As originally discussed, the contributing terms follow from basic symmetry conditions. In the case of C_{2v} symmetry, only those excited states that transform as the A_1 representation of the C_{2v}

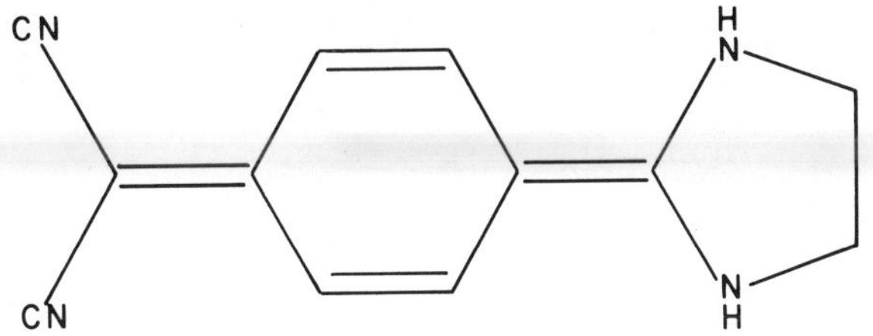

DCNQI

Figure 2: Molecular structure of DCNQI.

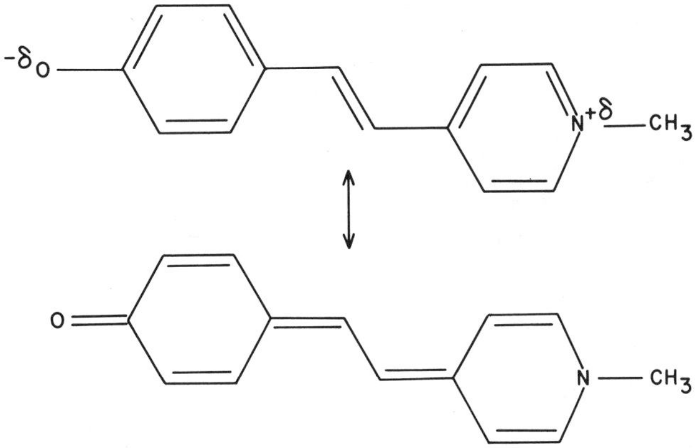

MPC

Figure 3: Molecular structure of MPC.

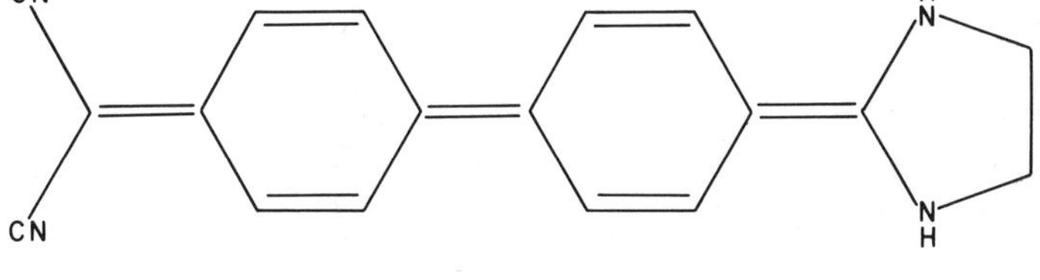

DPHQ

Figure 4: Molecular structure of DPHQ.

point group are symmetry allowed in the determination of β_{xxx}. The x operator is isomorphic to the A_1 representation of the group C_{2v}. Thus, dipole coupling may occur between states of the same symmetry since A_1 is the totally symmetric representation. In the perturbation expansion for β_{ijk}, every term consists of a product in which at least one factor is a dipole coupling to the ground state which is also totally symmetric. Hence, any two excited states, n and n', in an arbitrary term for β_{xxx} must both be of A_1 symmetry.

The remaining components β_{xyy}, β_{yyx}, β_{xzz}, and β_{zzx} contained in β_x are smaller than β_{xxx} because the dipole moment difference between ground and excited state, $\Delta r_{n'}$, which is an important quantity in determining β_{ijk} has no y or z component due to C_{2v} symmetry. Only off-diagonal dipole moment matrix elements and transition moments remain contributing to the components of β_{ijk} involving y or z. Inspection of the intermediate results of the calculation for each system reveal that these off-diagonal terms are always relatively small. For example, at 0.65 eV (1.907 μm), the calculated gas phase β_x components in units of 10^{-30} esu are for DCNQI β_{xxx}: -46.4; β_{xyy}: 3.7; β_{xzz}: -0.1; β_{yyx}: 2.9; and β_{zzx}: 0.1; and thus, the major component of β_x for DCNQI is, in fact, β_{xxx}.

The calculated gas phase dispersion curves of β_x for DCNQI and MPC are given in Figures 5 and 6. The calculated and experimental absorption spectra of DCNQI compared in Figure 7 illustrate the characteristic large absorption maximum at low energy and spectral features found commonly in the singlet-singlet excitation spectrum calculated for each of the three cases. As seen in Figures 5 and 6 for DCNQI and MPC, the low frequency β_x smoothly increases with increased frequency to magnitudes greater than 200×10^{-30} esu in approaching the 2ω singularity whose frequency is half the value of the excitation energy of the first major absorption peak of the singlet-singlet excitation spectrum. Thus, each 2ω singularity at 1.16 and 0.96 eV for DCNQI and MPC, respectively, corresponds to the large oscillator strength singlet-singlet excitation to the respective first excited states at 2.31 and 1.92 eV.

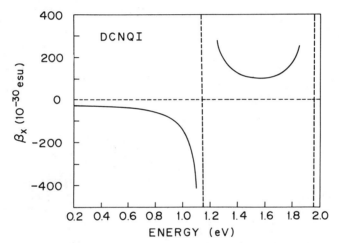

Figure 5: Calculated gas phase dispersion curve of β_x for DCNQI where E is the input photon energy. (7x8; 4x3 SDCI)

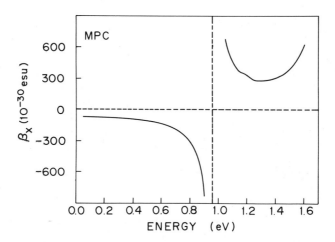

Figure 6: Calculated gas phase dispersion curve of β_x for MPC where E is the input photon energy. (4x4; 4x4 SDCI)

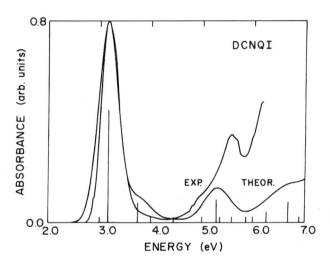

Figure 7: Calculated and experimental optical absorption spectra of DCNQI after accounting for DMSO solvent shift.

There are both diagonal β_{xxx}^{nn} and off-diagonal $\beta_{xxx}^{nn'}$

contributions to β_{xxx}. The percentage contributions to β_x
from β_{xxx}^{nn} and $\beta_{xxx}^{nn'}$ are, for example, at 0.65 eV for DCNQI 44%
and 56%, respectively, and for MPC 55% and 45%, respectively.
At 0.65 eV, the calculated β_x components in units of $\cdot 10^{-30}$
esu are for MPC β_{xxx}: -158; β_{xyy}: 1.1; β_{xzz}: 0.3; β_{yyx}: 0.4
and β_{zzx}: 0. The diagonal component β_{xxx}^{nn} is given by the
expression

$$\beta_{xxx}^{nn} = \frac{3}{2\hbar^2} \sum_n \mu_{ng}^x \mu_{ng}^x \Delta\mu_n^x \omega_{ng}^2 (\omega_{ng}^2 - \omega^2)^{-1} (\omega_{hg}^2 - 4\omega^2)^{-1}$$

where $\mu_{ng}^x = -er_{ng}^x$ is the x-component of the transition moment

between the ground (g) and excited (n) states; $\Delta\mu_{ng}^x = -e\Delta r_{ng}^x$
is the x-component of the dipole moment difference between
the ground and excited states; and the sum is over all
states. In DCNQI and MPC, several major singlet excited
states contribute to β_{xxx}^{nn}. In DCNQI, major terms from states
I and III contribute with calculated gas phase values,
respectively, for ω_{ng} of 2.31 and 3.92 eV; μ_{ng}^x of 9.0 and 3.1

D; and $\Delta\mu_{ng}^x$ of -4.1 and -6.8 D. The calculated ground state

dipole moment μ_g of 17.0D agrees well with the experimental
value of 17.5 ±0.8D[31-32]. In MPC, the largest terms are due
to states I, V and XI with values, respectively, for ω_{ng}:
1.92, 3.69 and 5.36 eV; μ_{ng}^x: 12.0, 4.7 and 1.8D; $\Delta\mu_{ng}^x$: -4.4, -

8.3 and -7.7D. The calculated μ_g of 21.4D compares reasonably
with the recently obtained experimental result of
17.4±2.5D[33]. Thus, for both DCNQI and MPC, the state I
contribution to β_{xxx}^{nn} is largest because in each case, both

values of μ_{Ig}^{x} and $\Delta\mu_{I}^{x}$ are relatively large and the excitation

energy ω_{Ig} is lowest.

The off diagonal component $\beta_{xxx}^{nn'}$ is given by the
expression

$$\beta_{xxx}^{nn'} = \frac{1}{2\hbar^2} \sum_{\substack{n\neq n' \\ n\neq g \\ n'\neq g}} (\mu_{gn}^{x}\mu_{nn'}^{x}\mu_{n'g}^{x}) \times [(\omega_{ng}-\omega)^{-1}(\omega_{n'g}-2\omega)^{-1}$$
$$+ (\omega_{ng}-\omega)^{-1}(\omega_{n'g}+\omega)^{-1} + (\omega_{ng}+\omega)^{-1}(\omega_{n'g}+2\omega)^{-1}]$$

where $\mu_{nn'}^{x} = -er_{nn'}^{x}$ is the x-component of the transition

moment between excited states n and n'. Major terms
contribute to $\beta_{xxx}^{nn'}$ from several singlet excited states in

DCNQI and MPC, especially transition moments involving the
first excited state I in each system. For example, the
calculated gas phase values for $\mu_{nn'}^{x}$ are for DCNQI: $\mu_{I}^{x}{}_{III}$:

4.1; $\mu_{I}^{x}{}_{VIII}$: 6.9; and $\mu_{I}^{x}{}_{XI}$: 5.0D; and for MPC: $\mu_{I}^{x}{}_{V}$: 8.0;

and $\mu_{I}^{x}{}_{XI}$: 11.6D. Thus, for both β_{xxx}^{nn} and $\beta_{xxx}^{nn'}$, several

singlet excited states provide major contributing terms.

Common to each system, the calculated and experimental
β_{x} at low frequencies is negative. This results from the

dipole moment diference $\Delta\mu_{ng}^{x}$ being negative in the β_{xxx}^{nn}

component. The negative dipole moment difference $\Delta\mu_{n}^{x}$ is due

to a virtual charge redistribution in the excited state
opposing the ground state charge density so that the dipole
moment of the excited state is smaller than that of the
ground state. In addition to DC SHG measurements, the
negative values for $\Delta\mu_{n}^{x}$ of DCNQI and MPC are consistent with

blue shifts observed for their solution absorption spectra
with increased solvent polarity. In each case, the
difference between the calculated gas phase and measured
solution phase transition energies is due to the solvent
shifts.

On the high frequency side of the 2ω singularity, the diagonal term β_{xxx}^{nn} in each case is larger than the off-diagonal term $\beta_{xxx}^{nn'}$, and with the change in sign of the energy denominator of β_{xxx}^{nn}, β_x becomes large and positive.

For a given low frequency, the magnitude of β_x for MPC is larger than the value for DCNQI because the state I transition moment μ_{Ig}^x of MPC (12.0D) is larger than that of DCNQI (9.0D). In the calculation of β_x, the product term $(\mu_{Ig})^2 \, \Delta\mu_I^x$ occurs in β_{xxx}^{nn} as a major contributing term. Although $\Delta\mu_I^x$ and ω_{Ig} are comparable for each case, the term is dominated by the larger μ_{Ig} of MPC.

At frequencies above the first 2ω singularity, each β_x dispersion curve (Figures 5 and 6) exhibits a second 2ω singularity located at 1.96 eV for DCNQI and at 1.84 eV for MPC . In each case, the singularity arises from virtual excitations to the next higher lying π-electron excited state above state I that possesses a large transition moment μ_{ng}^x. This excited state in DCNQI is state III with ω_{ng} of 3.92 eV and μ_{ng}^x of 3.1D, and in MPC state V with ω_{ng} of 3.69 eV and μ_{ng}^x of 4.7 D. In contrast to the low frequency 2ω singularity, even though the principal dipole moment difference $\Delta\mu_n^x$ is negative in sign, β_x below the high frequency 2ω singularity is positive. The main reason is that the positive off-diagonal $\beta_{xxx}^{nn'}$ components of state III in DCNQI and of state V in MPC are larger than their corresponding negative diagonal β_{xxx}^{nn} components. In this frequency range, the positive off diagonal term together with the remaining positive contribution from the first 2ω singularity results in the total β_x being positive and large.

These important features are conveniently illustrated by density matrices of the state functions. The density matrices in the form of contour diagrams provide a compact diagrammatic representation of the microscopic nature and characteristic features of second order nonlinear optical processes for molecular structures. Terms involving the dipole moment difference $\Delta\mu_n^x$ are represented in the form of the difference density function $\rho_n-\rho_g$ between the excited and ground state functions

$$\langle\Delta\mu_n\rangle = -e \int r(\rho_n - \rho_g)\, dr$$

where ρ_n is the first order reduced density matrix. The transition moment $\mu_{nn'}^i$ is expressed in terms of the transition density matrix $\rho_{nn'}(r,r')$ by the relation

$$\langle\mu_{nn'}\rangle = -e \int_{r=r'} r\, \rho_{nn'}(r,r')\, dr.$$

As illustrative examples, contour diagrams of $\rho_I-\rho_g$ important to terms involving $\Delta\mu_n^x$ are given in Figure 8 and 9 for the respective principal excited state I of DCNQI and MPC, where the solid and dashed lines correspond to increased and decreased electron density, respectively. In each case, upon virtual excitation in the second order process, large asymmetric redistribution of electron density occurs along the x dipolar axis causing the dipole moment difference $\Delta\mu_I^x$.

The origin of the negative sign of $\Delta\mu_I^x$ and the low frequency β_x is clearly evident in each diagram by the decreased electron density at sites near the electron acceptor group and increased density in the region of the electron donor group which opposes the natural ground state electron density distribution. The excited state dipole moment μ_I^x is thus smaller than the ground state dipole moment resulting in $\Delta\mu_I^x$ and β_x being negative. The contour diagrams further illustrate the highly separated, localized nature of the

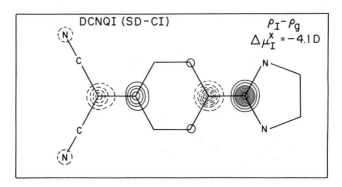

Figure 8: Contour diagram of difference density function $\rho_I - \rho_g$ of DCNQI for the first singlet excited state I.

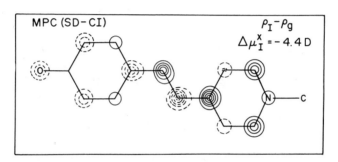

Figure 9: Contour diagram of difference density function $\rho_I-\rho_g$ of MPC for the first singlet excited state I.

virtual electron density distribution of DCNQI along the dipolar x direction compared to the modulated distribution in MPC. The net result is that although the molecular length along the dipolar axis is much larger in MPC than in DCNQI, the respective values of -4.4 and -4.1 D for the dipole moment difference $\Delta\mu_I^x$ are essentially the same in both cases.

The results for β_x and the highly charge correlated excited states of DCNQI motivated the theoretical analysis of the second order properties of the related extended ring structure DPHQ. The calculated gas phase dispersion curve for β_x of DPHQ is given in Figure 10. At 0.65 eV, the calculated gas phase components of β_x for DPHQ in units of 10^{-30} esu are β_{xxx}: -1413; β_{xyy}: 5.6; β_{xzz}: 0; β_{yyx}: 1.9; and β_{zzx}: 0 where β_{xxx} is the largest component. Although the general features of the β_x dispersion curve of DPHQ are qualitatively similar to those of DCNQI, two major features are pronounced. First, the magnitude of the measured β_x is unusually large, achieving magnitudes greater than 1000×10^{-30} esu. Second, the principal first excited state I is sufficiently separated from the upper excited states that the two singularities at 0.77 and 1.54 eV are, respectively, the 2ω and ω resonances of state I with ω_{Ig} of 1.54 eV.

The origin of the magnitude, sign, and dispersion of β_{xxx} resides in the electron density redistribution accompanying second order virtual processes in the dominant excited state I. At 0.65 eV, nearly the entire diagonal component β_{xxx}^{nn}, which accounts for 72% of β_{xxx}, results from the first excited state I with μ_{Ig} and $\Delta\mu_{Ig}^x$ vlaues of 14.4 and -7.7 D, respectively. The off-diagonal $\beta_{xxx}^{nn'}$ contribution of 28% is mostly due to virtual excitations between state I and three higher lying excited states.

The characteristic features of the principal excited state I are illustrated in the ρ_I-ρ_g contour diagram of Figure 11. Along the dipolar x axis, the electron density distribution exhibits the same type of highly asymmetric, correlated features and associated negative $\Delta\mu_I^x$ as found in the DCNQI case but on an enlarged scale. Upon virtual excitation, electron density redistributes in a localized

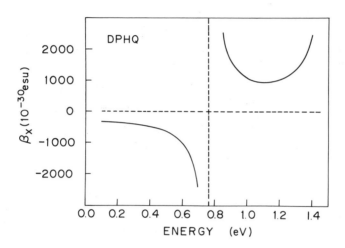

Figure 10: Calculated gas phase dispersion curve of β_x for DPHQ where E is the input photon energy. (3x3; 3x3 SDCI)

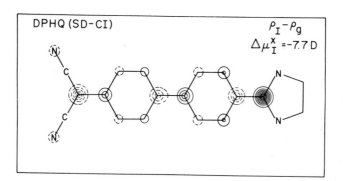

Figure 11: Contour diagram of difference density function $\rho_I - \rho_g$ of DPHQ for the first singlet excited state I.

fashion, nearly exclusively on the exocyclic carbon sites, and the correlation length spans the long molecular axis of the basic quinoid structure. Similar dramatic features are observed in contour diagrams of the transition density associated with the relatively large μ_{Ig}^{x} value of 14.4D.

Finally, at frequencies immediately below the ω singularity, the DPHQ β_x is almost entirely due to β_{xxx} whose β_{xxx}^{nn} and $\beta_{xxx}^{nn'}$ contributions are 39% and 61%, respectively.

The frequency interval between the 2ω and ω singularities is particularly significant because the value of β_x is unusually large at greater than 1000×10^{-30} esu over a broad frequency range.

   In summary, we reviewed fundamental theory for the microscopic second order susceptibility that is generally applicable to molecular and polymer structures. Recent results were summarized of combined theoretical and experimental studies of the frequency dependent microscopic second harmonic susceptibilities for the important class of noncentrosymmetric conjugated cyclic polyene and quinoid structures. The results provide a many-electron description of nonresonant second order processes as being determined by symmetry controlled, highly charge correlated, virtual excitations within the π-electron states for each structure. As a major illustration, such descriptions allow theoretical calculations of new structures such as DPHQ which show second order responses greater than 1000×10^{-30} esu over a broad frequency range.

ACKNOWLEDGEMENTS

   We gratefully acknowledge many stimulating discussions with colleagues Drs. A. Buckley and E.W. Choe of Hoechst-Celanese, and K.D. Singer and J. Sohn of AT&T. This research was generously supported by AFOSR and DARPA, F49620-85-C-0105 and NSF/MRL, DMR-85-19059.

REFERENCES

1. A.F. Garito and K.D. Singer, Laser Focus <u>18</u>, 59 (1982).

2. Nonlinear Optical Properties of Organic and Polymeric Materials; D. Williams, ed. (ACS Symp. Series, Plenum Press, New York, 1983).

3. A.F. Garito, C.C. Teng, K.Y. Wong and O. Zamani-Khamiri, Mol. Cryst. Liq. Cryst. <u>106</u>, 219 (1984).

4. D.J. Williams, Angew. Chem. Intl. Ed. Eng. <u>23</u>, 690 (1984).

5. A.F. Garito, SPIE Adv. in Mats. For Active Optics <u>567</u>, 51 (1985).

6. J. Zyss, J. Molec. Electron. <u>1</u>, 25 (1985).

7. A.F. Garito, SPIE Nonlinear Optics and Applications <u>613</u> (1986).

8. <u>Molecular and Polymeric Optoelectronic Materials: Fundamental and Applications</u>, G. Khanarian, ed. SPIE <u>682</u> (1986).

9. B.F. Levine and C.G. Bethea, J. Chem. Phys. <u>63</u>, 2666 (1975).

10. Y.R. Shen, Rev. Mod. Phys. <u>48</u>, 1 (1976).

11. G.F. Lipscomb, R.S. Narang, and A.F. Garito, Appl. Phys. Lett. <u>38</u>, 663 (1981); J. Chem. Phys. <u>75</u>, 1509 (1981).

12. K.D. Singer and A.F. Garito, J. Chem. Phys. <u>75</u>, 3572 (1981).

13. I.R. Girling, N.A. Cade, P.V. Kolinsky, and C.M. Montgomery, Electron. Lett. <u>21</u>, 169 (1985).

14. S.J. Lalama, J.E. Sohn, and K.D. Singer, SPIE Integrated Optical Circuit Engineering II. <u>578</u>, 168 (1985).

15. D.N. Rao, J. Swiatkiewicz, P. Chopra, S.K. Ghoshal, and P.N. Prasad, Appl. Phys. Lett. <u>48</u>, 1187 (1986).

16. K.D. Singer, J.E. Sohn, and S.J. Lalama, Appl. Phys. Lett. <u>49</u>, 248 (1986).

17. S.K. Saha and G.K. Wong, Appl. Phys. Lett. <u>34</u>, 423 (1979).

18. K.Y. Wong and A.F. Garito, Phys. Rev. A34, 5051 (1986).

19. I.C. Khoo and Y.R. Shen, Opt. Eng. 24, 579 (1985).

20. N.C. Kothari and T. Kobayashi, IEEE J. Quant. Electron. QE-20, 418 (1984).

21. Crystallographically Ordered Polymers, Daniel J. Sandman, Ed. (ACS Symp. Series, Washington, DC 1987).

22. H.M. Gibbs, Optical Bistability: Controlling Light with Light (Academic Press, New York, 1985).

23. R.W. Hellwarth, J. Opt. Soc. Am. 67, 1 (1977).

24. N.H. Farhat, D. Psaltis, A. Prata and E. Pack, Appl. Optics 24, 1469 (1985).

25. S.J. Lalama and A.F. Garito, Phys. Rev. A20, 1179 (1979).

26. A.F. Garito, K.D. Singer, and C.C. Teng, in Nonlinear Optical Properties of Organic and Polymeric Materials, D. Williams, Ed. (ACS Symp. Series, Plenum Press, New York, 1983).

27. C.C. Teng and A.F. Garito, Phys. Rev. Lett. 50, 350 (1983); Phys. Rev. B28, 6766 (1983).

28. A.F. Garito, Y.M. Cai, H.T. Man and O. Zamani-Khamiri, Nonlinear Optics: Organic and Polymeric Systems in Crystallographically Ordered Polymers; Daniel J. Sandman, ed. (ACS Symp. Series, 1986).

29. Y.M. Cai, H.T. Man, C.C. Teng, O. Zamani-Khamiri and A.F. Garito, J. Opt. Soc. Am. B3, 84 (1986).

30. A.F. Garito and K.Y. Wong, Polym. J. 19, 51 (1987).

31. K.Y. Wong and A.F. Garito, J. Opt. Soc. Am. B3, 184 (1986).

32. Y.M. Cai, H.T. Man, O. Zamani-Khamiri, and A.F. Garito, submitted.

33. Y.M. Cai, H.T. Man, O. Zamani-Khamiri, A.F. Garito, K. D. Singer, and J. Sohn, submitted.

DESIGN, ULTRASTRUCTURE, AND DYNAMICS OF NONLINEAR OPTICAL

EFFECTS IN POLYMERIC THIN FILMS

Paras N. Prasad

Department of Chemistry
State University of New York at Buffalo
Buffalo, New York 14214

INTRODUCTION

The topic of nonlinear optical effects is currently at the forefront of research because of its potential application to optical signal processing and optical computing.[1] Organic systems can be expected to play a significant role because of their relatively large nonresonant optical nonlinearity and subpicosecond response time.[2,3] This paper presents a review of recent work conducted in the author's laboratory and also reports some new previously unpublished results. Our research program on organic nonlinear optics is a comprehensive one which covers the following aspects: (i) Microscopic understanding of optical nonlinearities, (ii) Syntheses of novel nonlinear organic polymers, (iii) Design and molecular engineering of polymeric thin films, (iv) Ultrastructure determination for these films, and (v) Study of nonlinear optical effects.

The focus of our work is on the third order nonlinear optical effect which is characterized by a microscopic second hyperpolarizability, γ, and a corresponding bulk susceptibility $\chi^{(3)}$. The third order nonlinear optical effect can best be understood in the general picture of four wave mixing (FWM).[4] Three input waves $E(\omega_1)$, $E(\omega_2)$, and $E(\omega_3)$ interact in a medium to generate a coherent phase-matched output at $E(\omega_4)$ where

$$E(\omega_4) \; \alpha \; \chi^{(3)}(-\omega_4; \; \omega_1, \; \omega_2, \; \omega_3) \; E(\omega_1) \; E(\omega_2) \; E(\omega_3) \tag{1}$$

The measurement of $E(\omega_4)$ at known input fields $E(\omega_1)$, $E(\omega_2)$ and $E(\omega_3)$ and known polarization yields information about $\chi^{(3)}$, a fourth rank frequency dependent tensor with 81 components. Under the symmetry of the medium, the number of independent nonzero tensor components of $\chi^{(3)}$ can be significantly reduced[4] and probed by polarization selective experiments. Manifestations of the third-order nonlinear optical interactions include a rich variety of phenomena that can be observed. Important third-order effects are: (i) third harmonic generation (THG) and (ii) self-action. In third harmonic generation, $\omega_1 = \omega_2 = \omega_3$ and $\omega_4 = 3\omega$. In other words, one passes a beam of frequency ω through the medium and interaction of three photon fields from the same beam generates a coherent output $E(3\omega)$. The measurement of third harmonic generation can yield $\chi^{(3)}(-3\omega; \omega, \omega, \omega)$. In self-action, all fields are at the same frequency. Two examples of this type of effect are degenerate four wave mixing (DFWM) and intensity dependent refractive index n_2. In DFWM experiments, one generally uses two counterpropagating beams at $E_1(\omega)$ and $E_2^*(-\omega)$; a third beam $E_3(\omega)$ incident at a small angle with respect to $E_1(\omega)$ generates a phase conjugate $E_4^*(-\omega)$ which counterpropagates with respect to $E_3(\omega)$. $E_4^*(-\omega)$ is proportional to $\chi^{(3)}(-\omega; \omega, -\omega, \omega)$. For intensity dependent refractive index, one can visualize that two fields $E(\omega)$ are derived from the same beam to produce a refractive index change $n_2(\omega)$ $E^2(\omega)$ so that

$$n(\omega) = n_0 + n_2(\omega)E^2(\omega) = n_0 + n_2(\omega)I(\omega) \tag{2}$$

and

$$n_2(\omega) = \frac{16\pi^2}{n_0^2 c} \chi^{(3)}(-\omega; \omega, -\omega, \omega) \tag{3}$$

The THG method of determining $\chi^{(3)}$ has the advantage that the measurement is not complicated by other effects. In contrast, $\chi^{(3)}$ measurement by DFWM or nonlinear refractive index determination can be complicated by contributions due to resonant excitations which can create excited state species, thermal effects, and density changes.[5,6] However, the advantage of DFWM is that it allows one to readily obtain the time response of the optical nonlinearity and from that to resolve different contributions as shall be discussed later. A complicating factor in THG is phase matching whereas DFWM is a self-phase matched process. The intensity dependent refractive index measurement has the merit that it also yields the sign of $\chi^{(3)}$. It should be noted, however, that THG yields $\chi^{(3)}(3\omega; \omega, \omega, \omega)$ where as in a self-action effect one measures $\chi^{(3)}(-\omega; \omega, -\omega, \omega)$. If there is a sufficient dispersion induced in $\chi^{(3)}$ due to resonances, $\chi^{(3)}$ obtained by THG and that measured by DFWM may not be the same. Furthermore, in view of the tensor

42

property of $\chi^{(3)}$, it also depends on which component of the tensor is measured.

Compared to second order nonlinear optical effects, the origin of third order optical nonlinearity is not well understood. A simple one dimensional pseudopotential model has been used to explain the large π-electron contribution to $\chi^{(3)}$ for conjugated polymeric systems.[7] In this model, the second hyperpolarizability Y for a conjugated polymer is the largest along the polymer chain direction (z) and is given by

$$Y_{zzzz} \; \alpha \; \frac{e^{10}}{\sigma} \; (\frac{a_o}{d})^3 \; \frac{1}{E_g^6} \tag{4}$$

In this equation, a_o = Bohr's radius, d = average C-C distance, σ = cross-sectional area per chain, and E_g = band gap. The larger is the effective π-electron conjugation, the smaller is the band gap E_g resulting in an increased value for Y_{zzzz}.

For weak intermolecular coupling, one can use an oriented gas model and relate the microscopic nonlinearity Y with the bulk nonlinearity $\chi^{(3)}$ using the Lorentz approximation for the local field correction.[4] In such a case

$$\chi^{(3)}(-\omega_4; \; \omega_1, \; \omega_2, \; \omega_3) \; = \; F(\omega_1)F(\omega_2)F(\omega_3)F(\omega_4)\sum_n \langle Y^n(\theta,\phi)\rangle \tag{5}$$

In this equation $F(\omega_1)$ represents the local field at frequency ω_1. The summation n runs over all molecular sites to give orientational averaging of $\langle Y \rangle$. The two limits are (i) all polymer chains aligned in the same direction in which case $\sum_n \langle Y^n(\theta,\phi) \rangle = NY_{zzzz}$ and (ii) a completely amorphous polymer for which $\sum_n \langle Y^n(\theta,\phi) \rangle = \frac{1}{5} NY_{zzzz}$. In either case, a large value of Y produces a large value for $\chi^{(3)}$ and, therefore, the third order effects are determined primarily by the microscopic nonlinearity provided local field effects are described by the oriented gas model. For conjugated polymeric systems where no strong bulk polarization exists, it may be a good approximation. Interest centers, therefore, on enhancing the microscopic nonlinearity Y. In this regard, our effort has been directed to refining our understanding of microscopic nonlinearities and achieving larger Y by synthesis of novel polymeric systems. We have started a program for ab initio calculation of Y. To investigate the structure-property relationship leading to improved understanding of the relation between Y and $\chi^{(3)}$, our approach has been to design and molecularly engineer novel molecular assemblies. The molecular and microscopic structures of these assemblies as well as their energy level structures are important parameters in understanding the nonlinear optical properties. We probe these features by a variety of spectroscopic techniques. Finally, towards the goal of device application, we are involved in testing the concept of optical switching.

In this article we present some initial results of our ab initio calculation of molecular hyperpolarizability Y. Then our Langmuir-Blodgett work on design of polymeric thin films will be discussed. In relation to ultrastructure determination, only the surface plasmon technique will be discussed. The application of four wave mixing in the determination of $\chi^{(3)}$ will be discussed and results will be presented from our picosecond and subpicosecond FWM experiments. Application of surface plasmon nonlinear optics in determining the intensity dependent refractive index will be discussed. Finally, some recent results on optical switching at a nonlinear interface will be presented.

AB INITIO CALCULATIONS OF SECOND HYPERPOLARIZABILITY

We are evaluating nonlinear second hyperpolarizabilities by ab initio methods. The method involves solution of the non-relativistic time-independent, Schrodinger equation for the molecules in the presence of an applied electric field, utilizing the self-consistent field (SCF) formalism. A brief outline of this method is presented below.

The time independent Hamiltonian for a molecular system in the presence of an electric field $\underset{\sim}{F}$ can be written as

$$\hat{H}(\underset{\sim}{F}) = \sum_{I>J} \frac{Z_I Z_J}{R_{IJ}} - \sum_I Z_I R_I \cdot \underset{\sim}{F} - \sum_i \left[\frac{1}{2} \nabla_i^2 + \sum_I \frac{Z_I}{R_{iI}} - \underset{\sim}{F} \cdot \underset{\sim}{r}_i \right] + \sum_{i>j} \frac{1}{r_{ij}} \qquad (6)$$

where i and j label electrons and I and J label nuclei; Z_I are the nuclear charges; R_I and r_i positions of nucleus I and electron i respectively.

In the presence of an electric field and ignoring configuration interaction, the electronic configuration can be expressed as an antisymmetrized function (a single Slater determinant) of molecular orbitals $\phi_k(\underset{\sim}{F})$

$$\Phi(\underset{\sim}{F}) = A[\pi \, \phi_k(\underset{\sim}{F})] \qquad (7)$$

The molecular orbitals $\phi_k(F)$ are so chosen as to minimize the total energy of the system and can be expressed in terms of a linear combination of nuclear centered atomic orbital basis functions χ_σ as follows

$$\phi_k(\underset{\sim}{F}) = \Sigma_\sigma \chi_\sigma C_{\sigma k}(\underset{\sim}{F}) \tag{8}$$

The self-consistent field method is now used to optimize these molecular orbitals to obtain the coefficients $C_{\sigma k}(\underset{\sim}{F})$. The molecular energy $\varepsilon(\underset{\sim}{F}) = \langle \Phi(\underset{\sim}{F}) | \hat{H}(\underset{\sim}{F}) | \Phi(\underset{\sim}{F}) \rangle$ and dipole moment are calculated as a function of the applied electric field, $\underset{\sim}{F}$. For low field strengths, the energy and the dipole moment can be described by a Taylor series expansion in the electric field:

$$\varepsilon(\underset{\sim}{F}) = \varepsilon^o - \Sigma_i \mu_i^o F_i - \frac{1}{2} \Sigma_{ij} \alpha_{ij} F_i F_j - \frac{1}{3} \Sigma_{ijkj} \beta_{ijk} F_i F_j F_k - \frac{1}{4} \Sigma_{ijkl} \gamma_{ijkl} F_i F_j F_k F_l \tag{9}$$

and

$$\mu_i = \mu_i^o + \Sigma_j \alpha_{ij} F_j + \Sigma_{jk} \beta_{ijk} F_j F_k + \Sigma_{jkl} \gamma_{ijkl} F_j F_k F_l \tag{10}$$

For a symmetrically substituted molecule, μ_o and β drop out. Polarizabilities (α) and second order hyperpolarizabilities (γ) are then calculated as derivatives of the field dependent energy and dipole moment.

Two approaches can be followed:

(i) Apply a finite field, solve the SCF equation, and obtain $\varepsilon(F)$ and $\mu(F)$; by finite differencing of numerical derivatives obtain α and γ.

(ii) Analytically compute the polarizability, this involves solutions of first order coupled perturbed H-F (CPHF) equations. Hyperpolarizability calculations require solving higher order CPHF equations.

We have initially used the first approach. The details of the calculation will be published elsewhere.[8] Here only the preliminary results of calculations using minimal basis sets, STO-3G ($1s_H$, $1s_C$, $2s_C$, $2p_C$), will be presented for the series $[-C \equiv C-]_n$, which in the infinite limit would yield the conjugated polymer trans-polyacetylene. The results of the calculation show that γ term is highly anisotropic, the largest component being along the chain direction (γ_{zzzz}). The magnitude of γ_{zzzz} increases rapidly in comparison to α as the chain length increases. The values are

MOLECULE	γ_{zzzz}			
C_2H_4	-83 au	=	-4.2×10^{-38}	esu
C_4H_6	$+530$ au	=	$+2.9 \times 10^{-37}$	esu
C_6H_8	$+5000$ au	=	$+2.6 \times 10^{-36}$	esu
C_8H_{10}	$+18000$ au	=	$+9.0 \times 10^{-36}$	esu
$C_{10}H_{12}$	$+52000$ au	=	$+2.8 \times 10^{-35}$	esu

As one notices, there is a change of sign from negative to positive γ with an increase in conjugation length.

The closed shell quantum mechanical wavefunction is invariant with respect to unitary transformations within the closed and virtual molecular orbital spaces. We define an appropriate transformation such that a given electric field mixes each occupied orbital with one and only one virtual orbital. These are called corresponding orbital pairs. The total molecular polarization is then expressed exactly as a sum of independent contributions from each pair. In this way one can separate out the contributions of the sigma and π orbitals, or any subsets of these. Our results of orbital transformation analysis show that the π-orbitals make dominant contribution to $\chi^{(3)}$ for conjugated π-systems. As the effective conjugation increases, the π contribution becomes increasingly more important. The calculations described above yield the static hyperpolarizabilities i.e., a frequency independent γ. This value should qualitatively represent the nonresonant behavior of γ.

DESIGN AND MOLECULAR ENGINEERING OF POLYMERIC THIN FILMS

For understanding structure-property relationship, an important approach is to create molecular assemblies in which the order and orientation of molecules can be varied. One of the approaches used in our laboratory is the Langmuir-Blodgett method by which one can accomplish the variation of molecular assemblies with a monomolecular resolution. This method also offers one the opportunity to build multilayer films for integrated optical device applications. Polydiacetylene films have been formed using this method.[9] In general, we have used an amphiphilic diacetylene monomer, and compressed the monolayer film to obtain a condensed film.[10] The film was then polymerized by a U.V. light and subsequently transferred to a solid substrate. Alternatively, the monomer film was transferred to the solid substrate and subsequently polymerized. In our experience, these films were not uniform because of the strain developed during polymerization. For this reason we extended our Langmuir-Blodgett work to soluble polydiacetylenes. We found that some of these soluble symmetric polydiacetylenes also formed rigid monolayer films.[11] Furthermore, these films showed conformational transitions in the monolayer state. These results will be discussed below.

The specific soluble polydiacetylenes reported here are commonly known as poly-3-BCMU and poly-4-BCMU. These are urethane substituted symmetric polydiacetylenes; poly-3-BCMU has one methylene shorter side group chain than poly-4-BCMU. Both polymers form condensed rigid monolayers. The surface pressure to molecular area per residue isotherms for poly-3-BCMU and poly-4-BCMU are shown in Figure 1. In the low surface pressure region the film is yellow. But when compressed to an area per residue of ~100A^2, a conformational transition from a less conjugated yellow form to a more conjugated red form in the case of poly-4-BCMU and blue form in the case of poly-3-BCMU occurs. This conformational transition is represented by a plateau. Our surface plasmon study to be discussed latter indicates that the less conjugated form is a monolayer film whereas the more extended form is actually a bilayer film. The third harmonic generation from this monolayer film of poly-4-BCMU as a function of the surface pressure shows a rapid increase in the third harmonic signal in going from a less conjugated yellow form to a more conjugated red form.[12] This result confirms the strong dependence of the third order nonlinearity on the π-electron conjugation. This work is described in detail in the article by Berkovic and Shen. The value of bulk $\chi^{(3)}$ calculated from the microscopic nonlinearity γ obtained from this monolayer study is in agreement with the measurement on the bulk system lending support to the assumption of an oriented gas model.

Films have successfully been transferred onto a solid substrate by a horizontal dipping procedure. These films in the monolayer and bilayer states were characterized by various spectroscopic techniques and shown to differ in the effective π-electron conjugation. Attempts to make multilayer

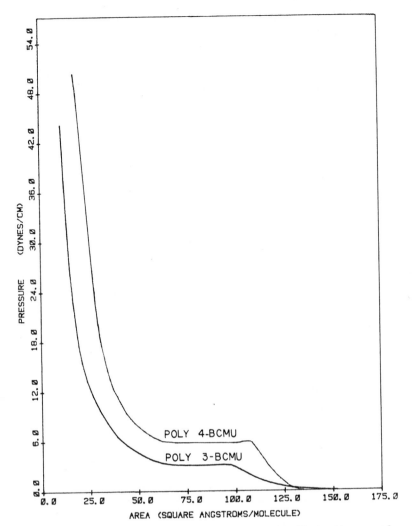

Fig. 1. Surface Pressure-Area per repeat unit isotherms of
poly-3-BCMU and poly-4-BCMU polydiacetylenes at 20°C.
A plateau in each case indicates a phase transition.

yellow films didn't succeed. Films transferred before the plateau (Figure
1) were yellow only in the monolayer (single dip) form. A successively
dipped multilayer film produced a red form for poly-4-BCMU and a blue form
for poly-3-BCMU. In our laboratory, we have deposited a 400 layer thick
film and examined its uniformity and optical quality. At low surface
pressure (< 10 dynes/cm) the transferred multilayer films are of excellent
optical quality. We conclude that this excellent quality results from
directly using the soluble polymer to form the Langmuir-Blodgett films as

opposed to using a monomer and then polymerizing it. We are in the process of testing the multilayer films of poly-3-BCMU and poly-4-BCMU for integrated optics studies.

ULTRASTRUCTURE CHARACTERIZATION

An important prerequisite for any study of nonlinear optical effects in polymeric films is characterization of the molecular structure, conformation and domain structure of the film. We use a variety of spectroscopic and structural techniques to accomplish this goal. In this paper only the surface plasmon method will be discussed. This method can be used to determine linear and nonlinear refractive index of monolayer and multilayer films. Furthermore, the method can provide useful information on the chemical structure and conformation of organic films deposited on a metal. Practical electronic devices would involve deposition of organic films on metals. However, an organic molecule (especially a polymeric unit) may change its structure and conformation on a metal surface due to a specific chemical interaction between the metal and the organic structure. In the surface plasmon geometry, we have successfully studied vibrational Raman spectra of monolayer and multilayer L-B films of polydiacetylenes deposited on metals.

Surface plasmons are electromagnetic waves which propagate along the interface between a metal and a dielectric material (such as a Langmuir-Blodgett film of an organic material).[13] Since the surface plasmons propagate in the frequency and wave vector ranges for which no propagation is allowed in either of the two media, no direct excitation of surface plasmons is possible. The most commonly used method to generate a surface plasmon wave is known as attenuated total reflection.[13]

In our laboratory, the Kretschmann configuration of attenuated total reflection has been used to couple surface plasmons. This configuration is shown in Figure 2. A microscopic slide is coated with a thin film of metal (usually 400-500 A thick silver film by vacuum deposition). Then the L-B film is coated on the metal surface. The microscopic slide now is coupled to a prism through an index matching fluid. For linear effects (nonlinear effects described latter) a low-power laser beam (HeNe) is incident at the prism at an angle larger than the critical angle. The total attenuated reflection of the laser beam is monitored. At a certain angle $\theta_{s.p.}$, the

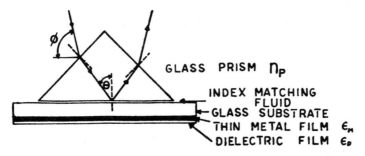

GLASS PRISM n_p

INDEX MATCHING FLUID

GLASS SUBSTRATE

THIN METAL FILM ϵ_m

DIELECTRIC FILM ϵ_D

Fig. 2.   The experimental arrangement for surface plasmon
experiments using Kretchmann configuration.

electromagnetic wave couples to the interface as a surface plasmon.   At this
angle the ATR signal drops.   The angle is determined by the relationship[13]

$$k_{s.p.} = k n_p \sin \theta_{sp} \tag{11}$$

In this equation $k_{s.p.}$ is the wave vector of the surface plasmon, k is the
wave vector of the bulk electromagnetic wave and n_p is the refractive index
of the prism.   The surface plasmon wave vector $k_{s.p.}$ is given by[13]

$$k_{s.p.} = \frac{\omega}{c} \left(\frac{\epsilon_m \epsilon_d}{\epsilon_m + \epsilon_d} \right)^{1/2} \tag{12}$$

where ω is the optical frequency, c, the speed of light, ϵ_m and ϵ_d are the
dielectric constant of the metal and the dielectric respectively.   In the
case of a bare silver film, ϵ_d is the dielectric constant of air and the dip
in reflectivity occurs at one angle.   In the case of silver coated with an
L-B film this angle shifts.   The result for a bare silver film and that for
a silver film coated with a monolayer L-B film of a polydiacetylene, poly-4-
BCMU, obtained in our experimental arrangement is shown in Figure 3.   A
significant shift in the resonance dip caused by the L-B film deposition
clearly demonstrates monolayer sensitivity of this method.   In this
experiment one measures the angle for the reflectivity minimum, the minimum
value of reflectivity, and the width of the resonance curves.   We use these
observables for a computer fit of the resonance curve using a least squares
fitting procedure with the Fresnel reflection formulas which yields three
parameters: real and imaginary part of the refractive index and the film

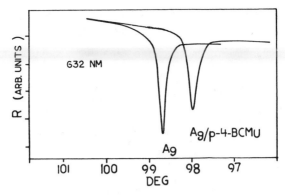

Fig. 3. The surface plasmon resonance dip in the
reflectivity obtained for a bare silver film
and that for a silver film coated with a monolayer L-B film of poly-4-BCMU are compared.

thickness. The surface plasmon study was conducted[14] for the poly-3-BCMU
and poly-4-BCMU Langmuir-Blodgett films deposited at two different points on
their isotherm curves shown in Figure 1. The first film was transferred at
a surface pressure before the plateau and the other film was transferred at
a surface pressure at the end of the plateau (where a change of conformation
is complete). For poly-4-BCMU, the plateau involves a change of
conformation from the yellow form to the red form. The surface plasmon
results for the yellow film transferred below the plateau yields 12Å for the
thickness and 1.5010 for the refractive index (real). For the film
transferred at the end of the plateau, the results are fitted by the
assumption that the film is bilayer. The first layer is in the yellow
conformation with thickness ~12Å and refractive index 1.5010; the second
layer is in the red conformation with thickness ~12Å and refractive index
1.5412. Our u.v. visible spectral study also confirms that the bilayer film
is actually a combination of the yellow and red forms.

In order to study if the films were uniform, the surface plasmon
resonance was studied as a function of micrometer translation of the film.
We didn't observe any shift in the angle for the resonance dip which
confirmed that within the resolution of the beam spot size (~20μ), the films
were uniform.

We have also successfully used the surface plasmon configuration to obtain Raman spectra of monolayer and multilayer L-B films. The experimental arrangement is the same as above except that the scattered light from the film is collected and analyzed by a Raman spectrometer. Enhanced Raman scattering is observed at an angle at which the electromagnetic wave propagates as surface plasmons (dip in reflectivity). The Raman spectra of poly-4-BCMU in the region of the double bond are shown in Figure 4. The spectrum of the yellow (monolayer) form deposited before the plateau has the -C=C- vibrational frequency of 1529 cm^{-1}. The vibrational frequency of the film transferred after the plateau is shifted to a lower value confirming the presence of a more conjugated (red) form.

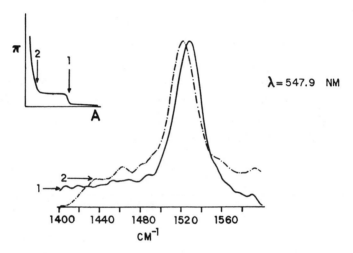

Fig. 4.   The Raman spectra of poly-4-BCMU L-B films transferred at two distinct points (1 and 2) of the isotherm (shown at left top) are obtained for the -C=C- region using the surface plasmon geometry.

FOUR WAVE MIXING STUDIES OF $\chi^{(3)}$

The four wave mixing process can be best described by using the general picture of a laser-induced dynamic grating. Two coherent beams $I_1(\omega_1)$ and $I_3(\omega_3)$ crossing at an angle θ set up an intensity grating in the material.

As a result of various resonant and/or nonresonant processes occuring in the material, a refractive index grating results. A time-delayed beam $I_2(\omega_2)$ is diffracted from this grating. The three input beams (I_1, I_2, and I_3) and the Bragg diffracted output beam, $I_4(\omega_4)$, constitute the four wave mixing process. This process is described by the grating efficiency $\eta = I_4/I_3$ which is given as[6]

$$\eta^{(\lambda)} \alpha \; (\frac{\partial n}{\partial x})^2 \; g(\lambda)^2 \; I^{2\beta}(\lambda)f^2(t,\theta) \tag{13}$$

For a resonant process producing material excitation (excited states, charge carriers etc.), x relates to the excitation density and $g(\lambda)$ is the quantum yield (photogeneration efficiency at wavelength λ. The term f is a function of the crossing angle θ and time delay t between the pump and probe pulses. The exponent β depends on the nature of the photon process. If the resonant excitation is generated by a one-photon process, $\beta=1$ for $I_1=I_2$. Therefore, determination of β from the dependence of diffraction efficiency on the pulse power I, gives information on the nature of photon processes creating material excitation. The time evolution of the grating provides information on the build up and decay of the excitation. The decay due to spatial migration of excitation will be dependent on the grating spacing and, therefore, on the crossing angle θ. Therefore, an angle dependence study of the grating decay will provide information on the nature of excitation migration (dispersive vs diffusive, etc).

For nonresonant processes, $\frac{\partial n}{\partial x}$ simply relates to the intensity dependent refractive index n_2 as discussed above. The intensity dependent refractive index n_2, derived from the π-electron nonlinearity in conjugated polymers, makes the largest non-resonant electronic contributions and has the fastest response time. In liquid crystals, an orientational nonlinearity can also make a significant non-resonant contribution, but its response is considerably slow.

In the case of resonant processes, local nonradiative relaxations produce local heating which results in change of temperature and also launches density waves leading to counterpropagating ultrasonic waves of wave length equal to the grating spacing.[6] Because of these two effects, diffracted signals are also observed where $\frac{\partial n}{\partial x}$ is $\frac{\partial n}{\partial T}$ and $\frac{\partial n}{\partial \rho}$. Table I summarizes various resonant and non-resonant processes which give rise to four wave mixing. Listed also are the time-response of these processes. In order to separate these contributions, one requires ultrashort pulses of several picoseconds or less. In our laboratory we can select pulses of

TABLE I

Different Mechanism of Four Wave Mixing

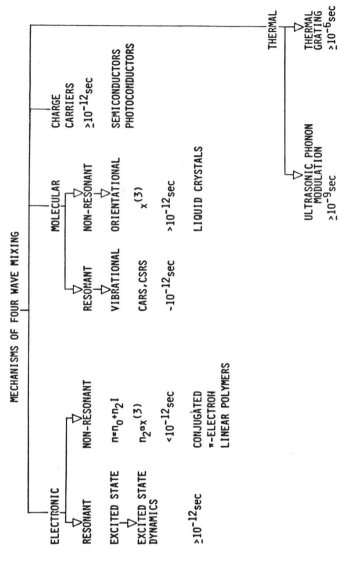

MECHANISMS OF FOUR WAVE MIXING

ELECTRONIC

RESONANT — NON-RESONANT

EXCITED STATE
EXCITED STATE DYNAMICS

$n = n_0 + n_2 I$

$n_2 \alpha \chi^{(3)}$

$>10^{-12}$ sec $<10^{-12}$ sec

CONJUGATED
π-ELECTRON
LINEAR POLYMERS

MOLECULAR

RESONANT — NON-RESONANT

VIBRATIONAL ORIENTATIONAL

CARS, CSRS $\chi^{(3)}$

$\sim 10^{-12}$ sec $>10^{-12}$ sec

LIQUID CRYSTALS

CHARGE CARRIERS

$\geq 10^{-12}$ sec

SEMICONDUCTORS
PHOTOCONDUCTORS

THERMAL

ULTRASONIC PHONON MODULATION

$\geq 10^{-9}$ sec

THERMAL GRATING

$\geq 10^{-6}$ sec

width varying from 40 picosecond to 600 femtoseconds. Two separate experimental arrangements are used. In one arrangement, a CW mode-locked Nd:Yag laser (Spectra-Physics Model 3000) is frequency-doubled and used to sync-pump two dye lasers (Spectra-Physics Model 375). The pulses from the dye lasers are subsequently amplified in two separate 3 stage amplifiers (Spectra-Physics Model PDA) which are pumped by the frequency doubled output of a 30 Hz pulsed Nd:Yag laser (model DCR-2A). The resulting pulses are typically 4 ps wide with pulse energy of ~0.5 mJ. For degenerate four wave mixing (DFWM), the output from only one pulse dye amplifier is split in three ways and used in the backward wave geometry as shown in Figure 5. For

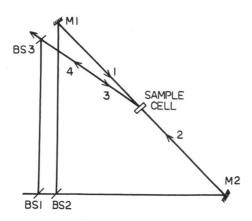

Fig. 5. Schematics of degenerate four wave mixing (DFWM)
in a backward wave geometry are displayed.

subpicosecond work, a saturable absorber is added to the dye in the sync-pumped dye laser. Pulses produced are now shortened to ~600 femtoseconds. To minimize broadening in the amplifier, a saturable absorber jet is now inserted in the place of a spatial filter between the second preamplifier stage and the final amplifier stage in the PDA amplifier. The best resulting pulses are around 700 femtoseconds. For coherent vibrational Raman spectroscopy (CARS and CSRS), two different frequencies ω_1 and ω_2 are required such that $\omega_1 - \omega_2$ is equal to a vibrational frequency ω_R. This experiment requires output from both amplifiers.

For experiments which do not require extremely short pulses and very high peak power, we use a different experimental arrangement in which a CW mode-locked Q-switched Nd:Yag laser with the frequency doubled output pumps dye lasers which are cavity dumped. The resulting output consists of ~40 ps pulses of energy 15 uJ at a repetition rate of ~500 Hz (the Q-switching rate). The advantage of this system is that it is more stable than the amplified system and that its high repetition rate allows a more rapid data collection.

Using DFWM, we have investigated a wide variety of polymers. First, a brief review of the published work will be presented. Then some recent results will be discussed. The DFWM experiments were performed in a backward wave geometry as discussed above. Unless otherwise indicated, we measured only the $\chi_{1111}^{(3)}$ components of the $\chi^{(3)}$ tensor where all the laser beams were vertically polarized. The $\chi^{(3)}$ measurements are made in relation to CS_2 as the reference material. Under the condition that the signal I_4 is much less than the beam intensity I_3, we have[5]

$$ I_4 \; \alpha \; G^3 \; \frac{I_1 I_2 I_3}{n_o^4} \; 1^2 [\chi^{(3)}]^2 \tag{14} $$

In the above equation, 1 is the interaction length in the sample, n_o is the linear refractive index, and G is a factor which corrects for reflection and scattering losses in the sample.

The soluble polydiacetylenes, specifically poly-4-BCMU were investigated by DFWM using wavelengths of 585 and 605 nm to obtain the $\chi^{(3)}$ values. This study provided the following important results:[3] (i) the magnitude of $\chi^{(3)}$ changed from 10^{-10} esu in the red form to 2.5×10^{-11} esu in the yellow form, establishing a strong dependence of $\chi^{(3)}$ on the π-electron conjugation; (ii) the time response was found to be in subpicoseconds, limited only by the pulse width indicating that the measured $\chi^{(3)}$ value was for a non-resonant electronic process; (iii) upon rotation of the film, the $\chi^{(3)}$ value did not change indicating that the film behaved isotropically.

Another class of polymer investigated is a conjugated aromatic heterocyclic polymer, poly-p-phenylene benzobisthiazole (PBT)[15] which has a very high mechanical strength due to its rigid rod conformation. Furthermore this polymer is also environmentally extremely stable and has a very high laser damage threshold. A 10μ thick as-spun biaxial film of PBT was studied by DFWM at wavelengths of 585 nm and 605 nm. The measured value of $\chi_{1111}^{(3)}$ is ~10^{-11} esu at both 585 nm and 604 nm. Because the PBT film is biaxial, the DFWM signal (I_4) was found to be dependent on the orientation of the PBT film. This anisotropic behavior has also been observed for other biaxial polymers and will be discussed below in detail. Also for PBT, a subpicosecond response time limited by the laser pulse width is observed. Another group of polymer is a polyimide which also has a very high mechanical stability. One specific polymer, commonly known as LARC-TPI, has the following molecular structure.

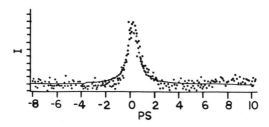

LARC-TPI

This polymer does not have a conjugated π-electron structure. It is yellow in color having a band gap close to that of PBT. However, the lowest lying transition in LARC-TPI corresponds to an intermolecular charge transfer band where as that in PBT corresponds to the π-π* transition (conjugation effect). Consequently, the value of $\chi^{(3)}_{1111}$ for LARC-TPI is found to be more than an order of magnitude less than that of PBT. This result again emphasizes the importance of π-electron conjugation. The DFWM signal observed for LARC-TPI as a function of delay of the backward beam is shown in Figure 6. Again, a subpicosecond response limited by the laser pulse width is observed. The $\chi^{(3)}$ value obtained on a biaxial film of LARC-TPI is again found to be dependent on the orientation of the film. This orientational anisotropy of $\chi^{(3)}$ is shown in Figure 7. This observed anisotropy can be explained by the fourth rank tensor behavior of $\chi^{(3)}$ as discussed below. Here we assume the highest symmetry for a biaxial system i.e. orthorhombic. The third order optical susceptibility tensor $\chi^{(3)}$ in the laboratory framework (defined by the polarization vectors of the radiation field) is related to the film based $\chi^{(3)}$ as follows:

Fig. 6. The degenerate four wave mixing signal observed for LARC-TPI is displaced as a function of the backward beam delay.

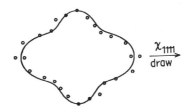

Fig. 7. The observed orientational anisotropy of
$\chi^{(3)}$ for LARC-TPI. The circles are the
observed data point; the solid curve is
the theoretical fit.

$$\chi_{ijk\ell}^{'(3)} = \Sigma A_{im} A_{jn} A_{kp} A_{\ell q} \chi_{mnpq}^{(3)} \tag{15}$$

In equation (15), A's are the elements of the transformation matrix.
In the case of all vertical polarizations, the field $E_4^{(1)}$ is given as:

$$E_4^{(1)} \, \alpha \, \chi_{1111}^{'(3)} E_1^{(1)} E_2^{(1)} E_3^{(1)} \tag{16}$$

For the orthorhombic in-plane symmetry one obtains:

$$\chi_{1111}^{'(3)} = \chi_{1111}^{(3)} \cos^4 \theta + \chi_{2222}^{(3)} \sin^4 \theta$$

$$+ (\chi_{1122}^{(3)} + \chi_{1212}^{(3)} + \chi_{1221}^{(3)} + \chi_{2122}^{(3)} + \chi_{2211}^{(3)} + \chi_{2112}^{(3)}) \sin^2\theta \cos^2\theta \tag{17}$$

In the above equation, θ is the angle of rotation of the film. For data
fitting, equation (17) can be rewritten as:

$$\chi^{'(3)} = A_1 \cos^4 \theta + A_2 \sin^4 \theta + A_3 \sin^2 \theta \cos^2 \theta \tag{18}$$

The observed anisotropy is well fitted by equation 18. The solid curve in
Figure 7 is the theoretical fit.

We have investigated the effect of both electronic and vibrational
resonances. For electronic resonance, an organic photoconductor namely
poly-N-vinyl carbazole (PVK): trinitrofluorenone (TNF) complex in various
stoichiometries was investigated. This organic polymeric photoconductor has

been a subject of considerable interest because of its application in photocopying.[16] The 602 nm laser line provides excitation into the tail of the charge-transfer band of this complex. The photophysics of the process can be visualized as the one creating a charge-transfer exciton which thermally relaxes to produce an electron-hole pair separated by the thermalization length. These carriers can subsequently move apart as free carriers or can undergo geminate recombination. The DFWM signal obtained for the complex with the stoichiometry 8PVK:TNF is shown in Figure 8. It can be seen that the signal has a long decay (~220 picoseconds). Furthermore, at longer delays the ultrasonic phonon modulation due to density wave excursion can be seen. The effective $\chi^{(3)}$ value calculated for this system is ~10^{-11} esu. A power dependence study of the zero time signal indicates the efficiency of the four wave mixing signal to be dependent on the square of the intensity i.e. $\beta=2$ for equation 13. Hence, the process involves a one-photon resonance. This work shows that even though a high $\chi^{(3)}$ can be achieved by electronic resonance, the response time is slow and further complication due to ultrasonic phonon generation can lead to long time ringing (over tens of nanoseconds as shown in Table I).

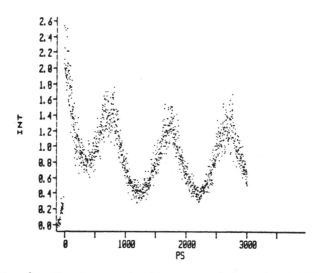

Fig. 8. The degenerate four wave mixing signal obtained
for the complex PVK:TNF with the stoichiometry
8:1 is displayed as a function of the background
beam delay.

Vibrational resonance can also increase $\chi^{(3)}$ of a system. Vibrational resonance of $\chi^{(3)}$ can be mapped out in a coherent Raman scattering experiments such as CARS and CSRS. Thermal and density effects generated by Raman scattering (vibrational resonance) can be expected to be much smaller than that observed for an electronic resonance. Time response of vibrationally resonance enhanced $\chi^{(3)}$ is determined by vibrational dephasing which can be measured by time-resolved coherent Raman spectroscopy experiments. Although, frequency resolved CARS and CSRS experiments have been conducted for polymers both in forms of films and solutions,[17,18] to our knowledge no time-resolved study of these processes in polymeric systems has been reported.

We have conducted both time-resolved and frequency domain CARS and CSRS studies of a polymeric system to investigate vibrational dephasing processes.[19] The system selected for this study was poly-4-BCMU. This conjugated polymeric system exhibits a large $\chi^{(3)}$ as determined by degenerate four wave mixing described above.

For CARS and CSRS experiments, pulses ω_1 (17076 cm^{-1}) and ω_2 (15544 cm^{-1}) were used from the two amplifiers. The signals were detected at $\omega_3 =$

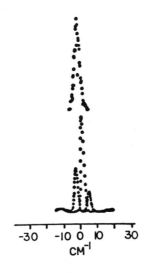

-30   -10 0 10   30
CM^{-1}

Fig. 9. The CARS signal for the C=C stretching band of poly-4-BCMU red films at room temperature is plotted as a function of the frequency detuning $\delta = \omega_R - (\omega_1 - \omega_2)$.

$2\omega_1 - \omega_2$ in case of CARS, and at $\omega_3 = 2\omega_2 - \omega_1$ in case of CSRS. For time-resolved experiments probing vibrational dephasing, the beam ω_1 was split in two for CARS; one was used as the pump beam and the other underwent a variable time delay to be used as the probe beam.

In Figure 9, the CARS signal is shown as a function of the detuning parameter $\delta = \omega_R - (\omega_1 - \omega_2)$ where ω_R is the Raman vibrational frequency for the -C=C- stretching mode in the poly-4-BCMU red form. At $\delta=0$, which corresponds to the vibrationally resonant four wave mixing, the signal is about two orders of magnitude larger than the non-resonant case ($\delta\neq0$). Figure 10 shows the time resolved studies of CARS spectra again for the -C=C- stretching mode of the red form of poly-4-BCMU at 4K. The decay of the vibrationally resonance enhanced four wave mixing signal is extremely fast even at 4K and is comparable to our cross correlation of the ω_1 and ω_2 pulses. A detailed analysis of the time-resolved and frequency resolved CARS and CSRS study of this vibration in poly-4-BCMU suggest that the vibrational dephasing is induced by an inhomogeneous mechanism due to the spread of the vibrational resonance resulting from a distribution of the π-electron conjugation.[19]

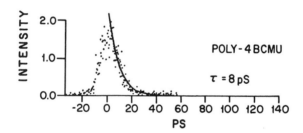

Fig. 10.   Decay of vibrational coherence for the -C=C-
stretching mode of poly-4-BCMU red form at
4K in the picosecond time-resolved CARS study.

SURFACE PLASMON NONLINEAR OPTICS

The surface plasmon coupling geometry discussed eariler can be used to investigate nonlinear optical effects in monolayer and multilayer L-B films. The $\chi^{(3)}$ values can be obtained from intensity dependence study of the refractive index. The experiment utilizes the study of angular shift (change in the reflectivity dip angle θ) and broadening of the surface plasmon resonance as a function of intensity of the incident laser beam. As discussed in the earlier section, the angle θ at which the dip in ATR signal occurs provides information on the dielectric constant, hence the refractive index, of the film. Therefore, intensity dependence of this angle gives the magnitude of intensity dependent refractive index n_2 and, consequently, $\chi^{(3)}$. An important advantage of this method is that one can also obtain the sign of n_2 (and hence $\chi^{(3)}$).

Combining equations (11) and (12) discussed above one can see that the change δθ in the surface plasmon resonance angle (angle corresponding to minimum reflectivity, for simplicity the subscript s.p. is dropped here) caused by changes $\delta\varepsilon_m$ and $\delta\varepsilon_d$ in the dielectric constants of the metal and film, respectively, is given by[20]

$$\cot\theta \; \delta\theta = \frac{1}{2\varepsilon_m \varepsilon_d (\varepsilon_m + \varepsilon_d)} (\varepsilon_m^2 \delta\varepsilon_d + \varepsilon_d^2 \; \delta\varepsilon_m) \tag{19}$$

Since $|\varepsilon_m| \gg |\varepsilon_d|$, the change in θ is much more sensitive to a change in ε_d than to a change in ε_m. Therefore, this method appears to be ideally suited to obtain $\delta\varepsilon_d$ as a function of laser intensity and hence obtain n_2. In actual experiment, however, utmost care is needed to interpret the data and obtain meaningful results. The reason is that for the visible wavelength region (~632 nm), the surface plasmon propagation is of the order of 10 μm in which the wave is completely damped. Therefore, even though the polymer films may be optically clear at this wavelength, the damping in the metal leads to heating. The $\delta\varepsilon_d$ and $\delta\varepsilon_m$ in such a situation have two contributions: one due to intrinsic $\chi^{(3)}$ and the other due to thermal effect ($\frac{\partial n}{\partial T}$ or $\frac{\partial \varepsilon}{\partial T}$). Further complication may arise if $\frac{\partial \varepsilon_d}{\partial T}$ and $\frac{\partial \varepsilon_m}{\partial T}$ have different signs. In such a case, depending on what contribution dominates, one may see a positive shift (δθ positive) or a negative shift as the laser beam intensity is increased. To determine the effect of $\delta\varepsilon_m$ itself, study was conducted with the slide coated with metal alone and with metal on which a thin film of polystyrene (having very small $\chi^{(3)}$) was deposited. Intensity range of the laser was selected for which very little shift was observed. Next, the role of $\frac{\partial \varepsilon_d}{\partial T}$ was determined by using longer pulses such as that from a CW chopped dye laser. The thermal effect is predominant when using low peak power longer pulses. For poly-4-BCMU, this shift was found to be in the negative direction indicating that $\frac{\partial n}{\partial T}$ is negative. Now we repeated our experiment on a spin coated poly-4-BCMU film using picosecond dye pulses (40 picosecond) from the Q-switched mode-locked Nd:Yag pumped dye laser. The observed shift $\Delta\theta_{sp}$ ($=\delta\theta$) as a function of the laser intensity is shown in Figure 11. One initially sees a positive shift but beyond

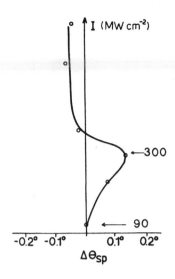

Fig. 11. The observed shift in the surface plasmon
resonance angle $\theta_{s.p.}$ for the poly-4-BCMU
film as a function of the laser intensity.

certain intensity level, the shift now moves to the negative direction.
This result emphasizes the complication due to thermal effects and that one
has to be extremely careful in the interpretation of the intensity dependent
surface plasmon experiment. Our tentative interpretation is that the
initial positive shift corresponds to the $\delta\varepsilon_d$ value derived from the
intrinsic $\chi^{(3)}$ of poly-4-BCMU. But with the increasing laser power, the
thermal effect takes over and the $\frac{\partial\varepsilon}{\partial T}$ effect produces a negative shift. For
the region where $\Delta\theta_{sp}$ is positive, we have calculated $\chi^{(3)}$ of poly-4-BCMU.
We find that calculated $\chi^{(3)} > 10^{-10}$ esu. This value is higher than what
has been determined by DFWM (discussed above) and by THG. We feel that a
portion of the positive shift $\Delta\theta_{sp}$ may be derived from the $\frac{\partial\varepsilon_m}{\partial T}$ term which
for silver is positive.[20]

OPTICAL SWITCHING

The concept of optical switching is based mainly on utilizing the
nonlinear refractive index. One kind of optical switching is optical

bistability which has drawn considerable attention for its promises in optical signal processing.[21] For optical bistability, one couples the nonlinear response with some mechanism of feedback to create two stable optical output states for the same input level within a given range of intensity. On increasing the input power to within this range, a sharp switching occurs at a given input level. However, in the reverse cycle i.e. on reducing the power level, a sharp switching occurs at a different input power level. Majority of optical bistability studies have been conducted with a nonlinear Fabry-Perot cavity.[21] The response time of such devices is limited by the cavity build up time.

In our approach we have used the concept of a planar leaky quasi-waveguide involving a nonlinear interface to study optical switching. A planar waveguide consists of a thin (~1 μm) film sandwiched between two cladding media (such as air and glass substrate) with the condition that the refractive index n_f of the film is higher than the refractive indices n_{c_1} and n_{c_2} of the two cladding media.[22] In such a case, the electromagnetic wave coupled into the wave guide is confined within it. Guided modes can also be supported even when $n_{c_1} < n_f < n_{c_2}$. However, in this case, their propagation is attenuated along the waveguide by leakage at the n_f and n_{c_2} interface. This type of wave guide is called a leaky quasi-wave guide.[23]

Modes of quasi-wave guides can be visually observed through so called m-lines. These lines result from in-plane scattering of light out of one excited modes into other directions of the same mode or into other modes. In the past, these m-lines of a leaky quasi-waveguide have been used to measure the refractive index and the thickness of thin films.[23] In our work we have used this type of wave guide to investigate optical switching.

A 1 μm thick film of polyphenylacetylene (PPA) was deposited on the base of a prism coupler of refractive index 1.646. The refractive index of PPA, determined by the m-line method, was found to be 1.639 at 590 nm (the wavelength chosen for this study). Therefore, this film at low power would work as a leaky quasi-waveguide, since the refractive index n_f of the film is lower than that of the prism (n_{c_2}; the other cladding medium is air for which $n_f > n_{c_1}$).

PPA is a nonlinear material. Therefore, $n_f = n_o + n_2 I$. If n_2 is positive, then increasing I would increase n_f (due to the evanescent field).

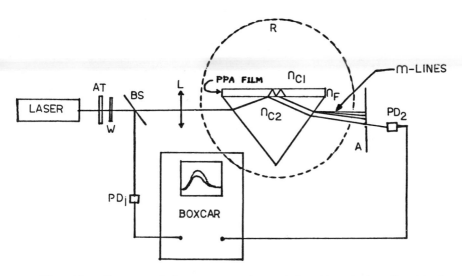

Fig. 12. Experimental arrangement used for the study of optical switching in a quasi-waveguide of the polyphenylacetylene film.

At some power level, $n_f > n_{c_2}$ at which point it would act as a waveguide and coupling of the electromagnetic wave is increased. Now in the reverse cycle, due to the beam confinement in the waveguide, the local field remains high and a reverse switching occurs at a lower intensity.

Our experimental arrangement is shown in Figure 12. For this experiment, a nanosecond laser source was used. It consists of a pulsed Nd:Yag laser (Quantel model 481) which pumps a dye laser (Quanta Ray). The dye output at 590 nm was used and injected into the film at a glancing angle of 5.32°, which is near the coupling angle for the m=2 line. Pulse shapes for the incident light, leaky mode (m=2 line), and the directly reflected light were monitored using photodiodes with a Boxcar integrator. From the pulse shape correlation, the relation between the output and input intensities were obtained. Figure 13 shows the intensity behavior of the directly reflected light. At higher input intensities, a decrease in the reflected light is indicative of increased coupling as a guided wave. In reducing the power, this switching occurs at a lower input level giving rise to hysteresis. The intensity of the leaky mode (m=2 line) shows a complementary behavior where the intensity of the line switches to a higher value at an input level at which the intensity of the reflected light drops,

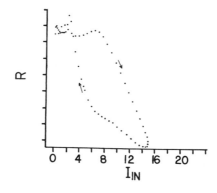

Fig. 13. The observed switching and hysteresis
for the directly reflected light from
the prism and polyphenylacetylene film
interface when plotted against the
input intensity.

again confirming an increased coupling as a guided wave. Also, for the
leaky mode, hysteresis is observed. However, the switching behavior in both
cases is not fast but shows a slow integrating behavior which is indicative
of an integrating nonlinearity derived from thermal effects. To separate
the contribution due to intrinsic nonlinearity from thermal effects,
experiments using picosecond pulses are planned.

ACKNOWLEDGEMENTS

The author thanks his collaborators Professors H. F. King and D. A.
Cadenhead; Drs. J. Swiatkiewicz, D. N. Rao, R. Burzynski, B. P. Singh and S.
K. Ghoshal; Mr. J. Biegajski, Mr. X. T. Huang, Ms. P. Chopra and Ms. X. Mi.
The author also thanks Dr. D. R. Ulrich and Dr. D. Williams for valuable
discussions. Interaction with Foster-Miller, Inc. is also acknowledged.
This work was supported by the Air Force Office of Scientific Research
contract number F49620855C0052. Dr. Rao was supported by a grant from
Eastman Kodak.

REFERENCES

1. B. Clymer and S. A. Collins, Jr., Opt. Eng. 24:74 (1985).

2.   D. J. Williams, "Nonlinear Optical Properties of Organic and Polymeric Materials" Am. Chem. Soc. (Washington, D.C. 1983).

3.   D. N. Rao, P. Chopra, S. K. Ghoshal, J. Swiatkiewicz, and P. N. Prasad, J. Chem. Phys. 84:7049 (1986).

4.   Y. R. Shen, "The Principles of Nonlinear Optics" Wiley & Sons (New York, 1984).

5.   R. A. Fisher, "Optical Phase Conjugation" Academic Press, Inc. (New York, 1983).

6.   H. Eichler, P. Gunter, and D. W. Pohl, "Laser-Induced Dynamic Gratings" Springer-Verlag (Berlin Heidelberg, 1986).

7.   C. Saukret, J. P. Hermann, R. Frey, F. Pradese, J. Ducuing, R. H. Baughman, and R. R. Chance, Phys. Rev. Lett. 36:956 (1976).

8.   P. Chopra, L. Carlacci, H. F. King and P. N. Prasad, to be published.

9.   G. G. Roberts, Adv. Phys. 34:475 (1985).

10. R. Burzynski, P. N. Prasad, J. Biegajski, and D. A. Cadenhead, Macromolecules 19:1059 (1986).

11. J. E. Biegajski, R. Burzynski, D. A. Cadenhead, and P. N. Prasad, Macromolecules 19:2457 (1986).

12. G. Berkovic, Y. R. Shen, and P. N. Prasad, J. Chem. Phys. (in press).

13. R. F. Wallis and G. I. Stegeman, "Electromagnetic Surface Excitations" Springer-Verlag (Berlin, 1986).

14. X. T. Huang, R. Burzynski, and P. N. Prasad, to be published.

15. D. N. Rao, J. Swiatkiewicz, P. Chopra, S. K. Ghoshal, and P. N. Prasad, Appl. Phys. Lett. 48:1187 (1986).

16. J. M. Pearson and M. Stalka, "Poly(N-Vinyl Carbazole" Gordon and Breach (New York, 1981).

17. W. M. Hetherington, III, N. E. VanWyck, E. W. Koenig, G. I. Stegeman, and R. M. Fortenberry, Opt. Lett. 9:88 (1984).

18. M. L. Shand and R. R. Chance "Nonlinear Optical Properties of Organic and Polymeric Materials" ed. D. J. Williams, ACS publication (Washington, D.C., 1983) p. 187.

19. J. Swiatkiewicz, X. Mi, P. Chopra, and P. N. Prasad, J. Chem. Phys. (in press).

20. J. M. Nunzi and D. Ricardo, Appl. Phys. B35:209 (1984).

21. H. M. Gibbs, "Optical Bistability" Academic Press (New York, 1986).

22. G. I. Stegeman and C. T. Seaton, J. Appl. Phys. 58:R57 (1985).

23. T. M. Ding and E. Garmire, Appl. Opt. 22:3177 (1983).

NEW NONLINEAR ORGANIC CRYSTALS FOR ULTRAFAST INFRA-RED OPTICAL PROCESSING

J. Zyss

CNET - Laboratoire de Bagneux (LA CNRS 250)

196 avenue Henri Ravera, 92220 Bagneux, France

INTRODUCTION

Based on now well tried "molecular engineering" groundrules /1, 2/, the "molecular diode", where an electon-donor and an electron-acceptor group are linked via a π-conjugated system, has emerged as a key-concept in quadratic nonlinear optics. The most compact packing of such microscopic units is achieved in molecular single crystals where molecules are arranged in translationnaly invariant non-centrosymetric lattices /3/. Such type of molecular organization played a seminal role at the early stage of research in nonlinear organic materials, providing ideal model systems to relate crystalline and molecular properties via an oriented gas type of description /4/. In terms of optimal active units packing density, the single crystalline structure, despite certain problems, still remains attractive as compared to other types of organization where a lesser statistical degree of orientation is achieved by doping or functionnalizing a given host structure and subsequent poling and stabilization /5,6/. However, in the latter approach, one may be willing to pay the price in terms of active units dilution, so as to take advantage of the possible qualities of the host system such as mechanical robustness in liquid crystalline polymers, or favourable optical propagation conditions in such simple polymers as PMMA. The possibility to shape materials in waveguiding nonlinear devices /7,8/ is required by the evolution towards integrated optical circuitry : single crystalline structures have been shown by a number of research groups to be compatible with such future technological trends : two-dimensionnal films /9/, crystal cored fibers (OCCF) /10,11/, organic crystal channel waveguides (OCCW) /12/ have been grown and tested. Furthermore, the relative mechanical softness and sensitivity to chemical or photochemical pollution of some molecular crystals may be significantly improved by the protective presence of sub-and/or superstrates and cladding-media. Keeping in mind this perspective, this work will focus on bulk single crystals where relatively poorly explored properties of nonlinear organic materials, such as their ultrafast (i.e. subpicosecond) reponse can be more readily studied and understood. The temporal compression resulting from mode-locking /13/ or colliding pulse mode (CPM) techniques /14/ together with the high nonlinear coefficients of the materials discussed here lead, as will be further described in this study, to gigantic parametric gains quite compatible with optical bulk propagation. However, as will be argued in concluding comments, extension of newly observed subpicosecond parametric effects in bulk crystals presently involving high power lasers, such as discussed here, on to integration-compatible semiconductor laser driven devices, will require waveguiding configurations.

In the first part of this article we shall recall and discuss the constant interplay berween theory and experiment which has helped set-up the guidelines towards new molecular crystals "optimized" in a quantitative sense at both molecular and crystalline levels. In particular, we propose a simple intrinsic definition of molecular efficiency, allowing to set-up a comparative scale between molecular species free of artificially enhancing or diminishing factors such as proximity to resonnances of experimentally used wavelength or molecular volume. Two new molecular crystals, namely N-(4-nitrophenyl)-L-prolinol (NPP) /14/ and N-(4-nitrophenyl)-N-methylamino-acetonitrile (NPAN) /11,15/ will be discussed at both molecular and crystalline levels so as to illustrate two strategies based on structural modifications of "molecular-diode" like microscopic entities so as to favour their organization in optimized /16/ crystalline lattices. In the second Part, various femtosecond timescale characterization techniques of NPP will be discussed such as second-harmonic generation, parametric amplification and emission processes, pump-probe cross-correlation experiments with two aims in view : evidencing the quasi-instantaneous parametric electronic response of organic crystals, within the limits of pump and probe pulse duration, and generating high gains for the detection of ultraweak infra-red signals. In the third Part, a new spectroscopic technique, termed PASS (after Parametric Amplification and Sampling Spectroscopy) /17/ will be seen to derive by a natural extension from the experimental set-up, as described in the second Part, used to characterized the NLO properties of NPP. This new technique is offering a unique way to time-resolve, below the picosecond, the infra-red luminescence of excited samples of interest. Finally, as already mentionned, combination of waveguiding structures and short pulse duration will be discussed in Conclusion as a most likely pathway towards future development of optical signal processing devices.

One can either compute independantly the 18 coefficients (27 reduce to 18 because $\beta_{ijk} = \beta_{ikj}$) of the $\beta(-2\omega ; \omega , \omega)$ tensor without the Kleinman symmetry assumption (which would further reduce to 10 the number of independant coefficients valid when 2 and/or are away from resonances)/15/ or measure by an appropriate experimental technique some combination of the tensor coefficients. In the latter case, the classical experimental tool is the electric-field induced second harmonic generation in solution (EFISH) /18/ leading to the determination of $\vec{\beta} \cdot \vec{\mu}$ where $\vec{\beta}$ is the so called vector part of tensor β , or, in irreducible group representation terminology the projection of tensor in the three dimensional $J = 1$ rotationnally invariant subspace $((\vec{\beta})_i = \Sigma \ \beta_{ijj})$ and $\vec{\mu}$ the ground state dipole moment. Three problems at least are associated with this technique : firstly one ends-up with a combination of coefficients assuming furthermore that the relative orientation of β and μ is known either from simple structural considerations (β in the case of a simple donor-acceptor system is generally along the charge transfer axis while μ can be reliably obtained from local bond dipole additivity considerations) or from molecular orbital calculations. Secondly, their may be solvent-solute interactions : Ledoux et. al. /19/ for example have shown significant alteration of results when changing the urea solvent while Singer et. al. /20/ in the case of paranitroaniline solutions have considered the wide dispersion of experimental results in the litterature and applied relevant local field corrections accounting for solvent-solute interactions. In any case, an additive model is always proposed where solvent and solute molecules are supposed to respond independantly and additively, which may not be the case when aggregates tend to form. The third problem is linked to the very nature of the experiment which leads in fact to $= \gamma_{e\ell} + \vec{\mu} \cdot \vec{\beta}$ /5 kT where $\gamma_e = \gamma_{e\ell}(-2\omega ; \omega , \omega , 0)$ is the scalar part (i.e. $J=0$ irreducible subspace projection) of the second-order molecular hyperpolarizability i.e. $\gamma_e = \frac{1}{5}\underset{ij}{\Sigma} \ \gamma_{iijj}$ when Kleinman symetry is valid and a more involved expression when it is not /15/. Extracting $\vec{\beta} \cdot \vec{\mu}$ from the experimentally determined overall may prove arduous and various methods have been proposed, some by-passing the problem rather than adressing it properly. Oudar /21/ has combined the four-wave mixing CARS and the EFISH techniques allowing for a two-level model to connect $\gamma(\omega_2 - 2\omega_1 ; \omega_1, \omega_1, -\omega_2)$ measured in the former experiment to $\gamma(-2\omega ; \omega , \omega , 0)$ needed in the latter one. Another more direct technique, more readily used in gas /22/ than in solutions where phase transitions impose drastic limitations, consist in varying the temperature at which the EFISH experiment is performed and equating $\vec{\mu} \cdot \vec{\beta}$ to the slope of the supposedly linear variation of   in terms of 1/T. We propose, in a forthcoming publication /23/ to measure the third-harmonic generation (THG) in similar solutions and concentrations as were used in the EFISH technique and relate $\gamma(-3\omega; \omega, \omega, \omega)$ to $\gamma(-2\omega; \omega, \omega, 0)$ by a two-level model which is more direct than such as applied in /21/. In particular, we have explored a number of wavelength such as   = 1.89 μm (first Stockes emission of a H_2 Raman shifted YAG : Nd^{3+}) for THG and $\omega_2 = 1.32$ μm (YAG : Nd^{3+} emission) in the EFISH generation so that 2 ω_2 comes as   close as possible to 3 ω_1. In these experiments one has to make room for the phase or imaginary parts of both γ and β owing to the proximity to resonnances of one of the interacting frequencies. This has been recently considered by Kajzar et al. /24/ in EFISH measurements at $\lambda = 1.06$ μm of red so-called pTS-12 polydiacetylene solutions or by Chance et al /25/ in CARS and resonant Raman experiments in solution involving the so-called 3-and 4-BCMU polydiacetylene moieties.

As for the theoretical modeling, two methods logical approaches may be followed. One is the so-called finite-field method /26/ first applied to calculations in Refs./27, 28, 29/, the second being the so-called summ-over-states (SOS) perturbationnal approach /30,31,32,33,34/. Both methods may allow for the introduction of electronic correlations and start from ab-initio or semi-empirical levels. In the finite field approach, a perturbating dipolar interaction monoelectronic potential - $\vec{\mu} \cdot \vec{E}$ is introduced in the Hamiltonian, leading to field-dependent eigenvalues and wavefunctions ψ (E). Differentiation of the expectation value of the dipole operator $< \psi(E)/\vec{\mu}/ \psi(E) >$ with respect to field components at $\vec{E} = \vec{0}$ leads to the so-called finite-field values of the molecular hyperpolarizabilities for example :

Difficulties such as raised by the trade-off between maximization of the field amplitude so as to diminish the numerical error and minimizing it so as to diminish the mathemical error and prevent divergence of the Hartree-Fock self consistent procedure as may result from too strong a perturbation are adressed in Refs. /28, 29, 35/. When no further refinement is introduced, this method leads to static susceptibilities while a quantum two level dispersion model has been introduced in Ref./28/ to allow for frequency dependance of the susceptibilities. Combination of finite-field and interaction of configuration methods has two our knowledge only been undertaken /34/ in the case of smaller molecules of no relevance to our present work. The other so-called SOS approach consists in filling in, at whichever level of approximation is required, the full perturbationnal expression of β as follows :

$$\beta^{2\omega}_{ijk} + \beta^{2\omega}_{ikj} = \frac{-e^3}{4\hbar^2} \sum_{n,n'} x$$

$$\{[\Gamma^i_{gn}\,\Gamma^i_{n'n}\,\Gamma^k_{n'ng} + \Gamma^k_{gn}\,\Gamma^i_{n'n}\,\Gamma^j_{n'ng}][((\omega_{n'g}-\omega)(\omega_{ng}+\omega))^{-1} + ((\omega_{n'g}+\omega)(\omega_{ng}-\omega))^{-1}]$$

$$+[\Gamma^i_{gn'}\Gamma^j_{n'n}\,\Gamma^k_{n'ng} + \Gamma^i_{gn'}\Gamma^k_{n'n}\,\Gamma^j_{n'ng}][((\omega_{n'g}+2\omega)(\omega_{ng}+\omega))^{-1} + ((\omega_{n'g}-2\omega)(\omega_{ng}-\omega))^{-1}]$$

$$+[\Gamma^j_{gn'}\Gamma^k_{n'n}\,\Gamma^i_{n'ng} + \Gamma^k_{gn'}\,\Gamma^j_{n'n}\,\Gamma^i_{n'ng}][((\omega_{n'g}2\omega)(\omega_{ng}-2\omega))^{-1} + ((\omega_{n'g}+\omega)(\omega_{ng}+2\omega))^{-1}]\}$$

Again the validity of this expression, in the absence of imaginary damping term or linewidth, is restricted to off-resonnant cases. An example of comparison between finite field and SOS calculations, can be found in Ref /15/ where, among others, NPAN and NPP have been specifically considered (see Fig. 1). Results are to be found in Table 1 where the static SOS and finite field values are seen to satisfactorily agree, especially for the dominant β_{xxx} coefficient (see definition of the molecular framework axis in the Table caption). Furthermore, such tensorial coefficients as β_{xyy} and β_{yxy} were computed independently and shown, in all cases, to be equal within computational error margins, in agreement with the assumption of the validity of Kleinman relations for static polarizabilities. Table 1 will be further invoked when discussing the role of the nitrile group in the structural and electronic properties of NPAN (see Fig. 2 for the struc-

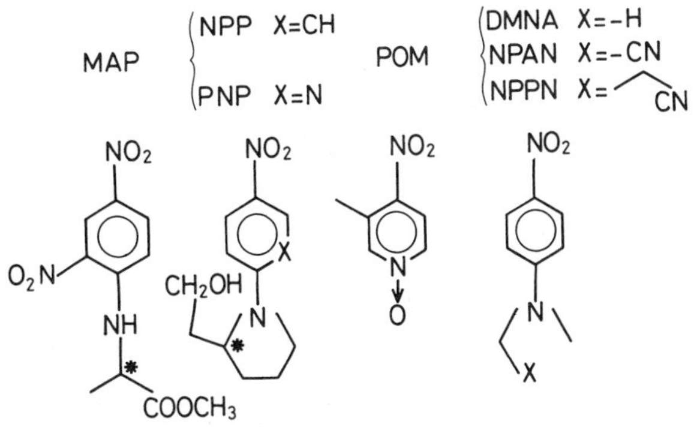

Fig 1 Structure of various "molecular-diode" like structures mentionnal in this paper, each illustrating specific molecular modefications meant to act on the transparency and/or the packing of the corresponding crystals.

Table 1. Calculated β Components of NPAN and NPP

$\beta^{2\omega}_{ijk}$ 10^{-30} e.s.u.	F.F. method (NPAN) $\hbar\omega = 0$	C N D O V S B $\hbar\omega = 0$	method (NPAN) $\hbar\omega = 1.17$ e.V.	F.F. method $\hbar\omega = 0$ (NPP)	C N D O V S B $\hbar\omega = 0$	method (NPP) $\hbar\omega = 1.17$ e.V.
β_{xxx}	$- 11.9 \pm 0.3$	$- 11.56$	$- 20.16$	$- 17.2 \pm 0.3$	$- 15.49$	$- 29.46$
β_{yyy}	$- 0.7 \pm 0.3$	0.06	0.04	0.3 ± 0.2	0.08	0.16
β_{zzz}	$- 0.1 \pm 0.3$	$- 0.05$	$- 0.06$	$- 0.2 \pm 0.3$	$- 0.01$	$- 0.01$
$\beta_{xyy} = \beta_{yyx}$	2.3 ± 0.3 2.2 ± 0.2	1.59	2.73 1.76	2.3 ± 0.3 2.3 ± 0.2	1.95	3.74 2.35
$\beta_{yxx} = \beta_{xyx}$	0.4 ± 0.2 0.1 ± 0.3	0.002	0.07 0.02	0.15 ± 0.3 0.2 ± 0.3	0.07	0.005 0.14
$\beta_{xzz} = \beta_{zzx}$	0.1 ± 0.3 0.2 ± 0.3	0.04	0.04 0.05	0.3 ± 0.3 0.3 ± 0.3	0.03	0.04 0.04
$\beta_{zxx} = \beta_{xzx}$	0.7 ± 0.3 0.7 ± 0.3	0.14	0.19 0.25	0.2 ± 0.3 0.2 ± 0.3	$- 0.12$	$- 0.32$ $- 0.24$
$\beta_{yzz} = \beta_{zzy}$	0.6 ± 0.2 0.1 ± 0.2	0.09	0.11 0.11	0.1 ± 0.2 0.2 ± 0.3	0.01	0.01 0.01
$\beta_{zyy} = \beta_{yzy}$	$0. \pm 0.2$ 0.05 ± 0.2	$- 0.12$	$- 0.04$ $- 0.05$	0.1 ± 0.3 0.1 ± 0.2	0.05	0.09 0.07
$\beta_{xyz} = \beta_{xzy}$ $\beta_{yxz} = \beta_{yzx}$ $\beta_{zxy} = \beta_{zyx}$	$- 0.05 \pm 0.2$ $- 0.4 \pm 0.2$ $- 0.1 \pm 0.2$	$- 0.03$	$- 0.04$ $- 0.03$ $- 0.03$	$- 0.1 \pm 0.2$ $- 0.1 \pm 0.2$ $- 0.1 \pm 0.3$	$- 0.05$	$- 0.06$ $- 0.07$ $- 0.07$

Molecular axis are defined similarly for NPAN and NPP, x is from the nitrogene of the donating group towards the nitrogene of the nitro accepting group, y is in the mean molecular plane perpendicular to x (see Ref. /15/, β xxx is the donating component of an essentially 1-D tensor for both NPP and NPAN, β xyy is the only otherwise small but non-negligible component for symmetry reasons discussed in text. The Finite-Field result (F.F.) reported herein refers to the method developped in /31/ while CNDOVSB results originate from the method developped in /27/.

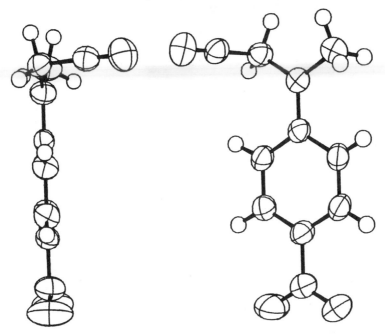

Fig. 2 Profile and front ORTEP II drawings of NPAN as from Ref /11/.
The nitrile group is shown to be stretching out the mean molecular aromatic plane.

ture of NPAN). NPP is shown to be more nonlinear (higher β_{xxx}) than NPAN at
λ = 1.06 μm, a difference which can be ascribed to two factors : one is
linked to intrinsic electronic properties of the molecules such as the intensity of the electron donor character of the prolinol group as compared
to that of the dimethylamino-group such as can be deduced from static calculation, the other to the resonnant enhancement stronger for NPP than for
NPAN in view of the relative positions of the absorption spectra maximum in
solutions from Fig. 3. However, from Table 1 it is seen that, at the wavelength of interest here, the intrinsitic contribution prevails over the
lesser dispersion contribution as, λ /2 = 530 nm is, for both molecules,
significantly remote from the absorbtion bands leading to an essentially
non-resonnant SHG process in both cases. However, in order to set-up a
meaningful comparative scale of SHG intensities, at the molecular powder or
crystalline levels it is necessary to define β or $\chi^{(2)}$ coefficients where
such parameters as molecular volume and possible resonnant enhancement are
accounted-for. Molecular volume is of major importance when the nonlinear
portion of the molecule, generally restricted to the π -conjugated system,
is dilute in an inactive environment of possible relevance towards the engineering of the macroscopic structure but of no relevance with respect to
the nonlinearity of the molecule itself. Such considerations should be given attention too, especially in guest-host system such as cyclodextrine in-

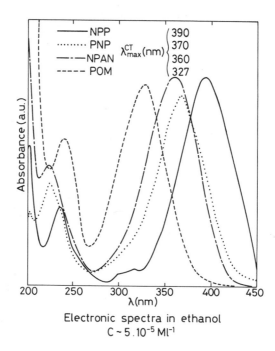

Electronic spectra in ethanol
$C \sim 5 . 10^{-5} Ml^{-1}$

Fig. 3 The significant maximum absorbtion shifts shown here can
be predictively ascribed to specific features of their molecular structures as discussed in text.

clusion compounds /36/, but even in simpler systems where, regardless of
the crystalline structure at the moment, preference, at the molecular level,
should be given to molecular systems where active and inactive parts coîn-
cide as closely as possible. NPAN offers a striking demonstration of this
concept : it may be seen from Ref. /15/ that the β tensor of DMNA (see
Fig. 1) a molecule fully similar to NPAN except for the substitution of a
methyl group to the nitrile group, is identical to that of NPAN, within
computational error margins (DMNA is refered to as NPAM in Ref. /37/). This
is a strong indication that the nitrile group plays no part in the nonlinear mechanism, as could be expected from the intermediary CH_2 link, screening the nitrile group from the rest of the conjugated system. Furthermore,
the molecular symetry of the nonlinear portion of NPAN is that of DMNA,
(i.e.mm2) which accounts for the quasi concellation of β_{yxx} as compared
to a relatively larger β_{xyy} coefficient where x is along the "quasi-"
twofold axis and y perpendicular to it within the mean molecular plane. Ho-

wever the volume of NPAN is slightly larger than that of NPAM, a price
which is willingfully paid for the benefit of the part played by the nitrile group in the optimization of the crystalline structure as will be further discussed.

Resonnant enhancement can be conventionnally accounted for, in most
charge transfer systems, by a two-level quantum model, essentially originating from the reduction of the summation in Eq . 1 to ground-and charge-
transfer states /38, 39/. The reduced Eq. 1 is given for a one dimensionnal
non-linearity by :

$$\beta = \frac{3e^2\hbar^2}{2m} \; f\Delta\mu \; \; F(W,\omega)$$ Eq.2

where the dispersion term $F(W,\omega)$ is defined as :

$$F(W,\omega) = W \Big/ \big[W^2 - (2\hbar\omega)^2\big] \big[(W^2 - (\hbar\omega)^2\big]$$ Eq.3

W (resp. $\Delta\mu$) is the energy (resp. dipole) difference between the two
levels, is the fundamental photon energy and f the oscillator strength
of the transition. A proper scaling of molecular efficiency would thus result from the following criterion :

$$\rho = \beta \; / \; V \; F \; (W,\omega) = \frac{3e^2\hbar^2 f\Delta\mu}{2\,m\,V}$$ Eq.4

where V is the molecular volume and β is, in one-dimensional systems, the
major hyperpolarizability coefficient. This criterion can be applied to
trans-4-amino-4'-nitrostilbene (DANS) as compared to paranitroaniline (pNA)
with a striking change in perspective from the usual point of view : it is
generally recognized that the β value of DANS surpasses that of pNA by over
one order of magnitude. However, when expressed in terms of ρ the normalized ratio decreases by less than a factor of two as can be seen from Table2.
Eq .1 can be generalized from the case of SHG on to any three-wave mixing
parametric interaction process involving frequencies ω_1, ω_2 and ω_3 where
$\omega_3 = \omega_1 + \omega_2$ using Eq.7 in Ref. /41/.

$$\rho(-\omega_3; \omega_1, \omega_2) = \beta(-\omega_3; \omega_1, \omega_2)/VF \; (W,\omega_1,\omega_2,\omega_2)$$ Eq.5

Table 2   Normalized nonlinear efficiency for pNA and DANS as defined by Eq.4.

	V (A^3)	$\hbar\omega$ (eV)	W (eV)	F (eV^{-3})	β (10^{-30} esu)	ρ (10^{-42} esu)
pNA	145.7	1.17	3.875	0.02976	34.5	32.59
DANS	240	1.17	3.062	0.0981	950	53.41

The dimensionnality of   is $[Q^3]$ $[M]$ $[L^2]$ $[T^{-2}]$.   is expressed in units of 10^{-42} e.s.u.. V is obtained from Ref. /31/   from Ref.38 and W in ethanol from Ref. /40/.

where :

$$\beta \; (-\omega_3 ; \omega_1 , \omega_2) = \frac{3e^2\hbar^2}{2m} \; F \; (W, \omega_1 , \omega_2 , \omega_3) \qquad\qquad \text{Eq.6}$$

and the dispersion factor F is given by :

$$F(W, \omega_1 , \omega_2 , \omega_3) = \frac{W\left[W + (\hbar^2 \; \omega_1 \; \omega_2 - \hbar^2\omega_3^2)/3\right]}{\left[W^2 - (\hbar\omega_1)^2\right]\left[W^2 - (\hbar\omega_2)^2\right]\left[W - (\hbar\omega_3)^2\right]} \qquad\qquad \text{Eq.7}$$

The latter Expression includes the case of electrooptic modulation where $\omega_3 = \omega_1$ and $\omega_2 = 0$ or corresponds to a low frequency externally ap-plied voltage.

Molecules represented in Fig. 1 illustrate different strategies meant to act, at the molecular level, on the transparency versus efficiency tra-de-off, and furthermore, to influence the crystalline packing, although it still remains a challenge to fully predict the crystalline structure in view of the sole molecular structure. MAP, NPP and PNP all contain a chiral carbon and are derivative of purely left-handed optically active natural amino-acids. These chiral carbons are part of the electron donating group of the molecule which is an ester derivative of alanine in the case of MAP and an alcohol derivative of proline in the case of NPP. The implementation of a chiral group in a molecule is a purely geometrical means to preclude a center of inversion and may end-up in a disappointing result : the depar-ture from centrosymetry may be weak or only that portion of the molecule which is not connected to the conjugated system may be arranged in a non centro-symmetric lattice while the π -electronic systems correspond to each other via a center of inversion. Here again, it may be useful to proceed, at the crystalline level and distinguish between the actual physical latti-ce, composed of the full molecular entities, and a virtual sub-lattice com-posed of only that portion of the molecules which take part in the nonli-near effect.

The "full lattice" may be non-centrosymmetric while the optically nonlinear
relevant sublattice may be close to centrosymmetric. MAP gives room to sub-
tler considerations (see Ref. /39/) whereby it has been shown that the
crystal packing favours the weaker ortho-charge transfer at the expense of
the stronger para charge transfer : it has been shown that an order of
magnitude could be gained if an hypothetical "optimized" geometry as defined in Ref. /16/, could be reached by proper modification of a 2-4 dini-
troaniline basic molecule. A number of statistical studies gathered in
Ref./37/ point-out that a minority (less than 30% at most) of organic crystals are reportedly non-centrosymmetric. More than other intermolecular
forces underlying a given crystalline structure, dipole-dipole interactions
are responsible for this trend towards centrosymmetry. Besides, or in addition to chirality, ways to influence dipolar interactions must therefore be
sought, through proper molecular modifications, without alteration of the
basic properties accounting for the molecular nonlinearity itself. One way
is to try and combine intramolecular charge-transfer (i.e. a significant)
and a vanishing ground state dipole moment, which is achieved in 3-methyl-
4-nitropyridine-1-oxydo (POM) /42/ (See Fig. 1) : the dipole moment of the
pyridine-N-oxyde molecule cancels-out that of nitrobenzene while the weakly
dipolar methyl group is just meant to act sterically on the structure (un-
substituted 4-nitro-pyridine-1-oxyde would be centrosymmetric). Owing to
its so-called "push-pull" character /43/, the N-oxyde group is capable of
either electron accepting or donating behaviour depending on the nature of
the substituent in para-position. The nitro-group being essentially an
electron attractor, the N-oxyde group will behave as an electron donating
group essentially via the conjugated p_z -lone pair of nitrogene. Another
strategy, illustrated by NPP is meant to surpass the dipolar interaction
forces by more energetic intermolecular hydrogen bonding which will not favour centrosymmetry : for this purpose an alcohol derivate of proline, namely prolinol is used as the electron donating portion of NPP, endowing it
with intermolecular hydrogen bonding potential via the hydroxylic group.
Finally NPAN, a derivate of the model DMNA molecule giving raise to an almost centrosymmetric crystal, is made to contain a highly dipolar group,
namely nitrile (see Fig. 2 for more detailed view) : it is expected that
strong local dipole-dipole interactions will couple adjacent nitrile group
/11, 15/, most probably in a quasi-centrosymmetric arrangement, but letting
free the rest of the molecule to set up a more adapted structure. The nitrile group may be seen to be decoupled, by a CH_2-link, from the conjugated
nonlinear system, as was already mentionned and is in fact stretching-out
by an angle close to 100° from the aromatic plane (see Ref. /11/). Figs. 4
and 5 display the crystalline packing of NPP and NPAN respectively.

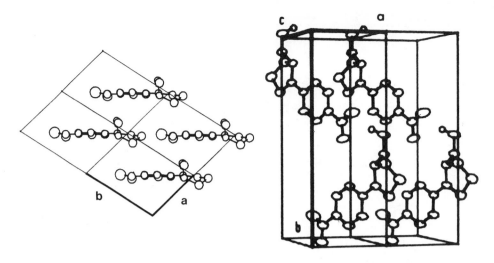

Fig. 4   Top and side views of the crystalline packing of the
NPP unit cell as from Ref. /14/ evidencing a quasi
lamellar structure and intermolecular hydrogen bonding from the alcohol group of a molecule to the nitro group of the next one. As discussed in text, the
angle beween the molecular charge transfer axis and
the crystalline two-fold axis is close to optimal /16/.

The lattice of NPP is highly lamellar with the with the mean molecular plane coinciding with the $< 101 >$ crystallographic plane, the structure being
monoclinic ($P2_1$) with two molecules per unit cell. Most noteworthy, in addition to the lamellar structure, is the angle θ of 58.6° between the two
nitrogene atoms linking through the aromatic ring the electron donating and
attracting groups. This angle comes close to the ideal angle of 54.3°, optimizing the projection factor $\cos^2\theta \sin^2\theta$linking the coefficient of   the
essentially one-dimensionnal β tensor of NPP to the phase-matchable d_{yxx}
(or d_{21} coefficient) as defined in Ref./16/. Within the framework of an o-
riented gas description of the crystal i.e. :

$$d_{21} = N f^3 \cos\theta \sin^2\theta \, \beta \qquad\qquad\qquad Eq.8$$

where N is the number of molecules per unit volume and f^3 accounts for local field corrections. It may be seen from Fig.4 that the hydrogen bonds
effectively account for the crystalline structure of NPP as the oxygene
atom of the alcohol group of a molecule is linked to the two oxygenes of
the nitro group of the neighbouring molecule within the same $< 101 >$ crystallographic plane which tends to be a cleavage plane.

   The packing of NPAN is also close to "optimal" : the cystal structure

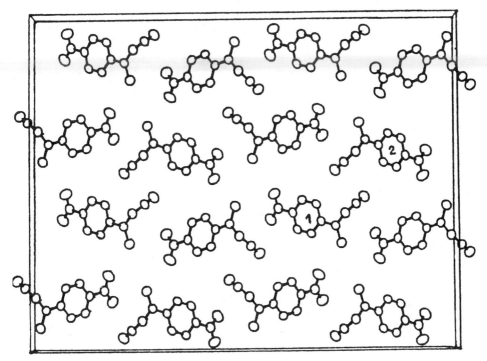

Fig.5 View of the crystalline molecular packing of NPAN
parallel to the c two-fold axis. Molecules are closely stacked perpendicular to the c axis. A typical
couple of adjacent molecules are singled-out in this
figure (molecules 1 and 2) and shown to be strongly
interacting via the antiparallel arrangement of their
highly dipolar nitrile ends.

is Fdd_2 (see Ref./11/) with 16 molecules per unit cell. As can be seen
from Fig.5 a projection of the unit cell on the < 001 > plane along the
c axis, the molecules are densely stacked along the c axis, the short distance between two consecutive aromatic planes (3.56 Å) coming close to the
contact distance of aromatic rings. The nitrile dipoles of molecule 1 and 2
projected on the < 001 > plane, are antiparallel and are both located parallel to the < 110 > plane. Furthermore, the angle betwen the charge
transfer axis, very much like in the case of NPP, is almost ideally oriented with respect to the c two-fold axis, by an angle of 58.6°. The main
difference with NPP is a much less pronounced tendency towards cleavage
when it pertains to crystal growth and processing questions.

As previously mentionned, the transparency of the molecule and,
therefrom, assuming a similar excitonic red-shift from molecules to crystals, that of the crystal can be tailored by proper definition of the mole-

cular structure. Such is the case for PNP where introduction of a nitrogene in the aromatic ring at a so-called "starred" position (se Ref./37/) shifts the absorbtion band of PNP, as compared to NPP, towards the blue by an amount of 20 mn (see Fig.3). However, surprisingly enough, the structure of PNP has been shown in Ref./44/ to be very close to that of NPP, regardless of the significant structural difference between the two molecules. Therefore, PNP is also "optimized", further detailed studies at the molecular and crystalline levels being still required to compare both materials. NPAN has been optimized, at the molecular level in terms of transparency-efficiency trade-off : one may consider, NPAN and NPPN as members of the following generic family of molecules : $A'-(CH_2)_n-D-$ $-A$ where ρ stands for a benzene ring, A(resp. A') is the electron accepting nitro (resp. nitrile) group while D is the electron donating NCH_3 group, n = 1 for NPAN and n = 2 for NPPN while A'= H and n = 1 for DMNA. Owing to the short-range so-called inductive effects the electron attracting nature of A', in the cases of NPAN and NPPN will tend to diminish the donating nature of D. This tendency will be stronger when A' is closer to D raising the blue shift in solution, taking DMNA for comparison, from 13 Å in the case of NPPN to 29 Å in the case of NPAN (see Ref./15/). NPAN is thus significantly more transparent than PNP by 10 Å while retaining the desired optimal crystalline structure.

Eqs. 4 and 5 can be readily extended from molecular values, on to semi-quantitative powder SHG efficiencies or full crystalline three-wave parametric mixing experiments in order to rationalize the comparison between materials at either powder or crystalline levels, regardless of the wavelength of interest, leading to :

$$\rho = A/F \qquad\qquad\qquad Eq.9$$

where F is defined as in Eqs. 3 or 7, the molecular volume V being of no more relevance here. A is a normalized second-harmonic intensity in the case of powders and is the effective phase-matched crystalline susceptibility in the case of crystals (i.e. with projection cosine factors corresponding to phase-matching wavelength which for consistence, are also used in F). Various methods have been utilized to measure or compute the crystalline susceptibility of NPP. In the next section, we will emphasize parametric gain measurements performed at the femtosecond time-scale, as an original means to obtain χ^2 or d values. Other methods have been employed: such as transverse SHG measurements through crystalline NPP samples thinner than a coherence length /9/, or the combination of EFISH value measurements with an oriented gas description of the solid, based on our knowledge of the detailed crystalline structure of NPP /15/. This latter method has also been applied to NPAN where full crystalline measurements will soon be available. CNDOVSB or finite field theoretical values of can equivalently be used. For both materials, estimated values of the highest phase-matchable coefficient (d_{21} for NPP and d_{31} for NPAN) coincide around 150.10^{-9} esu which qualifies them as the most efficient quadratic nonlinear molecular crystals reported so far. For comparison, the d_{21} coefficient of 3-methyl-4-nitroaline (MNA) is of the order of $75 \ 10^{-9}$ e.s.u. /45/. Furthermore, a non-critically phase-matchable three-wave mixing configuration will be shown in the next section, to take place in NPP at a fundamental wavelength of 1.15 μm resulting in $d_{eff} = d_{21}$.

ULTRAFAST SUBPICOSECOND HIGH-GAIN PARAMETRIC MIXING IN NPP

The purpose of the study reported in this Section is essentially two-fold : firstly, directly characterize the phase-matching configurations of NPP for both second-harmonic generation and parametric amplification process. Secondly, probe the supposedly quasi-instantaneous time-response of non-resonnant electron-driven parametric interaction processes in molecular crystals when shined by ultra-short (ie. subpicosecond) laser pulses. It turns-out that both aims are intimately related as they call upon a common experimental set-up described in Fig. 6. As for the first direction, the the approach described here (see also Refs. /17 ,46, 47, 48/ together with that reported for POM in Ref./49/ make both use of tunable I.R. sources : a double-stage $1i1O_3$ parametric system (pumped by a frequency doubled Q-switched mode-locked Nd^{3+} : YAG laser) emitting in the 10 picoseconds time-

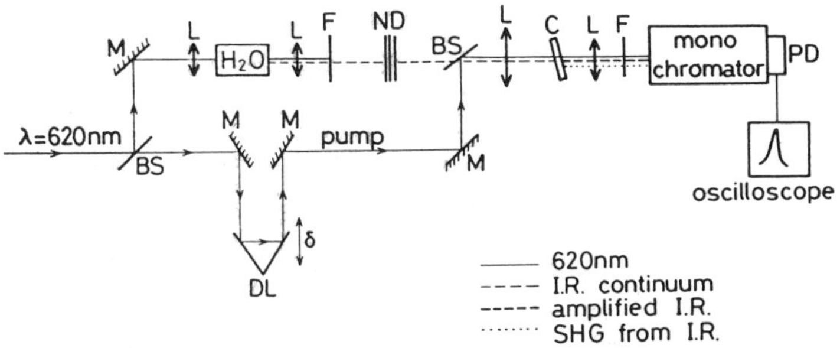

Fig. 6 Experimental set-up for various three-wave parametric interaction configurations. M's, mirrors . BS's beam splitters ; L's, lenses ; F's, filters ; ND, neutral density ; DL, delay line ; PD ; photodiode ; C non-linear crystal.

scale in Ref /49/ and a colliding pulse mode dye laser generating, by self-phase modulation, an I.R. continuum in the 100 fs duration range. Such systems systems allow for both angular and time characterization of samples of interest as will be further described. Previous approaches, where tunable I.R. sources were not available, relied on Sellmeir index dispersion based computational extrapollation of a fixed frequency experimental determination (see Ref. /42, 50/ for example) or on parametric fluorescence experiments (see Ref. /51/) which precludes direct parametric gain measurement. In addition, a number of benefits can be derived from going towards ever shorter pulse, although a number of previously irrelevant problems are now raised. Among the advantages are essentially the higher parametric gain and and increasing damage threshold, the cumulative nature of the latter relating more to energy than power considerations. It turns-out that the maximum power-density allowed in the visible to shine without damage para-nitroaniline-like crystals is of the order of a few tens of $MWcm^{-2}$ at the 10 ps time-scale, an a few $GW\ cm^{-2}$ at the 100 fs time-scale, allowing, as will be seen in this Section, to trigger gigantic I.R. gains in the latter case. However, brute extrapollation of experiments performed at relatively longer time-scales is not permitted when no attention is given to group-velocity matching considerations, remembering that, in vacuum the spatial extension of a 100 fs duration pulse is 30 μm. Maintaining sufficient space-time overlap over that kind of dimensions requires revisiting the classical phase-matching concept as follows.

Phase-and group-velocity matching considerations

Phase matching is an essential prerequisite towards efficient nonlinear three-wave interaction in crystals. A non-phase-matchable crystal, such as quartz, cannot give rise to efficient frequency conversion, since only a small fraction of the crystal length, the coherence length, will contribute to the nonlinear process. If phase matching-occures, the coherence length goes to infinity, and the entire length of the material cumulatively contributes to the nonlinear interaction. Other limiting factors must then be taken into account in the case of short pulses. If one wishes to collect as much IR signal as possible, angularly and spectrally, the nonlinear material must then be phase-matchable for angular and spectral mismatch factors $\Delta\theta$ and $\Delta\lambda$ being increased in order to process a maximal flux of incoming photons. When time-resolution is required, the maximization of the spectral acceptance $\Delta\lambda$ is essential, so that the short pulses travelling inside the material will not be time expanded at the price of a decrease of the time-resolution capability of the system.

In the plane-wave approximation, the scond-harmonic intensity is given

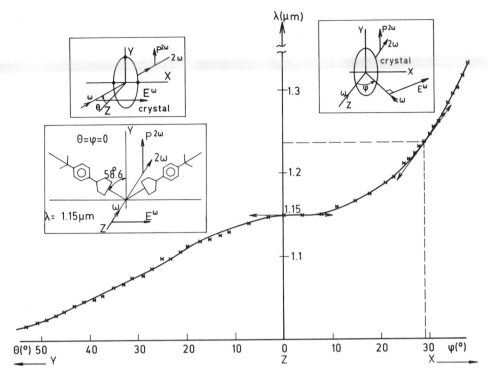

Fig. 7 SHG phase-matching curve of NPP as determined by the experiment sketched in Fig. 6. θ (resp. ρ) are external incidence angles of the IR beam with respect to Z within the
ZY (resp. ZX) plane. An angularly non-critical phase-matching configuration along Z (i.e. perpendicular to the
mean molecular planes) occures at 1.15 μ m while a marked spectrally non critical tendency can be noted at
1.24 μm from the rather steep slope of the phase-matching
curve at this point.

by a function of the type sinc^2 ($\Delta k l/2$), where k is the difference between the fundamental and harmonic wave vectors :

$$\Delta k = k_{2\omega} - 2k_{\omega} = 4\pi(n_{2\omega} - n_{\omega})/\lambda \qquad \text{Eq.10}$$

where l is the interaction length and n_{ω} ($n_{2\omega}$) is the refractive index of
the nonlinear crystal at the fundamental (harmonic) wavelength. The indices
are generally also a function of the angle between the phase-matched collinear beam wave vectors and a dielectric axis of the crystal, when the
propagation takes place in a principal dielectric plane. The angular and
spectral aperture conditions can be expressed as follows :

$$-1 < \Delta k(\theta,\lambda) l < 1 \qquad \text{Eq.11}$$

85

A first-order Taylor expansion of Δk (θ, λ) at the proximity of the exact phase-matching values (θ_0, λ_0) yields :

$$-1/1 < \left[\partial(\Delta k)/\partial\theta\right]_{\theta_0,\lambda_0} \Delta\theta + \left[\partial(\Delta k)/\partial\lambda\right]_{\theta_0,\lambda_0} \Delta\lambda < 1/1, \qquad \text{Eq.12}$$

where $\Delta\theta = \theta - \theta_0$ and $\Delta\lambda = \lambda - \lambda_0$ correspond, respectively, to the angular and spectral mismatches. The maximal spectral (angular) acceptance is obtained for $\left[\partial(\Delta k)/\partial\lambda\right]_{\theta_0,\lambda_0} = 0$ $\{\left[\partial(\Delta k)/\partial\theta\right]_{\theta_0,\lambda_0} = 0\}$

The SHG phase-matching curve being given by $\Delta k(\theta,\lambda)=0$, the condition $\left[\partial(\Delta k)/\partial\theta\right]_{\theta_0,\lambda_0} = 0$ shows that the phase-matching curve has a horizontal slope at (θ_0,λ_0) (see Fig. 7). It corresponds to the well-known noncritical phase-matching configuration and ensures the double advantage of an optimal aperture of the acceptance cone and the absence of walk-off. Let us turn next to broadband well-collimated signal, a situation symmetric to the previous one. The permutation of θ and λ leads to similar considerations for the spectral acceptance, the relation $\left[\partial(\Delta k)/\partial\lambda\right]_{\theta_0,\lambda_0} = 0$ corresponding to a vertical slope of the phase-matching curve. Such tendency towards a vertical slope appears on the right-hand side of Fig. 7.

It is then adequate to use the terms of θ-noncritical in the former case and λ-noncritical in the latter one, as this terminology points-out explicitly their respective advantages.

The occurrence of quasi-λ noncritical phase-matching conditions for a fundamental wave in the near-IR range seems to be a characteristic property of paranitroaniline derivatives : the condition $\left[\partial(\Delta k)/\partial\lambda\right]_{\theta_0,\lambda_0} = 0$ leads in the case of a type I, $e^{\omega} + e^{\omega} \rightarrow o^{2\omega}$ configuration to the following relationship between the refractive indices :

$$2(\partial n^e/\partial\lambda)_{\lambda_0} = (\partial n^0/\partial\lambda)_{\lambda_0/2} \qquad \text{Eq.13}$$

so that the slope of the dispersion curve of the ordinary refractive index at $\lambda_0/2$ in the visible range is twice that of the dispersion curve of the extraordinary refractive index at λ_0 in the near IR. This situation is freqently encountered in organic crystals, such as paranitroaniline derivatives which absorb in the near-UV range. This property can be explained by the proximity to $\lambda_0/2$ of a UV resonnance (charge-transfer-excited state around 350-400 nm), which greatly enhances the slope of the curve n^0 (λ). On the other hand, near the IR wavelength λ_0 , the refractive index values are less sensitive to the influence of UV resonnances, leading to a decrease of the slope of the dispersion curve n^e (λ).

The next relevant requirement involved in short-pulse nonlinear interaction is the matching of group velocities. In SHG, the useful crystal interaction length 1 is limited by the following condition :

$$1(u_{2\omega}^{-1} - u_{\omega}^{-1}) \leq \tau, \qquad \text{Eq.14}$$

where $u_\omega (u_{2\omega})$ refers to the fundamental (harmonic) wave group velocity and τ is the fundamental wave pulse duration. If $u_{2\omega}$ differs from u_ω , the interacting pulses will cease to overlap significantly beyond 1, and no further harmonic can be generated. Calling u_1 and u_2 the classical unit vector reference frame associated with θ (that is u_2 along the wavevectors of the interacting waves and u_1 perpendicular to it in the principal dielectric plane) it was shown in Ref. /52/ that the group velocity of a plane wave with frequency ω and walk-off angle ρ is given by :

$$u = \frac{c}{n + \lambda(\partial n/\partial \lambda)} (\tan \rho u_1 + u_2),$$
Eq.15

where n is the refractive index at   .

The phase-velocity-matching condition for an $e^\omega + e^\omega \to o$ configuration implies that $n^e_\omega = n^0_{2\omega}$; in a λ noncritical phase matching, Eq. 13 combined with Eq. 15 leads to

$$(u_{2\omega}^0 - u_\omega^0)u_2 = 0$$
Eq.16

Thus, in a λ noncritical configuration, the group velocities are matched along the wave-vector. The harmonic group velocity is then along the wave-vectors since there is no walk-off angle, and identical to the projection along that direction of the harmonic group velocity $\mathbf{u}_{2\omega}^0$. Quasi-group-velocity matching is then achieved. In a θ noncritical configuration, both group velocities are parallel to the wave-vector direction and to a principal dielectric axis, but their moduli are not matched.

There we have shown that materials exhibiting a λ -noncritical phase-matching configuration give rise to quasi-groupe-velocitiy matching, allowing for efficient nonlinear interactions of ultrashort laser pulses. Owing to the specific index-dispersion properties in the visible-near-IR spectral range resulting from the existence of near-UV charge-transfer absorption bands of the molecules, organic crystals made of paranitroaline-like molecules seem to be good candidates for efficient parametric interactions at the subpicosecond time scale, as will be shown in the following.

Parametric-mixing experiments in NPP

The high-power femtosecond source used as shown hereafter is based on the principle of pulse mode-locking. The ultrashort pulse (wavelength 620 nm, duration 100 fsec) built into the oscillator cavity is further amplified in a four-stage dye amplifier, pumped by a frequency-doubled Q-switched Nd:YAG laser, and separated by saturable-absorber jets. Output pulses of energy above 1 mJ are obtained for a pumping energy of 300 mJ at 532 nm with a pump duration of 6nsec at a 10-Hz repetition rate. The output beam is split in two parts : one of the beams is vertically polarized . It can be sent directly onto the nonlinear organic crystal to be studied and used as the pump beam in parametric amplification and emission processes. The other one, with a horizontal polarization, is focused into a 20 mm-long water cell to generate, by self-phase modulation, a white-light continuum ; the conversion efficiency of this process is close to 100%, giving rise to a reproducible and uniform continuum with a spectral range from 0.2 to 1.6 μm and a pulse duration of the order of 180fsec. In this experiment, wavelengths below 1 μm in the continuum are filtered-out by use of a Schott RG 1000 filter.

The various configurations depicted in Fig.6 correspond to the different three-wave interactions that occure in the organic crystal. To obtain SHG, the pump beam is removed and the IR part of the continuum, horizontally polarized, is used as a fundamental beam. In parametric emission experiments, the continuum is removed, and only the pump beam is focused onto the crystal. To perform parametric amplification, both the IR continuum and the visible pump beam are focused into the nonlinear crystal to let the beams overlap inside the crystal. The time delay between the two incident beams is adjusted to ensure simultaneous arrival of the two incident pulses in the crystal ; this can be achieved by monitoring the arrival time of the pump with a delay line consisting of mirrors mounted upon a translational stepping motor. Both beams are focused into the crystal with a beam diameter of the order of 1 mm. The energy of the pump is of the order of 1 μJ, corresponding to an incident intensity of 1 GW cm^{-2} ; the IR signal must be much weaker, especially when high parametric gains are involved, in order to avoid pump-saturation phenomena.

The beams emerging from the nonlinear crystal are collected onto a monochromator and then detected by a Silicon (in the visible range) or a Germanium (in the IR range) photodiode connected to a computer that controls the delay line (path δ).

In this study we used a 1.5-mm-thick NPP crystal, grown by a modified Bridgman-Stockbarger method and cleaved so that the Z dielectric axis is normal to the input face. The sample is placed onto a stage consisting of two independent rotating plates with their rotation axes perpendicular to each other and to the incident beam. The crystal is oriented so that the Y (twofold) axis is parallel to the vertical polarization of the pump beam. The angle of incidence can be varied by rotation either around the Y axis or around the X axis, depending on the phase-matching plane.

Preliminary tests were performed on both POM and NPP in order to estimate their respective damage-threshold intensities at 0.62 μm. POM crystals withstand intensities slightly higher than 1 TW cm^{-2} , whereas for NPP the maximum tolerable intensity lies around 10 GW cm^{-2} . It must be noted that, whereas in POM optical damage is rapidly destructive and disqualifies the crystal for further optical applications, damage appears progressively in

NPP, leading to a local shift of the colour of the sample from yellow to red or brown.

SHG phase-matching conditions were determined for various fundamental wavelengths ranging from 1.018 to 1.332 μm by using the continuum, and tuning the monochromator, for a given incidence angle on the crystal, to the wavelength maximizing the value of the second-harmonic intensity. The crystal acts as a frequency selector, each angular value corresponding, through the phase-matching condition, to a given value of the IR wavelength. It must be pointed out that , although the thickness of the sample is small (1.5 mm), the second -harmonic beam is clearly visible between 0.5 and 0.62 μm ; this feature is helpful for the initial determination of the phase-matching angles. We have found two phase-matching planes : the YZ plane for $\lambda > 1.15\mu$m, the corresponding curve being obtained by rotation of the sample around the horizontal axis X by the angle θ , and the ZX plane for $\lambda < 1.15$ μm, the sample being rotated around the vertical axis Y by the angle ρ . A θ noncritical phase-matching configuration has been found at $\lambda = 1.15$ μm for a propagation direction along the Z axis and perpendicular to the input faces of the sample (Fig. 7). When the phase-matching curve is departing from the Z axis in the ZX plane, and consequently when the wavelength is increasing, its slope strongly increases, leading to a quasi- noncritical phase-matching configuration as described in the previous subsection. SHG is then possible without a large increase of the harmonic pulse duration with respect to fundamental, because of the large spectral acceptance and quasi group-velocity-matching characteristic of a noncritical configuration. It must be pointed-out that the pump wavelength of the amplification process is 575 nm, falling within the emission domain of the currently used Rhodamine 6G dye laser. In this configuration $d_{eff}= d_{21}$, without any angular projection factor. The polarizations sketched in the lower inset of Fig. 6 are then all in the molecular planes as defined in Fig. 4, thus fully involving the π electrons of the conjugated system of the molecule : the polarization plane of the three interacting waves is then in coincidence with the mean plane of the charge-transfer electrons responsible for the nonlinear effect. Another advantage of this configuration lies in the absence of any walk-off, ensuring a perfect transverse overlap of the various interacting beams along the whole interaction length.

The evidence of a θ-noncritical phase-matching configuration in NPP, fully optimizing the effective nonlinear coefficient d_{eff} for near-IR wavelengths and occuring at a highly convenient incidence angle (normal incidence on a naturally cleaved sample, therefore avoiding such critical technological steps as cutting and polishing), makes this compound a good candidate in view of parametric experiments such as parametric oscillation at the nanosecond pump-pulse-duration range. However, the optical quality has still to be improved over that of currently available NPP samples. On a femtosecond time-scale, where only single pass parametric interactions are possible, satisfactory results are in-keeping even with a relatively poor crystal quality.

The parametric phase-matching curve (Fig. 8) was plotted by angular tracking of the phase-matching amplified IR spectrum. Only that portion of the continuum ($\lambda \pm \Delta\lambda$) emitted by the water cell that corresponds to the phase-matching angle $\rho \pm \Delta\rho$ is amplified in the crystal ($\Delta\rho$ being the acceptance). The experiment is initiated by generating the second-harmonic wave for $\lambda = 1.24 \ \mu$m corresponding to degeneracy ($\lambda_i = \lambda_s = 2\lambda_p$). The temporal delay τ is then adjusted to ensure a simultaneous arrival

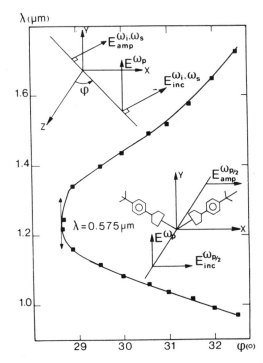

Fig 8   Parametric amplification phase-matching curve in NPP. The idler and signal wavelengths (λ_i and λ_s) are related to ρ (see caption of Fig 7) at the pump wavelength λ_p = 620 mm. Polarization orientations are represented in the upper inset while the lower inset depicts the θ -non critical degenerated parametric configuration at λ_p = 0.575 nm corresponding to Rhodamine 6G.

time of the pump and signal pulses inside the crystal, which corresponds to a maximal values of the amplified signal with respect to τ .The point on the phase-matching curve where the slope is vertical can then be located. For each signal wavelength λ_s , as indicated by the monochromator, the incidence angle on the crystal that maximizes the signal is measured. The whole phase-matching curve can then plotted for wavelength ranging from 0.95 to 1.7 μm. For a pump wavelength λ_p = 0.62 μm, the curve lies in the ZX plane, the crystal acting, for each position, as an amplifying filter for the frequency set (λ_i , λ_s). The quasi-vertical slope of the curve in the spectral domain studied here increases the spectral acceptance of the amplifier (800 - 1 000 Å around degeneracy) and consequently minimizes the group-velocity mismatch.

The maximum gain value obtained around degeneracy (λ = 1.24 μm) is of the order of 10^4 , corresponding to a maximum IR intensity of 100 MW cm^{-2} in a nonsaturated regime. Beyond this value, the amplified IR intensity is not negligible when compared with the pump intensity (up to 1 GW cm^{-2}), and therefore saturation of the gain must occur. In fact, these huge gain values make NPP a good candidate for amplification of small-intensity IR signals. Amplification carried-out in two successive NPP crystals, in which the amplified IR beam from the first sample is subsequently amplified in the second one, leads to a remarkable overall gain value of the order of 2×10^7 at λ_s = 1.1 μm. This result clearly demonstrates the ability of NPP crystals to efficiently amplify weak IR signals, with incident intensities of the order of a few tens of watts per square centimeter, corresponding to a few thousands photons per 100 fsec. NPP thus qualifies as a highly performing detector and processor of ultrashort near-IR laser pulses.

An estimate of d_{eff} can be deduced from the experimental gain value by using the following approximation /53/ :

$$\frac{\ln(4G)}{E_p} \left(\frac{n_i n_s}{\omega_1 \omega_2}\right)^{1/2} \frac{\varepsilon_0 c}{1} = d_{eff}, \qquad \text{Eq. 13}$$

where l is the interaction length, E_p the amplitude of the pump field, and $n_i (n_s)$ the refractive index of the idler (signal). In the experimental conditions mentioned above, we find that d_{eff} = (100 $\pm$ 20) x 10^{-9} e.s.u. at degeneracy. The d_{21} value, derived from d_{eff} by taking geometrical factors into account, is therefore (135 $\pm$ 30) x 10^{-9} e.s.u., a value that compares satisfactorily with that determined in Ref. /9/ by SHG at 1.064 μm in a thin crystalline layer. The relatively small discrepancy can be accounted for by three factors : first, the value determined here by parametric amplification is measured at a pump wavelength of 0.62 μm, a value that lies further away from electronic resonance than the 0.532 μm harmonic wavelength involved in the afore mentioned experiment ; secondly, the losses due to various defects, such as chemical impurities, dislocations, cleavages, and crystalline disorientations, are related to specific problems associated with the Bridgman-Stockbarger crystal growth technique, such as thermal decomposition of the material and relaxation of stresses. These defects can be held responsible for a phase-velocity mismatch of the interacting beams that results in a decrease of the interaction length compared with the full physical length. Third, group-velocity mismatch between the pump and the signal beams could be responsible for a non negligible shortening of the effective interaction length. However, this reduction must not be very significant, since group-velocity matching is approximately ensured, as explained in the previous subsection and confirmed by temporal analysis of the amplified signal (see next subsection). However, even a slight decrease (for example, from 1.5 to 1 mm) of the effective interaction length has a significant influence on the d_{eff} value determined from Eq.13. This experiment illustrates an important requirement for nonlinear materials applied to subpicosecond signal processing : their nonlinearitues must be high, because otherwise low nonlinear efficiency cannot be compensated by an increase of the crystal interaction length ; the ultimate limitation results from group-velocity-mismatch considerations which, in usual cases, lead to an effective interaction length of the order of 1 mm. Therefore only crystals, such as POM and NPP, that exhibit strong nonlinearity have been used so far in efficient parametric amplification and generation of tunable IR laser pulses.

It is also of interest to study the influence on gain values G of the pump intensity I_p. The pump is progressively attenuated by introducing neutral densities into the optical path. The behavior of G is linear with respect to $I_p^{1/2}$ for intermediary values of Ip as shown in Fig. 5 of Ref /47/ and corresponds to the following expression of G for high gain values /53/:

Eq 14

$$G = 1/2 \ \exp(2\Gamma 1),$$

with

$$\Gamma = \left(\frac{\mu_0}{\varepsilon_0}\right)^{3/4} \left(\frac{4\pi\omega_i\omega_s}{cn_p n_i n_s}\right)^{1/2} d_{eff}\sqrt{I_p},$$

Eq 15

where n_p is the refractive index at the pump wavelength.

For small gain values, $G = ch^2\Gamma 1$ and varies less steeply with respect to $\sqrt{I_p}$. For high pump intensities, huge values of the gain are reached, which are no longer compatible with the non depleted-pump approximation. Equation 14 is no longer valid because of the pump depletion. The full expession of the gain is a Jacobian elliptic function of the incident and signal pulse amplitudes, which saturates at higher pump intensity values.

It must be pointed-out that, although the parametric gain decreases

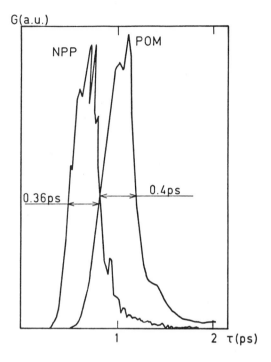

Fig 9   Cross-correlation plot of the pump 80 fs duration and
        signal (180 fs duration water continuum) by amplifica-
        tion in a 1.4 mm - thick NPP sample and a 4 mm - thick
        POM sample at degeneracy (λ = 1.24 μ m).

strongly when going toward 750 nm (see Fig. 9 in Ref. /47/), a value that is far removed from degeneracy (the corresponding idler wavelength is λ_i = 3.58μm). The width of the correlation (520 fsec) is significantly broader than in the degenerate case ; this can probably be accounted for by larger velocity mismatch among the three interacting waves. As the pump intensity was kept relatively small (less than 1 GW cm^{-2}), the gain is low (i.e., of the order of 20). The demonstration of the ability of NPP to perform parametric amplification at 750 nm confirms the potential of this material for ultrashort signal processing in a large spectral range.

PARAMETRIC AMPLIFICATION AND SAMPLING SPECTROSCOPY

The specific nonlinear properties of NPP crystals at the subpicosecond time scale can be taken advantage of to develop a new technique for time resolution of any near-IR emission, such as signals originating from parametric interaction (amplification, emission) or from luminescence processes. The method described further here has definite advantages over other types of techniques such as Streak Cameras (lesser time resolution and limited spectral response in the I.R.) as well as Kerr spectroscopy or summmixing spectroscopy, both of poor yield as opposed to PASS (or difference mixing) which involves a gain. It is in principle possible to substitute for the water cell that generates the continuum a nonlinear crystal emitting IR pulses by parametric emission or a sample giving rise to IR luminescence after excitation. Frequency mixing in NPP with a variable delay of such IR signals with the pump pulse allows for temporal sampling and amplification of the incident signal. In fact, the cross-correlation experiments described above represent a specific application of this more general configuration. Temporal analysis is now possible with IR signals of much longer duration than that of the sampling pulse. Therefore NPP can be used in a new spectroscopic technique, termed Parametric Amplification and Sampling Spectroscopy (PASS) /17, 47, 48/ which is of particular relevance in the temporal resolution of weak luminescence signals in the near IR. The signal is sent onto the nonlinear crystal and collected, after propagation through the nonlinear medium and a monochromator, on a photodiode, giving a value proportional to the total number N of collected photons. By varying the time delay between the pump pulse and the onset of the signal to be analyzed, one can follow the time evolution of the signal at the selected wavelength with a time resolution limited by the pump-pulse duration and with a satisfactory contrast owing to the high parametric gain. The actual experimental set-up is shown in Fig 10, while its principle is illustrated in Fig 11. However, if the signal (such as luminescence emission) is much longer than the pump pulse, the signal-to-noise ratio decreases strongly in contrast with the case of interacting beams of similar duration. The contrast is defined by:

$$C = \frac{N + S}{N}$$

Eq. 16

where the number of photons is.

$$N = \langle n \rangle \tau_1$$

Eq. 17

94

and the number of amplified signal photons is

Eq. 18

$$S = <n> \tau_p G$$

$\tau_1 (\tau_p)$ being the duration of the luminescence signal (pump beam), $<n>$ the number of photons per time unit, and G the parametric gain.

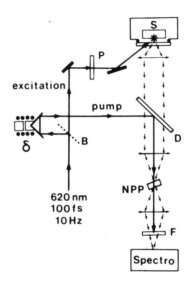

Fig.10   Sketch of the actual PASS spectroscopy /17/ experimental configuration where   stands for any I.R. luminescent sample of interest. δ , delay-line ; B, beamsplitter ; D, dichroic mirror ; F, filter ; P, polarizer.

Taking for τ_ℓ the typical values (100 psec), τ_p = 100 fsec, G = 10^4, we deduce from Eqs (16)-(18) that :

$$C = 1 + G \frac{\tau_p}{\tau_1} \sim 10$$

Eq. 19

Therefore, using nonlinear crystals with large gains over small interaction lengths is an efficient means to increase the contrast of long luminescence

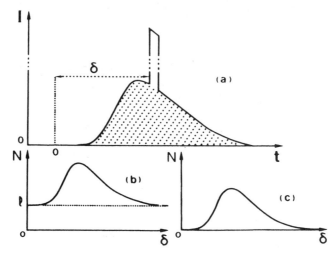

Fig. 11   Principle of PASS spectroscopy /17/. (a) : outcoming partially amplified I.R. intensity I with respect to time t.
An additionnal advantage of PASS lies in the possibility
to locate precisely, as described in text, the reference
time 0 when the exciting pump pulse intially reaches the
sample ; (b) : integrated signal energy (or photon number)
N as measured by the photodetector with respect to the
pump-luminescence variable δ ; (c) : corresponding idler
photons with no baseline.

signals while keeping an excellent time resolution. Note that a better contrast may be obtained by detecting the idler frequency ω in the absence of
incoming photons at the same frequency as shown in case (c) of Fig.11 (for
each amplified photon at ω_s , an idler photon is emitted to satisfy the
energy-conservation relation $\omega_p = \omega_i + \omega_s$). One additional major advantage in using the idler frequency consists in the eventual increase of the
detected frequency : if the signal to be studied lies at a large wavelength
($\lambda_s > 1.4$ μ m), for which the currently available detectors have a poor
sensitivity, observation at the idler wavelength $\lambda_i < 1.1$ μ m becomes easier
since the sensitivity of the detectors improves when the wavelength goes
down.

Determination of the arrival time of the pump beam on the sample is of
essential physical interest if such parameter as the delay in the onset of

the luminescence is studied. It can be reached by replacing the sample S
(see Fig. 10) at the same location, by a reflecting surface shined, at the
luminescence wavelength of interest by a filtered portion of the water continuum at the same wavelength.

The infrared luminescence of various samples was studied by this method. The samples were excited by 0.62 μm ultra-short pulses originating
from the same laser as the pump pulse. The infrared luminescence was collected through a large-aperture lens and focused into the NPP crystal so as
to undergo parametric amplification by interaction with the pump pulse. Time-resolved luminescence of an IR dye dissolved in methanol and of a multiple -quantum-well structure (MQWS) AlAs/InAlAs was investigated. The wavelengths of luminescence were 1.3 μm for the dye and 1.44 μm for the MQWS.
From the time-resolved spectra by PASS it was possible to infer interesting
features of the optical properties of the sample. For example, the time-
resolved luminescence of MQWS yields the delay between the arrival of the
excitation pulse at the sample and the rise of the photoluminescence signal
leading to information about the migration of the photocarriers through the
buffer layer down to their trapping sites in the quantum wells, as shown in
Fig. 12.

The highly interesting configuration refered to as case (c) in Fig. 11
where idler rather than signal photons, are detected so as to increase the
signal-to-noise ratio is described in Fig. 13. The feasibility of this expe-

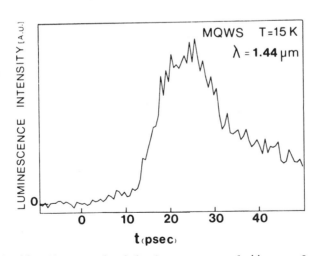

Fig 12   Time-resolved luminescence at 1.44 μ m of an
          InGaAs-InAlAs MQWS at 15 K excited by a
          0.62 μm pump beam.

riment is clearly demonstrated here, by amplifing in a NPP crystal the filtered portion at 1.62 μ m of the water continuum. The idler frequency signal at 1.01 μm is easily detected and the PASS spectrum, as expected,
shows a minimal baseline in the absence of incoming photons at this frequency.

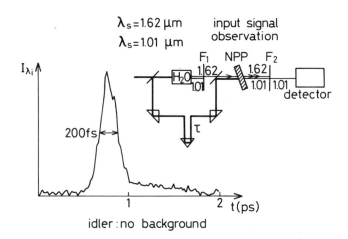

idler : no background

Fig, 13   Illustration of PASS spectroscopy in case (c) of Fig 11
where idler photons at 1.01 μ m are detected, corresponding to an incoming filtered water continuum signal at
1.62 μ m. F_1 and F_2, filters.

CONCLUSION

Time-compression has been called-upon, in the previously discussed
mixing configurations, in order to increase the gain (provided, as explained here, that group velocity matching and spectral acceptance requirements
are dealt with), and to increase the time resolution of the I.R optical

signal sampling and amplification process of interest here. It has been shown at the subpicosecond regime that a bulk crystal like NPP, or of similar efficiency such as NPAN qualifies for such an application as well as for ultrafast tunable emission, or highly sensitive detection of low energy C.W. I.R. signals. If the goal is to amplify randomly incoming short pulses such as would be of interest towards a parametric amplification device in an optical digital communication system, the point of view has to be drastically different : triggering a pulsed gain makes no more sense if only because synchronism between the pump and the signal cannot be ensured and the relatively lower repetition rate of a short pulse pump is not in keeping with the ever increasing transmission data rate of modern communication systems. A high gain must be ensured at CW powers of the order of 100 mW for a visible pump which requires, in order to reach ordes of magnitude such as reported here, spatial confinement of the interaction signal over transverse dimensions of the order of a fraction of the wavelength by some nonlinear waveguiding configuration /8/. As shown in another section of this book /54/, single mode organic crystalline cored fibers (OCCF) made of materials comparable to those reported here are theoretically capable of up-converting by second-harmonic generation CW semiconductor laser emission with almost 100% yield over interaction lengths of the order of a few mm. It is therefore of crucial importance to pursue further research in the domains of growth of high-quality OCCF's and their nonlinear optical characterization under short-pulse as well as continuous conditions of operation. Fig. 14 from Ref. /11, 12/ displays the respective advantages and drawbacks of NPP and NPAN, with respect to OCCF types of configuration at this stage of the research. As shown in Ref. /11/ where an up-to-date summary of past and present work in the field of OCCF's can be found, low symmetry crystals, such as monoclinic NPP, tend to grow with their symmetry elements, be it a plane for MNA/54/ or a two-fold axis for NPP, transversely oriented with respect to the cladding, while materials belonging to higher symmetry classes such as orthorhombic m-NA or NPAN tend to exhibit a structure where molecules are symetrically oriented with respect to the growth axis. The former structure leads to a favourable non-vanishing radiating tranverse component of the nonlinear polarization which conversely cancels-out in the latter one. Therefore, future research in this field should aim towards two directions : firstly try to modify the relative orientation of the crys-

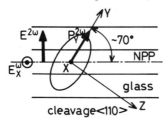

NPP waveguide (P2₁)

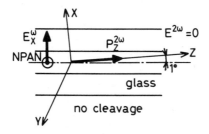

NPAN waveguide (Fdd₂)

Fig 14   NPP and NPAN waveguides from /12/ and /11/. The low
          symmetry monoclinic NPP crystalline structure leads
          to a favourable transverse orientation of the charge-
          transfer dipole moment although a marked tendency
          towards cleavage along the molecular plane (see Fig.14)
          ultimately limits the future of this material in OCCF
          structures. The quality of NPAN OCCF's is satisfactory
          but the higher symmetry of this material leads to an
          almost longitudinal unfavorable position of the charge-
          transfer axis.

talline line core with respect to the cladding by appropriate modifications
of the growth condition, a more realistic goal than try to act on the natural intrinsic crystalline structure of a given material ; secondly, work-
out a material which is capable of sustaining high CW laser power without
damage in compliance with the previouly exposed requirements of continuous
parametric amplification. Other types of configuration such as 2-D (and 1-D
structures derived therefrom by appropriate treatments) films are of great

interest : among the various film formation techniques, the Langmuir-Blodgett /2/ mono-and multilayer deposition technique is being investigated in
connection with quadratic nonlinear optics by a number of groups. As initially propsed in Ref. /8/, non-centrosymmetric bilayers composed of so –
called "inverted dyes" are in-keeping with the natural tendency of hydro-
drophobic molecular ends of successive layers to couple in an antiparallel
lattice and can be formed to remain stable. After the initial experiment by
Aktsipetrov et all. /55/, various group /56,57/ have investigated the SHG
efficiency of L-B layers. In Ref./57/, as shown in Fig.15, the transverse
conversion efficiency, surface macroscopic nonlinear susceptibility $\chi^{(2)}$,
molecular susceptibility β and relative geometry of molecules with respect
to substrates can be reached by the interference patterns between the
front-and back-layers emitted nonlinear waves. This method and the resulting fringing pattern was first proposed and demonstrated in Ref. /58/, in
the case of third-harmonic generation and polydiacetylene L-B layers. The
very high value measured here, of the order of 10^{-27} e.s.u. can be accoun-

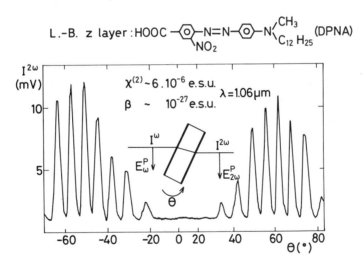

Fig.15 Interference fringes of second-harmonic generation by
a DPNA L-B layers coated silica plate following a measurement technique developped in the case of third-harmonic generating polydiacetylene L-B layers in Ref ./57/.

ted-for by resonnance between the harmonic wave at 530 nm and the absorption peak of the diazo-stilbene DPAN molecule. Waveguiding parallel to the layers is presently being performed in a configuration proposed in Ref. /8/ (case of Fig. 2 in this reference) and more thoroughly investigated in Ref. /52/ (see section C and Fig. 49 in this reference, where phase-matching, birefringency, and transverse mode overlap are being considered). An adequate trade-off between absorption and conversion efficiency will have to be defined and actually implemented by playing in such degrees of freedom as passive and active film thicknesses and refractive index difference.

In conclusion, statistically oriented media such as L-B layers, or doped polymer guests /5,6,9/ provided there optical quality is compatible with the stringent demands of nonlinear waveguides optics as summarized in Ref. /8/, are of interest in view of a convenient macroscopic structural flexibility of which single crystalline structures are deprived. However, single crystals in various configurations, bulk as mainly discussed her, films /9/, or fibers /11,12/ have merits of their own, essentially strict ordering and maximal packing density, which is bound to stirp-up further research.

ACKNOWLEDGEMENTS

Research results reported herein have benefited from various contributors : J. BADAN, M. BARZOUKAS, D. JOSSE, I. LEDOUX, P. FREMAUX, P. VIDAKOVIC from CNET for bulk or fiber crystal growth and NLO characterization; J. MORLEY from ICI for some theoretical results; J.F. NICOUD from ESPCI for molecular engineering and synthesis ; A. ETCHEPARE G. GRILLON, D. HULIN and A. MIGUS from Ecole Polytechnique-ENSTA for femtosecond time-scale NLO experiments. The author is grateful to F. Herin for the typing work. Partial support from the EEC through ESPRIT Project 443 is gratefully acknowledged.

REFERENCES

/1/ "Nonlinear Optical Properties of Organic Molecules and Polymers", ACS Symp.Ser.233 Edit. D.J. Williams (A.C.S., Washington, 1983)
/2/ "Nonlinear Optical Properties of Organic Molecules and Crystals",

2 vols., series : Quantum Electronics, Principles and Applications.
Edits. D.S. Chemla and J. Zyss. (Academic Press, Orlando, 1987).

/3/ "Molecular Crystals and Molecules" A.I. Kitaigorodsky (Academic Press, Orlando, 1973).

/4/ D.S. Chemla, J.L. Oudar and J. Jerphagnon, Phys. Rev.B12, 4534 (1975).

/5/ "Nonlinear Optical Process in Organic Materials" feature issue of J. Opt. Soc. Am. B 4(6) june 1987. See for example Singer et al. for doped PMMA and Tompkin et al. for dye containing glass.

/6/ G.R. Meredith et al. p.109 in Ref. /1/ and P. le Barny, G. Ravaux, J.C. Dubois, J.P. Parneix, R. Njeumo, C. Legrand and A.M. Levelut Proceeding of SPIE "Molecular and Polymeric Materials, Fundamentals and Applications", vol. 682, 56 (1986).

/7/ G.I. Stegeman et al. p. 31 in "Nonlinear Optics : Materials Devices", Springer Proceeding in Physics 7, Edits. C. Flytzanis and J.L. Oudar (Springer Verlag, Berlin, 1986).

/8/ J. Zyss, J. Mol. Electron 1, 25 (1985).

/9/ I. Ledoux, D. Josse, P. Vidakovic and J. Zyss, Opt.Eng. 25, 202 (1985)

/10/ B.K. Nayar p. 142 in Ref. /7/.

/11/ P.V. Vidakovic, M. Coquillay and F. Salin in Ref. /5/.

/12/ P.V. Vidakovic, J. Badan R. Hierle and J. Zyss, IQEC-84, Postdeadline Commun. PD-C5-1.

/13/ "Solid-State Laser Engineering", W. Koechner, Springer Series in Optical Sciences (Springer Verlag, New-York, 1976).

/14/ J. Zyss, J.F. Nicoud and M. Coquillay, J. Chem.Phys. (81), 4160 (1984)

/15/ M. Barzoukas, D. Josse, P. Fremaux, J. Zyss, J.F. Nicoud, and J.O. Morley in Ref. /5/.

/16/ J. Zyss and JL. Oudar, Phys. Rev. A26, 2028 (1982).

/17/ D. Hulin, A. Migus, A. Antonetti, I. Ledoux, J. Badan, J.L. Oudar and J. Zyss, Appl. Phys. lett 49 (13), 761 (1986).

/18/ G. Hauchercorne, F. Kerherve and G. Mayer, J. Phys. 32, 47 (1971).

/19/ I. Ledoux and J. Zyss, Chem. Phys. 73, 203 (1982) see also J. Zyss and G. Berthier, J. Chem Phys. 77, 365 (1982).

/20/ K.D. Singer and F. Garito, J. Chem. Phys. 75, 3752 (1981).

/21/ J.L. Oudar, J. Chem. Phys. 67, 446 (1977).

/22/ J.F. Ward and C.K. Miller, Phys. Rev A 19, 826 (1979).

/23/ M. Barzoukas, D. Josse and J. Zyss and F. Kajzar to appear in the Proceedings of the MRS fall meeting (dec. 1987).

/24/ F. Kajzar, I. Ledoux and J. Zyss, to be published in Phys. Rev. A.

/25/ M.L. Shand and R.R. Chance p. 187 in Ref. /1/ and R.R. Chance in this book.

/26/ H.D. Cohen and C.C.J. Roothan, J. Chem. Phys. 43, 534 (1965).

/27/ J. Zyss, J. Chem. Phys. 70, 3333 (1979).

/28/ J. Zyss, J. Chem. Phys. 70, 3341 (1979).

/29/ J. Zyss, J. Chem. Phys. 71, 909 (1979).

/30/ S.J. Lalama and A.F. Garito, Phys. Rev. A20, 1179 (1979).

/31/ V.J. Docherty, D. Pugh and J.O. Morley J. Chem. Soc. Faraday Trans. 2 81, 1179 (1985).

J.O. Morley, V.J. Docherty and D. Pugh to be published in J. Chem. Soc., Perkin Trans. in the press (Part I, Part II, Part III).

/32/ E.F. Mc Intyre and H.F. Hameka, J. Chem.Phys. 69, 4814 (1978) and references therein.

/33/ C.W. Dirk, R.J. Twieg and G. Wagnière, J. Am. Chem. Soc. 108, 5387 (1986).

/34/ H. Sekino and R.J. Bartlett, J. Chem. Phys. 85, 976 (1986).

/35/ J.A. Pople, J.W. McIver and N.S. Ostlund, J. Chem. Phys.49, 2960 (1986).

/36/ S. Tomaru, S. Zembutsu, M. Kawachi and M. Kobayashi, J. Chem. Soc., Chem. Commun., 1207 (1984).

/37/ J.F. Nicoud and R.J. Twieg in Ref. /2/ p.227 in vol. 1, pp. 221 and 255 in vol. 2.

/38/ J.L. Oudar, J. Chem. Phys. 67, 446 (1977).

/39/ J.L. Oudar and J. Zyss, Phys. Rev. A26, 2076 (1982).

/40/ "Atlas of Spectral Data and Physical Constants for Organic Compounds" 2nd Edition, eds. J.G. Grasselli and W.M. Ritchey (CRC Press, Cleveland, 1975).

/41/ M. Sigelle and Hierle, J. Appl. Phys. 52, 4199 (1981).

/42/ J. Zyss, D.S. Chemla and J.F. Nicoud, J. Chem. Phys. 74, 4800 (1981).

/43/ A.R. Katritzky, E.W. Randall and L.E. Sutton, J.Chem.Soc., 1769 (1957)

/44/ R.J. Twieg and C.W. Dirk, J. Chem. Phys. 85, 3537 (1986).

/45/ B.F. Levine, C.G. Bethea, C.G. Thurmond, C.D. Lynch and J.L. Bernstein, J.Appl. Phys. 50, 2523 (1979).

/46/ I. Ledoux, J. Zyss, A. Migus, J. Etchepare, G. Grillon and A. Antonetti, Appl. Phys. Lett. 48, 1564 (1986).

/47/ I. Ledoux, J. Badan, J. Zyss, A. Migus, D. Hulin, J. Etchepare, G. Grillon and A. Antonetti, in Ref. /5/.

/48/ J. Zyss, I. Ledoux, J. Badan and J.L. Oudar to be published in the "Revue de Physique Appliquée".

/49/ J. Zyss, I. Ledoux, R. Hierle, R. Raj, J.L. Oudar, IEEE J. Quantum Electron. QE 21, 1286 (1985).

/50/ J.L. Oudar and R. Hierle, J. Appl. Phys. 48, 2699 (1977).

/51/ J.L. Oudar and M. Sigelles private communication reported in Ref. /2/ vol. 1 p. 149 (Fig. 32).

/52/ J. Zyss and D.S. Chemla p. 23 in Ref. /2/, vol. 1.

/53/ "Quantum Electronics" A. Yariv (Wiley, New-York, 1975).

/54/ see B.K. Nayar in this book and private communication.

/55/ O.A. Aktsipetrov, N.N. Akhmediev, E.D. Mishina and V.R. Novak, JETP lett. 37, 207 (1983).

/56/ I.R. Girling, N.A. Cade, P.V. Kolinsky, J.D. Earls, G.H. Gross and I.R. Peterson, Thin Solid Films 132, 101 (1985) and Refs. therein.

/57/ I. Ledoux, D. Josse, P. Vidakovic, J. Zyss, R.A. Hann, P.F. Gordon, B.D. Bothwell, S.K. Gupta, S. Allen, P. Robin, E. Chastaing and J.C. Dubois, Europhys. Lett. 3, 803 (1987).

/58/ F. Kajzar, J. Messier, J. Zyss and I. Ledoux, Opt. Commun. 45, 133 (1983).

/59/ K.D. Singer, M.G. Kuzyk and J.E. Sohn in Ref. /5/ and K.D. Singer et al. in this book.

ELECTRIC-FIELD POLING OF NONLINEAR OPTICAL POLYMERS

C. S. Willand, S. E. Feth, M. Scozzafava, and D. J. Williams

Corporate Research Laboratories, Eastman Kodak Company
Rochester, NY

G. D. Green[*], J. I. Weinschenk, III, H. K. Hall, Jr., and
J. E. Mulvaney

C. S. Marvel Laboratories, Department of Chemistry
University of Arizona, Tucson, AZ

INTRODUCTION

Second-order nonlinear optical (NLO) processes, including harmonic generation and the electrooptic effect, are the result of a nonlinear response in the polarization P of a dielectric which is quadratic in the applied electric field of the light beams E

$$P = \chi^{(2)}:EE . \qquad (1)$$

The coupling coefficient $\chi^{(2)}$ is the second-order electric susceptibility tensor for the material. Maximizing the nonlinear susceptibility coefficients can result in large NLO effects. However, understanding the origins of $\chi^{(2)}$ often requires examination of these interactions on the microscopic level. Here, the process can be viewed via the molecular polarization p which is induced by an electric field e local to the molecular environment

$$p = \beta:ee , \qquad (2)$$

where β is referred to as the second-order molecular hyperpolarizability. Much theoretical and experimental research in the past has been directed toward understanding the molecular origins of β.[1-3] Of particular importance has been the discovery of the contributions of intramolecular charge-transfer in conjugated systems to the hyperpolarizability. This finding along with others has resulted in the discovery of numerous molecules with large hyperpolarizabilities.

A major obstacle in the development of new second-order NLO materials has been the incorporation of molecules with large β into condensed phases with appropriate symmetries for observing large values of $\chi^{(2)}$. For systems composed of noninteracting molecules (the "oriented gas" approximation), one can express the second-order susceptibility as a weighted orientational average of the hyperpolarizability

$$\chi_{IJK}^{(2)} = N \sum_{rst} \langle F_{rst} A_{Ir} A_{Js} A_{Kt} \rangle \beta_{rst} , \tag{3}$$

where $\langle \; \rangle$ denotes an orientational average over all possible molecular orientations and is weighted by the molecular orientational distribution function. The summations extend over molecular Cartesian coordinates. The direction cosine matrix A contains the information linking the lab-frame Cartesian coordinate system to the molecular frame. Finally, F is the local field factor tensor which relates the external electric fields to those present at the molecular site.

For isotropic or centrosymmetric materials, Eq. (3) reduces to the well known result of $\chi^{(2)} \equiv 0$, regardless of the magnitude of the molecular hyperpolarizability. Hence it is important to have systematic approaches for achieving optimal molecular orientation in condensed phases. Additional important criteria relating to photochemical and thermal stability, linear optical properties including transparency, birefringence, and uniformity, as well as other requirements for processing in appropriate formats determine the suitability of a material for applications in nonlinear optics. To this end various methods are being explored for producing such materials.[4-9] Approaches based on single crystal growth could lead to the highest achievable nonlinearities due to the high concentration of active chromophores and would have linear optical properties determined by the degree of perfection of the crystal.[8] Unfortunately, crystal symmetries are unpredictable and large crystals with reasonable mechanical properties are difficult to achieve. Hence a variety of alternative approaches have emerged for achieving well controlled orientation with some sacrifice in the ultimate nonlinear coefficient. These include monomolecular films,[5] poled polymer glasses,[7] poled polymeric liquid crystals,[9] and poled polar polymers. Monomolecular films rely on control of inter- and intralayer forces and bonding to control symmetry. Poled polymer glasses combine the useful properties of polymeric host media with those of the nonlinear molecule and use electric field poling to achieve partial removal of orientational averaging of the isotropic medium. Dilution of the chromophore concentration and the practical problems associated with high poling fields constitute the limitations on this approach. Polymeric liquid crystalline media have been shown to provide enhancement in the orientation of dopant molecules due to the contribution of orientational potential of the host to the guest polar alignment. In this paper we explore the possibility of an additional alignment enhancement mechanism in polymers due to cooperation of individual molecular dipole contributions in polar polymer chains. The discussion will be confined to second harmonic generation, but the results may easily be generalized to other second-order nonlinear optical processes.

ELECTRIC-FIELD POLING

In static electric-field poling, the interaction of an electric field and the permanent molecular dipole moment causes the dipoles to orient preferentially in the direction of the field. The orientational distribution function for this process can be written in terms of a Boltzmann distribution in an isotropic medium:

$$W = \exp(-\mu E_0 \cos \theta / kT) , \tag{4}$$

where θ is the angle between the permanent molecular dipole moment μ and the static electric field E_0; kT is the Boltzmann thermal energy. Using Eqs. (3,4) and neglecting contributions from third-order frequency mixing processes $(2\omega = \omega + \omega + 0)$ yields only two independent nonzero susceptibility tensor elements:[9]

$$\chi^{(2)}_{ZZZ} = N\, f_\omega^2\, f_{2\omega}\, L_3(f_0 E_0 \mu_z/kT)\beta_z \,, \tag{5}$$

$$\chi^{(2)}_{ZXX} = \frac{1}{2}\left[\frac{L_1(f_0 E_0 \mu_z/kT)}{L_3(f_0 E_0 \mu_z/kT)} - 1\right]\chi^{(2)}_{ZZZ} \,, \tag{6}$$

where

$$\beta_r = \frac{1}{3}\sum_i (\beta_{rii} + \beta_{iri} + \beta_{iir}) \,, \tag{7}$$

$$f_\omega = \frac{\eta_\omega^2 + 2}{3} \,, \tag{8}$$

$$f_0 = \frac{\epsilon_\infty + 2}{\epsilon_\infty + 2\epsilon_0}\, \epsilon_0 \,, \tag{9}$$

and N is the number density of nonlinear chromophores. The macroscopic Z-axis and molecular z-axis are defined to lie along and in the direction of the static electric field and the molecular dipole moment, respectively. The Onsager f_0 and Lorentz-Lorentz f_ω local field factors are assumed to be rotationally invariant, being only a function of the indices of refraction η_ω, the static dielectric constant ϵ_0, and the dielectric constant for frequencies above dipolar relaxation ϵ_∞. The nth-order Langevin functions $L_n(x)$ represent the orientational average of $\cos^n\theta$ and are defined as

$$L_1(x) = \coth(x) - \frac{1}{x} \tag{10}$$

$$= \frac{x}{3} - \frac{x^3}{45} + \cdots \,,$$

$$L_3(x) = L_1(x)\left(1 - \frac{6}{x^2}\right) - \frac{2}{x} \tag{11}$$

$$= \frac{x}{5} - \frac{x^3}{105} + \cdots \quad .$$

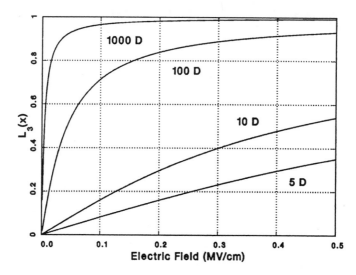

Fig. 1. $L_3(f_0E_0\mu_z/kT)$ versus electric field strength at T = 298 K. The various curves denote differing values for $(f_0\mu_z)$.

The electric-field dependence of the Langevin function $L_3(f_0E_0\mu_z/kT)$ is shown in Fig. 1 for various values of $(f_0\mu_z)$. It can be seen that the value of this function is significantly less than unity for typical molecular dipole moments; this is even true for field strengths approaching those nominally associated with dielectric breakdown. In contrast, molecules with large dipole moments display saturation (non-linear) behavior of $L_n(x)$ for very low field strengths. Therefore, one approach to increasing the overall alignment of the NLO chromophores, and hence the nonlinear susceptibility, would be to bond covalently the molecules into polar polymer main and side chain structures.

In the low field limit where $|\mu E_0| \ll kT$, the Langevin functions are linear and the susceptibility can be approximated by

$$\chi^{(2)}_{ZZZ} = N\left(\frac{f_\omega^2 f_{2\omega}f_0}{5kT}\right)\mu_z\beta_z \quad ,\tag{12}$$

$$\chi^{(2)}_{ZXX} = \frac{1}{3}\chi^{(2)}_{ZZZ} \quad .\tag{13}$$

For a polymer containing n monomer subunits, Eq. (12) can be rewritten as

$$\chi^{(2)}_{ZZZ} = N\left(\frac{f_\omega^2 f_{2\omega}f_0}{5kT}\right)\left(\frac{M_z B_z}{n}\right) \quad ,\tag{14}$$

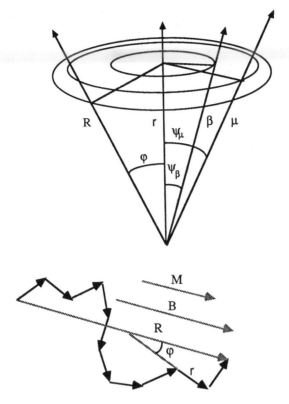

Fig. 2.   Schematic representation of a polymer chain showing the relationship between vectors M, B, R, and μ, β, r.

where N is the number density of monomer units, and M_z and B_z are the
polymer dipole moment and vector part of the hyperpolarizability,
respectively. The quantity ($M_z B_z/n$) is analogous to ($\mu_z \beta_z$) for the
single molecule case and can be interpreted as an "average monomer
susceptibility" in the polymer chain. The relationships between M, μ,
B, and β are illustrated in Fig. 2. Here hypothetical polymer chains
are drawn and characterized by an end to end vector R. The angle
between the planes defined by the vector pairs (r,β) and (r,μ), as well
as the angles Ψ_β and Ψ_μ, are determined by the chromophore. For the
purpose of this discussion it is assumed that there is a random
distribution in the angle between (r,R) and (r,μ). M and B are
therefore parallel to R and represent the vector sums

$$M = \sum_{i=1}^{n} \mu^i = n\mu \langle \cos \phi \rangle \cos \Psi_\mu \, \hat{z} \, , \tag{15}$$

$$B = \sum_{i=1}^{n} \beta^i = n\beta \langle \cos \phi \rangle \cos \Psi_\beta \, \hat{z} \, , \tag{16}$$

where $\langle \, \rangle$ indicates an average value over the polymer chain.

An enhancement factor G in the nonlinear susceptibility can be defined as

$$G = \frac{\chi^{(2)}_{ZZZ}(\text{polymer})}{\chi^{(2)}_{ZZZ}(\text{monomer})} = \frac{\left(\frac{M_z B_z}{n}\right)}{\mu\beta} = n \langle \cos\phi \rangle^2 \cos\Psi_\mu \cos\Psi_\beta . \quad (17)$$

From Eq. (17) it is clear that extended polymer chains with μ and β parallel to the backbone could have a large enhancement relative to an equivalent concentration of monomers, whereas a tightly coiled random conformation or a chain composed of monomers having either μ or β perpendicular to r could reduce the nonlinearity to near zero. It should be noted that $\langle\cos\phi\rangle$ could be field dependent if the chain conformation is perturbed by the field.

Earlier work of Levine and Bethea[10] on poly-γ-benzyl-L-glutamate ($n \simeq 2500$) in ethylene dichloride solutions showed that it is possible to achieve substantial enhancement of the average monomer susceptibility by incorporating NLO chromophores into a polymer. This increase was attributed to the polymer's large dipole moment (8000 D) as a result of the rigid α-helix structure. While these results are promising for those molecules which form rigid polymer structures, the question remains as to whether similar effects can be realized for flexible polymers. In a theoretical treatment of a prototype flexible polymer, Khanarian[11] used the methods of Flory to calculate the average monomer susceptibility for poly(oxyethylene dimethyl ether) as a function of the polymer chain length. Here, the magnitude of the monomer susceptibility fell off rapidly as the chain length increased from one to five; the value for chains of size $n \geq 20$ was approximately one-third that for a single molecule ($n = 1$). However, the model used was limited to polymers in the absence of electric fields and thereby ignored any field-induced conformational changes which occur in typical poling experiments. In most instances, these changes would be expected to increase the average monomer susceptibility.

ELECTRIC-FIELD INDUCED SECOND HARMONIC GENERATION FROM FLEXIBLE POLYMER SOLUTIONS

We have performed a series of electric-field induced second harmonic generation (EFISH)[2,12] measurements in order to compare the average monomer susceptibility for some single molecule and flexible main chain polymers containing NLO chromophores. The EFISH technique utilizes second harmonic generation to probe the nonlinear susceptibility of a solution induced by a pulsed electric field. Using properly selected field polarizations, the intensity of the second harmonic (SH) generated is related to the square of the nonlinear susceptibility $\chi^{(2)}_{ZZZ}$. The magnitude of $(\mu_z\beta_z)$ for the solute and solvent molecules can be determined using the concentration dependence of the susceptibility and Eq. (12). The sign of $(\mu_z\beta_z)$ is derived for the solute relative to that of the solvent.

A new group of nonlinear optical chromophores based on p-oxy-α-cyanocinnamate were synthesized for use in our investigations (see Table 1).[13] These molecules would be expected to display moderate second-order NLO properties on the basis of their conjugated electron donor-acceptor structures. In addition, their lowest electronic transitions lie in the ultraviolet region of the spectrum making them potentially useful for frequency doubling into the visible. All measurements

Table 1. EFISH Results for p-oxy-α-cyanocinnamate Derivatives

Molecule	$\mu_z\beta_z(10^{-48}$ esu$)$
I — benzene ring with OCH$_3$ (top), CH$_3$O— (left), CH$_3$O (bottom), bearing CH=C(CN)—COOCH$_3$	113
II — benzene ring with HO(CH$_2$)$_3$O— (left), CH$_3$O (bottom), bearing CH=C(CN)—COOCH$_3$	57
III — benzene ring with CH$_3$O— (left), CH$_3$O (bottom), bearing CH=C(CN)—COOCH$_3$	48
IV — benzene ring with CH$_3$O (top), CH$_3$O— (left), CH$_3$O (bottom), bearing CH=C(CN)—COOCH$_3$	-10

were performed on dilute solutions (<3% by weight) using chloroform as a solvent.

The experimental setup for the EFISH measurements is shown in Fig. 3. The solutions to be examined were placed in a specially designed cell where a pulsed electric field of approximately 6 kV/cm and 100 μsec in duration was applied. Synchronous to the high voltage pulses, the 1064 nm output beam from a 10 pps, Q-switched Nd:YAG laser was focussed by a long focal length lens into the sample cell. The polarization and intensity of the light reaching the sample were controlled by a half-wave plate and polarizer. The second harmonic generated in the cell was collected by a lens, focussed onto the slits of a monochromator, and detected by a photomultiplier tube. Optical filters were placed before the sample cell to absorb any spurious SH light and after the cell to decrease the amount of laser radiation reaching the detector. A portion of the initial laser beam was also redirected and focussed into a reference quartz crystal. The SH generated from the crystal was detected in a manner identical to that previously described. The sample and reference signals were integrated and divided by a gated integrator to yield a signal level normalized for fluctuations in the incident laser intensity. The nonlinear susceptibility was derived from the normalized SH intensity levels using standard Maker fringe techniques.[2,12]

The results of the EFISH measurements are shown in Table 1. The values listed are similar in magnitude to cinnamaldehyde derivatives previously reported.[14] Comparison of $(\mu_z \beta_z)$ for structures I and IV indicates a pronounced difference between meta and ortho positioning of the electron donating methoxy group on the phenyl ring relative to that of the electron acceptor. Only the ortho and para substituted forms have low lying, donor-to-acceptor charge-transfer states; they would therefore be expected to display larger hyperpolarizabilities than meta forms. Similar effects are also seen for nitroaniline.[15]

The flexible polymers studied in this investigation are a series of main-chain polyesters based on structure II.[13] These polymers all contain monomeric units which are oriented in the same direction along the polymer backbone. Homopolymers of molecule II were found to be only slightly soluble in common organic solvents. Hence, 1:1 random copolymers of molecule II and 12-hydroxydodecanoate were prepared which were

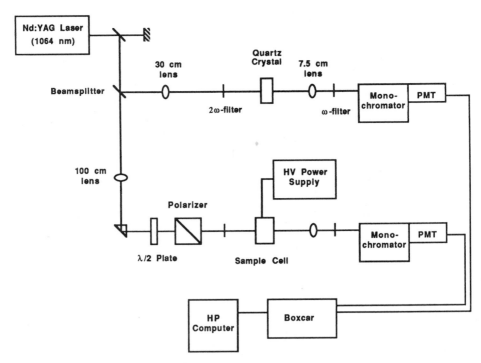

Fig. 3.   Experimental setup for the EFISH measurements.

readily soluble in chloroform. Absorption spectra for the copolymers and molecule II were nearly identical for the lowest excited, charge-transfer state (λ_{max} = 363 nm). Therefore, the chromophore electronic properties critical to the hyperpolarizability are assumed to be unperturbed by the polymerization; changes in the average molecular susceptibility are the result of orientational rather than chemical interactions.

The experimental data for two different length 1:1 copolymers, as well as the NLO chromophore, are given in Table 2. The polymers display enhancements as large as a factor of 20 for the average monomer susceptibility relative to that of the NLO chromophore. This is due to the average polymer chain conformation in solution which results in an effective chain dipole considerably larger than those of individual monomer units. The electric field induced SHG signal is therefore considerably larger for these correlated chains than for an equivalent concentration of monomeric units. The fact that the enhancement factor G is substantially less than n indicates that the conformation is far from extended.

PROPERTIES OF POLED POLYMERIC THIN FILMS

An advantage of polymers with regard to poling applications is that they can exist in various viscous and nonviscous states. Molecular orientation can be frozen into polymeric materials by using the following sequence:

fixed,	$\xrightarrow{\text{mobilize}}$	mobile,	$\xrightarrow{\text{field on}}$	mobile,	$\xrightarrow{\text{freeze}}$	fixed,
isotropic		isotropic		polar		polar

"Thermal-locking" of a polar orientational distribution can be obtained by heating the polymer above the glass transition temperature T_g (mobilizing the polymer chains) and poling with an electric field. The orientational order is then fixed by cooling to temperatures below the glass transition before removing the field. Previously published work has concentrated on the use of guest-host systems whereby monomeric NLO chromophores are doped into a host polymer matrix (such as poly(methyl

Table 2. Average Monomer Susceptibilities for
Various Molecular Weight Copolymers

Molecular Weight,[a] $\bar{M}_n$	Calculated Number of Monomer Units, n	$M_z B_z/n$ (10^{-48} esu)	Enhancement factor, G
NLO chromophore[b]	--	57	--
17,000	37	830	15
70,000	152	1140	20

[a]Determined by GPC relative to polystyrene.
[b]Structure II from Table 1.

methacrylate)). Nonlinear optical polymers would have the added advantage that no host matrix is required; the host matrix dilutes the nonlinear effect by lowering the number density of chromophores. However, the linear optical and physical properties of potential NLO materials must also be considered.

We have examined thermal-locking in thin films of the copolymers described in the previous section. The sample cell used in these experiments is shown in Fig. 4. The two electrodes were constructed by evaporating a thin layer of chromium onto a quartz optical flat separated by a 150 micron gap. Thin films were prepared by spin coating solutions of the $\bar{M}_n$ = 70,000 copolymer in chloroform onto the sample cell. The solvent was allowed to evaporate at ambient temperatures for about an hour. The films were then annealed at temperatures above 100°C for a period of at least 24 hours. Typical film thicknesses were measured to be on the order of one micron. The setup for the second harmonic generation experiments was similar to that in Fig. 3 except that a 7.5-cm lens was used to focus the laser beam onto the sample. The laser light was focussed onto the electrode gap at normal incidence and was polarized parallel to the static electric field. The incident laser pulse energy was maintained below 15 μJ/pulse to avoid any laser-induced thermal effects in the film. A DC power source was used to supply the poling voltage. Cooling of the sample was achieved by passing a stream of dry nitrogen gas through a liquid nitrogen heat exchanger and into a sealed chamber containing the sample cell. The temperature was monitored with a thermocouple. For all experiments the cooling and heating rates were less than 0.5 K/min to provide adequate thermal homogeneity of the film.

Differential scanning calorimetry (DSC) measurements revealed the glass transition temperature of the film to lie below room temperature.

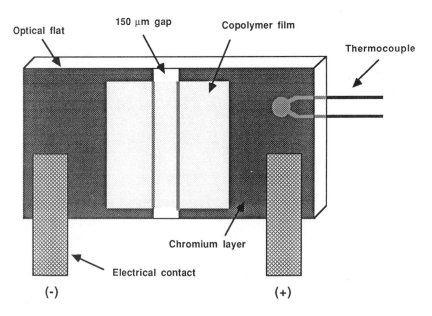

Fig. 4. Sample cell used in the thin film poling experiments.

Hence, the sample could be poled at ambient temperature. The maximum electric field strength used in our experiments was 0.2 MV/cm. At room temperature, the SH intensity from the poled film was found to increase and decrease by at least one order of magnitude when the static field was turned on and off, respectively. In addition, a sub-second time response of the molecular orientation was encountered with no slow processes occurring over the period of one to two hours. Next, the sample was cooled to approximately 220 K (well below T_g) before removal of the field. Here, the SH intensity from the poled sample appeared to be independent of the presence of the electric field. The signal was therefore assumed to be the result of a field-induced molecular reorientation with negligible contribution from third-order NLO mixing processes. In addition, the signal at 220 K was constant over a period of hours indicating the fixed polar molecular alignment to be stable.

Next, the sample was slowly heated to room temperature with the field off. Two distinct transition temperatures in the SH signal were found (see Fig. 5). At ~250 K a slow decline of the signal was noted followed by a steeper decline beginning at ~285 K. The heat capacity, as measured by DSC, for the copolymer is also displayed in Fig. 5. There appears to be excellent overlap in the glass transition region of the heat capacity curve and the steepest declining region of the SH intensity curve. In this region, one would expect to observe large-scale segmental motions in the polymer chains allowing for an orientational relaxation of the poled film and a subsequent loss in SH signal. The slow decline of the signal in the region between 250 K and the glass transition temperature is most likely a result of the loss of short-range ordering through motions on the distance scale of the size of bonds or single molecules. This type of relaxation would be expected to be thermally accessible at temperatures beneath T_g and would have less drastic effects on the SH intensity than the larger-scale relaxation present above the glass transition.

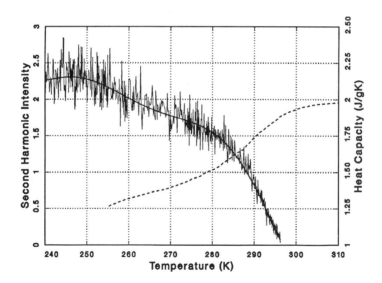

Fig. 5. SH intensity (——, original and smoothed) generated from the oriented film and heat capacity (---) of the copolymer versus temperature.

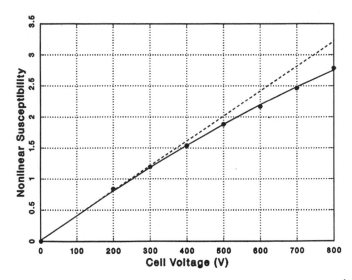

Fig. 6.   Experimental (*) and theoretical best fit (——) of $\chi_{ZZZ}^{(2)}$ as a function of poling voltage for the copolymer film. The dashed line represents the susceptibility expected using Eq. (14).

The electric-field dependence of $\chi_{ZZZ}^{(2)}$ for the film at room temperature is shown in Fig. 6. Nonlinear least-squares analysis fitting the data to a third-order Langevin function (see Eq. (5)) gives an effective dipole moment for the poling process of $(30 \pm 5\ D)/f_0$. This value is much less than expected in solution and is the equivalent of segmental alignment over the distance of only 3 or 4 monomer units. Measurements performed at 310 K and 360 K gave similar results suggesting that the polymer motion is constrained even at temperatures well above T_g.

The magnitude of $\chi_{ZZZ}^{(2)}$ for the film was measured to be 5×10^{-9} esu by comparison with quartz. Rewriting Eq. (5) for the polymer we obtain

$$B_z = \frac{n\ \chi_{ZZZ}^{(2)}}{N\left(\dfrac{\eta^2 + 2}{3}\right)^3 L_3\left(\dfrac{f_0 E_0 M_z}{kT}\right)} . \tag{18}$$

Using the measured values for $\chi_{ZZZ}^{(2)}$ and L_3 along with $n = 152$, $\epsilon = 2.6$, and $\eta = 1.6$, we obtain $B_z = 182 \times 10^{-30}$ esu. The enhancement factor G from Eq. (17) is therefore

$$\frac{24 \times 10^{-48}\ \text{esu}}{57 \times 10^{-48}\ \text{esu}} = 0.4 .$$

It is instructive to compare this to the solution value of

$$G = \frac{1140 \times 10^{-48} \text{ esu}}{57 \times 10^{-48} \text{ esu}} = 20 \ .$$

Although the enhancement factor is calculated relative to the solution value for the monomer, we feel the comparison is valid since monomeric species in viscous glassy media appear to follow the behavior predicted by Eq. (5).[7]

Some additional insights into the assumptions of this model, i.e., the relationship between μ and β, can be obtained by comparing the following ratios: $B_z/\beta_z \sim 18$ and $M_z/\mu_z \sim 4$ (assuming $\mu_z = 5$ D). Apparently the vectors μ and β are not strictly parallel since the vector summation of the hyperpolarizability B_z is considerably more effective than that for the dipole moment M_z (cf. Eq. (15,16)). In an optimized system the ratios would be much closer in value.

In conclusion, incorporation of the monomer into the polymer and poling under the conditions we investigated appears to reduce the achievable value of $\chi_{ZZZ}^{(2)}$ relative to an equivalent (hypothetical) concentration of monomeric species. The constraints of the viscous rubbery state clearly limit the response of this polymer chain to the poling field relative to chains in dilute solution.

SUMMARY

The second-order nonlinear optical susceptibilities were examined for a series of model monomer and copolymer compounds in dilute solution and thin film formats. Second harmonic generation experiments on poled solutions revealed substantial enhancements in the SH conversion efficiency (up to 400-fold increases) could be attained by incorporating the nonlinear optical chromophores into a polar main-chain polymer. This effect is the manifestation of overall increases in the field-induced molecular alignment through cooperative effects of the separate molecular dipoles. While investigations on thin films of the same copolymers showed that the poled molecular orientation could be frozen and maintained in the absence of the field, the degree of orientation in the thin films was found to be much less than that in the solutions but considerably more than an uncorrelated assembly. Finally, second harmonic generation was shown to be a sensitive probe for studying polymer dynamics in thin films.

ACKNOWLEDGMENTS

We wish to thank Dr. J. M. O'Reilly of Eastman Kodak Company for performing the differential scanning calorimetry measurements and helpful discussions.

REFERENCES

*Present address: Allied-Signal Engineered Materials Research Center, Des Plaines, Illinois 60016.

1. A. Dulcic and C. Flytzanis, A new class of conjugated molecules with large second order polarizability, Opt. Commun. 25:402 (1978).
2. J. L. Oudar, Optical nonlinearities of conjugated molecules. Stilbene derivatives and highly polar aromatic compounds, J. Chem. Phys. 67:446 (1977).

3. S. J. Lalama, K. D. Singer, A. F. Garito, and K. N. Desai, Exceptional second-order nonlinear optical susceptibilities of quinoid systems, Appl. Phys. Lett. 39:940 (1981).

4. "Nonlinear Optical Properties of Organic and Polymeric Materials," D. J. Williams, ed., ACS Symp. Series, Plenum Press, New York (1983).

5. I. R. Girling, P. V. Kolinsky, N. A. Cade, J. D. Earls, and I. R. Peterson, Second harmonic generation from alternating Langmuir-Blodgett films, Opt. Commun. 55:289 (1985).

6. B. K. Nayar, Optical fibres with organic crystalline cores, in: "Nonlinear Optics: Materials and Devices," C. Flytzanis and J. L. Oudar, eds., Springer-Verlag, New York (1986).

7. K. D. Singer, J. E. Sohn, and S. J. Lalama, Second harmonic generation in poled polymer films, Appl. Phys. Lett. 49:248 (1986).

8. J. Zyss, Nonlinear organic materials for integrated optics: a review, J. Mol. Electron. 1:25 (1985).

9. G. R. Meredith, J. G. VanDusen, and D. J. Williams, Optical and nonlinear optical characterization of molecularly doped thermotropic liquid crystalline polymers, Macromolecules 15:1385 (1982).

10. B. F. Levine and C. B. Bethea, Second order hyperpolarizability of a polypeptide α-helix: poly-γ-benzyl-L-glutamate, J. Chem. Phys. 65:1989 (1976).

11. G. Khanarian, Direct current electric field induced second harmonic generation in flexible molecules and polymers, J. Chem. Phys. 77:2684 (1982).

12. B. F. Levine and C. G. Bethea, Second and third order hyperpolarizabilities of organic molecules, J. Chem. Phys. 63:2666 (1975).

13. G. D. Green, J. I. Weinschenk, III, J. E. Mulvaney, and H. K. Hall, Jr., The synthesis of polyesters containing a nonrandomly placed highly polar repeating unit, Macromolecules (submitted for publication).

14. A. Dulcic, C. Flytzanis, C. L. Tang, D. Pepin, M. Fetizon, and Y. Hoppilliard, Length dependence of the second-order optical nonlinearity in conjugated hydrocarbons, J. Chem. Phys. 74:1559 (1981).

15. B. F. Levine, Donor-acceptor charge transfer contributions to the second order hyperpolarizability, Chem. Phys. Lett. 37:516 (1976).

FREQUENCY AND TEMPERATURE VARIATIONS OF CUBIC SUSCEPTIBILITY IN POLYDIACETYLENES

P.A. Chollet, F. Kajzar, and J. Messier

IRDI/D.LETI/DEIN-LERA
CEN Saclay – 91191 Gif sur Yvette Cedex, France

1. INTRODUCTION

Emergence of conjugated polymers with one dimensional π electron delocalization has opened new perspectives for getting ultra-past nonlinear optical devices with high damage threshold. Since first pioneering works, both theoretical[1-4] and experimental[5-6] by Ecole Polytechnique's group in France a great interest of these materials has been demonstrated. Before, organic materials have been known as having either a large second order hyperpolarizability $\chi^{(2)}$ due to big dipolar moment in fundamental and/or excited state in charge transfer molecules[7] or a large third order hyperpolarizability $\chi^{(3)}$ with a long response time (of the order of ms) in liquid crystals[8] due to the molecular reorganization under light illumination.

Very simple free electron models considering electrons as confined in one dimensional box and applied to cyanines, carotenes and polyenes have shown a strong dependence of linear polarizability $\chi^{(1)}$[9] ($\chi^{(1)} \sim L^3$) and maximum absorption wavelength λ_{max} on L[10-11]. These models recovered by Rustagi and Ducuing[1] predicted also a strong dependence of $\chi^{(3)}$ on L ($\chi^{(3)} \sim L^5$). More sophisticated calculations by Cojan et al.[3] and Agrawal et al.[4] within framework of Pariser-Parr-Pople approximation confirmed these predictions. Large values of $\chi^{(3)}$ have been found in β-carotene[5] and in polytoluene sulphonate (PTS) single crystals[6]. Recently a large variety of conjugated polymers (mostly of polydiacetylenes) has been studied by different techniques. These studies confirm large hyperpolarizability of these systems, fast, subpicosecond response time. They show also existence of multiphoton resonances in visible and near IR range.

In this paper, we review our recent results of nonlinear optical properties studies of thin films of polydiacetylenes with different side groups and obtained by various techniques. The thin film technology permits to do hyperpolarizability studies in the polymer absorption range (nonlinear spectroscopy) and is promising in integrated optics applications. Different thin film deposition techniques are reviewed and described as well as some recent results of wave dispersed third harmonic generation measurements at room and liquid helium temperature are reported.

2. THIN FILM DEPOSITION TECHNIQUES

There exists a large class of organic polymers with one dimensional π electron delocalization. Among them of special interest are polydiacetylenes with general formula $R - C - C \equiv C - C - R'$ (details on chemistry and physicochemical properties can be found in ref. 12). These polymers have a unique property to polymerize in solid state by UV, X-Ray or γ irradiation and in some cases simply by heating permitting to get polymer single crystals. Some of these polymers, like poly-mBCMU are soluble. This property is also useful in thin film deposition. Fundamentally there are two methods permitting to get a polymer thin film by:

A. Monomers film deposition and its subsequent polymerization

In this case the monomer thin film should be composed from sufficiently large crystallites ($\ell > 100$ Å) permitting to get polymers with enhanced nonlinear optical properties. On the other hand, it is imperative that the size of crystallites be smaller (or much bigger) than the laser wavelength in order to avoid the light scattering. A good compromise present crystallites with size $100 \text{ Å} \lesssim \ell \lesssim 300 \text{ Å}$ (or macroscopic single crystals). For monomer deposition we use two techniques:

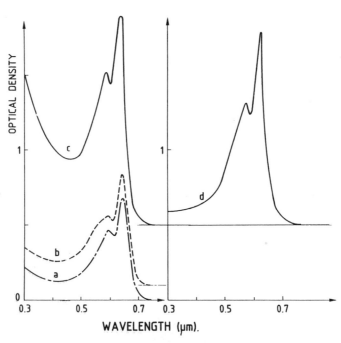

WAVELENGTH (μm).

Fig. 1 - Optical absorption spectra of a 71 layers thick Langmuir-Blodgett (a) 4500 Å (b) and 8700 Å (c) thick C_{16-8} and 5000 Å thick 4-BCMU (d) evaporated under vacuum polymer films. Horizontal lines denote corresponding zeroth density reference values. Blue form polymers are obtained by monomer thin film polymerization with a UV lamp.

1. Langmuir–Blodgett method

This technique is adequate for diacetylenes with aliphatic chain side groups ($R = (CH_2)_m-COOH$, $R' = (CH_2)_n-CH_3$). For sufficiently long aliphatic chains ($m = 0, 8$, $n = 10 \div 15$) one obtains stable monomolecular films on water subphase which can be transferred on a substrate and subsequently polymerized. Addition of Cd^{2+} ions to water stabilizes in some cases the monomolecular films giving in result Cd salts ($R = (CH_2)_m-COOC_d^{\frac{1}{2}}$). The Langmuir–Blodgett techniques give polymer films with controllable and well defined thickness (stacking of monolayers with about 30 Å individual layer thickness). Crystallites with about 60 Å thickness and a few microns wide are obtained in this technique with polymer chains parallel to the substrate. The technique is appropriate for very thin films realization.

2. Monomer evaporation in vacuum

Several diacetylene monomers can be deposited on a substrate by evaporation in high vacuum (10^{-6} T) and subsequently polymerized. We have used successfully this technique in thin film preparation from: PTS($R = R' = CH_2OSO_2\ C_6H_4CH_3$), DCH($R = R' = CH_2NC_{12}H_8$)[13], C_{16-8}($R = (CH_2)_{15}\ CH_3$, $R' = (CH_2)_8\ COOH$), 4-BCMU($R = R' = (CH_2)_4\ OCONHCH_2\ COOC_4H_9$). In Fig. 1, absorption spectra of Langmuir–Blodgett film (a) C_{16-8} PDA are compared with 4500 Å (b) and 8700 Å (c) thick films obtained by evaporation. One observes a significant decrease in transmission between 3000 Å and 4000 Å for the thickest film (c) due to an important light scattering. This is also seen in monomer films which are not translucent in visible light. In contrast, the thin films of poly-4BCMU scatter much less the light (Cf. Fig. 1d).

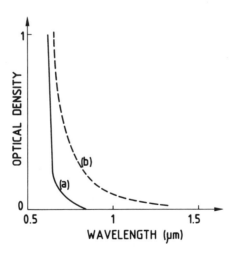

Fig. 2 – Tail of optical absorption spectrum of a 50 μm thick solution cast 4-BCMU film showing surface light scattering effect (b). This scattering decreases significantly when one presses slightly a silica plate on the free face of polymer at 400°K and under vacuum (a).

B. Thin film formation from polymer solution by solvent evaporation

Some polymers, e.g. n-BCMU(R = R' = $(CH_3)_n$OCONHCH$_2$ COOC$_4$H$_9$), PTS-12 (R = R' = $(CH_2)_4$OSO$_2$C$_6$H$_4$CH$_3$) are soluble. Using a 20 g/ℓ solution of 4-BCMU in chloroform, one can obtain thin films with thickness of a few tens of microns. A transmission spectrum of a 4-BCMU 50 μm thick film is shown Fig. 2. The increase of optical density between 0.8 μm and 1.5 μm is due to the light scattering by film surface. It disappears when one press slightly a silica plate on the free face of polymer at 400°K under vacuum. (Fig. 2a) With 4-BCMU polymers one can obtain also amorphous gels with a good optical quality.

Another possibility of thin film formation consists in a dipping of substrate in a polymer solution and its slow drawing out. Fig. 3 shows absorption spectra of two thin films of 3-BCMU. Regular and homogeneous films can be obtained in this way, whose thickness depends on polymer concentration on solvent and on drawing velocity (a preliminary filtering of polymer solution is advised).

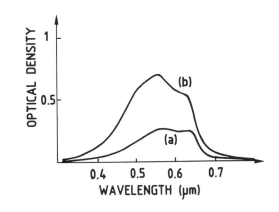

Fig. 3 — Absorption spectrum of a 800 Å (a) and 2150 Å (b) thick 3-BCMU film obtained by dipping technique from polymer solution in chloroform.

3. HARMONIC GENERATION AND NONLINEAR PROPAGATION IN ABSORBING THIN FILMS

We recall here fundamental formulas for third harmonic generation and nonlinear propagation in absorbing thin films (details can be found elsewhere[13-15]). For an incident plane wave with frequency ω:

$$E_\omega(\vec{r},t) = \tfrac{1}{2} [E_\omega(\vec{r})e^{-i\omega t} + E_\omega^+(\vec{r})e^{i\omega t}] \tag{1}$$

the amplitude of harmonic field generated in a nonlinear slab with thickness ℓ is given by

$$E_{3\omega}(\ell) = \frac{\pi \chi^{(3)}(3\omega;\omega,\omega,\omega)(E_\omega)^3}{\Delta\varepsilon} [e^{i\varphi_H} A_1(e^{i\Delta\varphi}-1) + A_2(e^{i(\varphi_F+\varphi_H)} - 1)]/A3 \tag{2}$$

where $\chi^{(3)}(-3\omega;\omega,\omega,\omega)$ is nonlinear susceptibility responsible for THG process, $\Delta\varepsilon$ is dieletric constant dispersion and

$$\Delta\varphi = \varphi_F - \varphi_H = 3\omega(n_\omega \cos\Theta_\omega - n_{3\omega}\cos\Theta_{3\omega})\ell/C \tag{3}$$

is phase mismatch between fundamental (F) and harmonic (H) wave inside the slab, Θ_ω and $\Theta_{3\omega}$ are propagation angles at ω and 3ω frequency, respectively. The coefficients $A_i(i = 1,3)$ arise from boundary conditions (for details, see ref. 14) and take account of multiple reflection effects of harmonic wave. In the case when refractive indices of nonlinear medium (polymer film) are not very different from those of surrounding media (substrate, air or vacuum) coefficient A_2 is close to zero and A_3 close to 1. Consequently Eq.(2) can be significantly simplified leading to:

$$E_{3\omega}(\ell) = \frac{\pi \chi^{(3)}(E_\omega)^3}{\Delta\varepsilon} A_1 e^{i\varphi_H} (e^{i\Delta\varphi} - 1) \tag{4}$$

For a thin film deposited on one or both sides of a substrate, the resultant harmonic field will be a sum of harmonic fields generated in thin films and substrate, respectively:

$$E_{3\omega}^R = T_s E_{3\omega}^s + T_{p_1} E_{3\omega}^{p_1} + T_{p_2} E_{3\omega}^{p_2} \tag{5}$$

where s and p refer to substrate and polymer film, respectively and T's are corresponding whole transmission factors. In many cases the harmonic field generated in thin film is much larger (due to large cubic susceptibility) than that generated in substrate itself and $E_{3\omega}^s$ in Eq.(5) can be neglected with respect to other fields, but not the phase mismatch, which in the case of harmonic generation in thin films deposited on both sides of substrate, leads to interference fringes similar to Maker fringes but different in origin (in general the thin film thickness is much smaller than the coherence length $\ell_c = \lambda_\omega/6\Delta n$ and harmonic intensity is directly proportional to ℓ^2). In the case of absorbing films at harmonic frequency (it is very hard to do harmonic generation experiment on films absorbing at fundamental frequency and we do not consider here this case) refractive index $n_{3\omega}$ is complex ($n_{3\omega} = \nu_{3\omega} + i K_{3\omega}$) and consequently complex are transmission factors, phase mismatch φ_H and A_1 coefficient. Thus for a thin film deposited on one side of substrate only the harmonic intensity is given by[16,17]

$$I_{3\omega} = \frac{64\pi^4}{c^2} | \chi^{(3)} A_1 |^2 I_\omega^3 f_a \tag{6}$$

where I_ω is fundamental beam intensity and f_a an absorption depending factor

$$f_a = \frac{(1 - \exp(-\alpha_{3\omega}\ell/2))^2 + (\Delta\varphi)^2\exp(-\alpha_{3\omega}\ell/2)}{[(\nu_{3\omega})^2 - (n_\omega)^2 - (K_{3\omega})^2]^2 + 4(\nu_{3\omega}K_{3\omega})^2} \tag{7}$$

where $\alpha_{3\omega} = 4\pi K_{3\omega}/\lambda_{3\omega}$ is linear absorption coefficient at harmonic wavelength.

As we mentioned before, Eq. (2) takes account multiple reflections of harmonic wave and does not those of the fundamental wave. The last may be most important than the former ones as the fundamental field contribute in power six to harmonic intensity. As discussed by Kajzar and Messier[14], multiple reflection effects are important only near normal incidence and lead to rapid oscillations of harmonic intensity for sufficiently thick substrate. In Fig. 4, we show third harmonic intensity in function of incidence angle for a 3-BCMU thin film obtained by dipping technique with a fused silica substrate and showing such oscillations around normal incidence. This figure shows importance of multiple reflections effects and necessity of checking it when harmonic generation measurements are carried out on thin films.

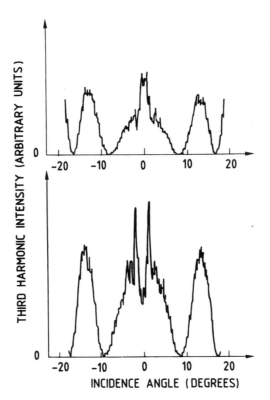

Fig. 4 - Multiple reflection effects observed from 3-BCMU thin films obtained by dipping technique. Polymer films are deposited on both side of a fused silica substrate rotated along an axis perpendicular to the beam propagation direction.

Fig. 5 shows also influence of multiple reflections on harmonic intensity for a very thin polymer film (~ 700 Å) deposited on one side of a sapphire substrate. Polymer film is translated in this case perpendicularly to the laser beam. A strong variation of harmonic intensity (Fabry-Perrot effect) is seen due only to the fact that substrate is slightly prismed and has a large refractive index (~ 1.75). Another origin of experimental artefacts is air contribution to harmonic intensity[14]. In general, for conjugated highly nonlinear polymer film, this contribution is

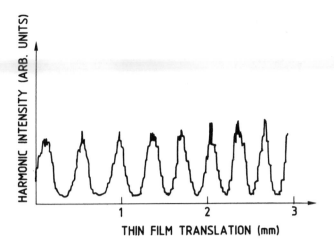

Fig. 5 - Multiple reflection effects observed from a polyacetylene thin
film deposited on one side of a slighten prismed sapphire substrate translated perpendicular to the beam propagation direction.

negligible, however it can be significant when harmonic generation measurements on very thin films are carried out and/or a comparison with a standard is made. This contribution can be avoided by doing experiments in
vacuum and with focused laser beams (in every case) or by introducing a
correction depending on laser wavelength and focal length[14].

3. TWO AND THREE PHOTON RESONANCES IN $\chi^{(3)}$

3.1 - Linear Absorption Spectra

Generally there exist two forms of solid polydiacetylene : blue
(unstable) with an absorption maximum wavelength $\lambda_{max} \sim 0.63$ μm and a red
one (stable) obtained by heating of blue polymer (or in same cases from
solution cast). The polydiacetylene absorption spectra were primarily
interpreted as polyacetylene band to band transition of a one dimensional
semiconductor[3-4]. However the fact that the conductivity is observed in
blue form at wavelength signifi cantly lower ($\lambda_c \sim 0.54$ μm) as well as
existence of anomalies around λ_c in electro-reflectivity spectra[19] imply
that λ_{max} corresponds to an excitonic transition and that band to band
transition lies higher in energy (around 0.54 μm). Thus the bonding energy
(0.4 eV) of such one dimensional exciton is much larger than that of a
bidimensional exciton in multiple quantum wells (few tens of meV in AsGa)
or a tridimensional exciton (a few meV in bulk AsGa). The photoconductivi-
ty measurements in DC and AC fields show[20] that very large electric fields
(up to 10^6 V/m) are necessary for electron separation in PDA's.

3.2 - Frequency Dependence of Cubic Susceptibility

The THG measurements have been performed in 0.9 ÷ 1.91 μm fundamental
wavelength range with use of apparatus described elsewhere[16]. The low temperature experiments have been done using an Oxford Instruments cryostat

(model CF 204). In all cases, we used thin films deposited on one side of substrate (fused silica) only. The data have been collected translating the polymer film over 3 mm distance perpendicular to the light propagation direction checking carefully for multiple reflections effects. Silica plate or silica wedge were used for calibration. In Fig. 6, the measured cubic susceptibility $\chi^{(3)}(-3\omega;\omega,\omega,\omega)$ is shown in function of fundamental

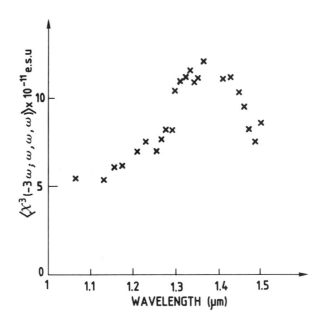

Fig. 6 – Measured average cubic susceptibility of a 71 layers thick Lang-muir-Blodgett film of C_{16-8} polydiacetylene (blue form).

wavelength for a 71 layers Langmuir-Blodgett film of C_{16-8} PDA. There are two resonance enhancements in $\chi^{(3)}$: one at 1.35 µm in the polymer transparency range (cf. Fig. 1) and the second one at 1.907 µm. In order to understand origin of these resonances we call the role of quantum mechanical formulas for cubic susceptibility derived by time dependent perturbation method which yields for molecules with localized excitations following expression[21-22]:

$$\chi^{(3)}_{xxxx}(-3\omega;\omega,\omega,\omega) \sim \sum_{u,g',u'} <g|x|u'><u'|x|g'><g'|x|u><u|x.g>$$

$$x\left[\frac{1}{(E_{u'}-3\omega-i\Gamma_{u'})(E_{g'}-2\omega-i\Gamma_{g'})(E_u-\omega-i\Gamma_u)} + \frac{1}{(E_{g'}-2\omega-i\Gamma_{g'})(E_u-\omega-i\Gamma_u)(E_{u'}+\omega+i\Gamma_{u'})} \right.$$

$$\left. + \frac{1}{(E_u-\omega-i\Gamma_u)(E_{u'}+\omega+i\Gamma_{u'})(E_{g'}+2\omega+i\Gamma_{g'})} + \frac{1}{(E_{u'}+3\omega+i\Gamma_{u'})(E_{g'}+2\omega+i\Gamma_{g'})(E_u+\omega+i\Gamma_u)} \right]$$

where g, u, g', u' are fondamental (g) and excited states with gerade (g'..) and ungerade (u,u'..) symmetry, <u|x|g> are dipolar moment transition matrix elements between u and g states and E_i (i = g',u,u'..) is energy difference between fundamental (g) and one of the corresponding excited states (i) in h units. The cubic susceptibility $\chi^{(3)}$ is particularly large for one dimensional wave functions for which the transition matrix elements <i|x|j> are enhanced. From Eq. (8), it is seen that one and multiphoton resonance will appear when the incident photon energy (or respectively its multiple) approaches the energy difference between fundamental and one of the excited states. From symmetry considerations the odd photon number transitions appear between states with opposite symmetry (g and u). Such transitions are allowed for one photon and are seen in optical absorption spectrum. Other transitions between states with the same symmetry (g and g') are forbidden for one photon but allowed for two photons (or even multiple). These transitions are not seen in optical absorption spectrum but may be observed in $\chi^{(3)}$ where according to Eq.(8) an enhancement will occur (two photon resonance (TPR)). Thus the observed enhancement in blue LB film at 1.35 μm (cf. Fig. 6) can be interpreted as two photon resonance whereas that at 1.907 as three photon resonance. We note here that the recent bleaching experiments 23 on blue PDA films demonstrate one photon resonance in optical absorption range. Thus we conclude that in LB PDA films the two photon level (g') located at 0.675 μm lies below one photon level (0.635 μm) inside the gap. Such situation is theoretically predicted for one dimensional systems with strong electronelectron correlations[24] and was also observed in polyenes by fluorescence measurements[25,26] and recently also in PDA oligomers by the same technique[27].

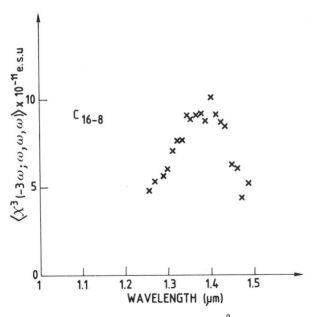

Fig. 7 — Average cubic susceptibility of 4000 Å thick evaporated C_{16-8} blue form PDA film.

For evaporated C_{16-8} PDA the two photon resonance (cf. Fig. 7) is slightly shifted to larger wavelength (1.39 µm). In one photon spectrum a small shift of maximum absorption is also observed (cf. Fig. 1). For the red form of LB film (cf. Fig. 8), the two photon resonance lies at about 1.22 µm. In the case of the red form of the C_{16-8} molecules where the absorption spectrum has two well defined peaks at 0.5 µ and 0.55 µ, a broad resonance is seen with two shoulders at 1.15 and 1.22 µ.

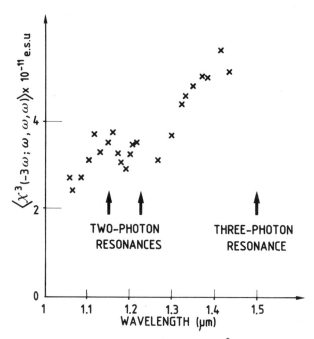

Fig. 8 - Average cubic susceptibility of 4000 Å thick evaporated C_{16-8} red form PDA film.

From Eq. (8), it is also seen that close to a resonance the cubic susceptibility $\chi^{(3)}$ is complex. Complex is also refractive index which is intensity dependent

$$n = n_o + n_2 \, I_\omega \tag{9}$$

where

$$n_2 = 12 \; \pi^2 \; \chi^{(3)}(-\omega;\omega,\omega,-\omega)/n_o^2 \; C \tag{10}$$

Although the Kerr susceptibility $\chi^{(3)}(-\omega;\omega,\omega,-\omega)$ is different from that responsible for THG process ($\chi^{(3)}(-3\omega;\omega,\omega,\omega)$) the one and two photon resonances will occur at the same frequency. Thus two photon resonance and two photon absorption will take place with an intensity dependent index of refraction. It follows also from Eq. (8) that exactly at resonance n_2 is purely imaginary (the real part of $\chi^{(3)}$ being equal to zero). In principle, it is possible to predict the sign of Re $\chi^{(3)}$. For example, for $E_\omega < E_u <$ 3ω and $E_{g'} > 2\omega$, Re $\chi^{(3)}$ will be negative. The measure of the phase of $\chi^{(3)}$ is possible by comparison with harmonic emission from a known standard (e.g. thin film and substrate if the generated harmonic fields are comparable[28]) or by measuring a concentration dependence of $\chi^{(3)}$ in solution. Using the last method, we have found the $\chi^{(3)}$ of yellow PTS-12 solution is complex at 1.064 µm with negative real and imaginary parts[29]. We note here that negative real part of $\chi^{(3)}(-\omega;\omega,\omega,-\omega)$ was also found in solution cast 4-BCMU films by surface polariton refractive index measurements[30].

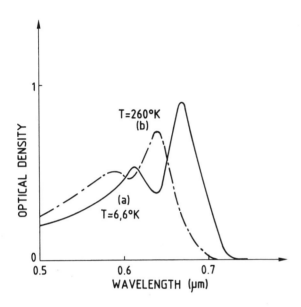

Fig. 9 – Optical absorption spectra of 71 layers thick LB film at 7K (a) and at room temperature (b).

4. TEMPERATURE VARIATION OF $\chi^{(3)}$

In Fig. 9 the linear absorption spectra at room temperature and at 7 K of blue form of an LB film are compared. One observes a small ($\sim$ 100 Å) shift of excitonic maximum absorption wavelength λ_{max} toward larger wavelengths at low temperature with a simultaneous narrowing of the absorption peak. The narrowing of absorption peak at low temperature may be well explained by temperature dependent interaction with vibronic bands (inhomogeneous broadening).

The low temperature (at 7 K) wave dispersed THG measurements on the same polymer film show a negligible shift in two photon resonance (cf. Figs. 6 and 10). With no significant variation in $\chi^{(3)}$ between room temperature and low temperature (7 K) is observed (cf. Fig. 11). Although the data were calibrated with a fused silica standard kept at the same temperature as polymer film, it is legitimate to assume that $\chi^{(3)}$ of silica is almost temperature independent; the Si-O bonds being saturated bonds.

131

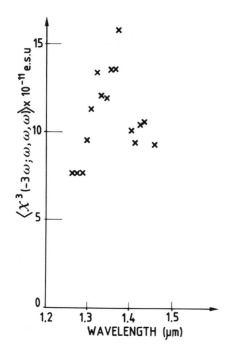

Fig. 10 - Average cubic susceptibility $\langle\chi^{(3)}(-3\omega;\omega,\omega,\omega)\rangle$ of a 71 layers thick LB film at 7 K.

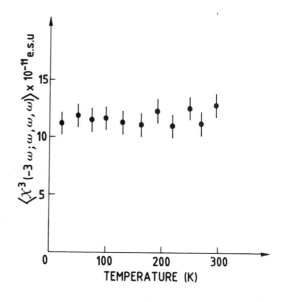

Fig.11 - Temperature variation of average cubic susceptibility $\langle\chi^{(3)}(-3\omega;\omega,\omega,\omega)\rangle$ of 71 LB layers film from DPA film at 1.35 μm fundamental wavelength.

Recently Schmitt-Rink et al.[31], in order to explain origin of large nonlinearities and their dynamics in multiple quantum wells considered three mechanisms contributing to it: phase space filling for excitons based on Pauli exclusion principle, exchange mechanism and free carriers contribution. They calculated temperature variation of saturation value for excitons density N_s (or free carriers density) for these mechanisms. The value of N_s depends on the extension of the excitonic wavefunction and consequently is related via Eq. (8) to the hyperpolarizability of the system. For two mechanisms (phase space filling and exchange) they found a temperature independent N_s for exciton bounding energy E_s < kT and a strong variation of N_s for free carriers. Although Schmitt-Rink et al.[31] consider two dimensional excitons and use the formalism for explanation of bleaching experiments, an analogy with polydiacetylenes, where one dimensional excitons are generated, can be seen. In fact, at two photon resonance the two photon level (g') is populated via two photon transition and the resonant $\chi^{(3)}$ value is related to the exciton number saturation value N_s. As we discussed before, the exciton binding energy E_s is very large in polydiacetylenes for one photon transitions ($E_s \simeq 400$ meV). We can expect the same for those created by two photon absorption. Thus the generated excitons are undissociable between room temperature and liquid helium temperature (thermal energy kT $\simeq$ 25 meV is significantly smaller than the exciton binding energy (~ 400 meV)). Thus the constancy of $\chi^{(3)}$ is function of temperature is in favor of excitonic origin of large $\chi^{(3)}$ in polydiacetylenes.

5. DISCUSSION

The discovery of two photon resonances in 1.35 μm ÷ 1.39 μm fundamental wavelength range in the polymer transparency range with two photon level lying below one photon state is of great importance from both fundamental and practical point of view. It shows that electron-electron correlations are important in these one dimensional model systems. The resonance wavelength falls exactly at the telecommunications window (maximum transmission range for optical fibers: 1.3 ÷ 1.5 μm). The resonant values of $\chi^{(3)}$ are very large ($\sim 10^{-10}$ e.s.u.). Although we measure $\chi^{(3)}(-3\omega; \omega, \omega, \omega)$ the two photon resonance will occur also in Kerr susceptibility $\chi^{(3)}(-\omega; \omega, \omega, -\omega)$ leading to large refractive index variation with change of sign of Re n_2 when crossing the two photon resonance (Re $\chi^{(3)}$ < 0 below and Re $\chi^{(3)}$ > 0 above two photon resonance). Although the response time for this two photon resonance is unknown one can expect that similarly as for one photon resonance where the recovery time is very fast (picoseconds or subpicoseconds[23,32]) it should be also rapid[33].

The temperature variation of linear absorption spectrum between room and liquid helium temperature (narrowing of excitonic absorption peak) is in favour of inhomogeneous broadening. The constancy of $\chi^{(3)}$ in function of temperature at two photon resonance confirms excitonic character of this resonance and a strong bounding of excitons. Although electric field induced second harmonic generation (EFISHG) experiments[34-35] evidence existence of internal polarization created via two photon absorption, the density of generated free carriers is very small. In fact a separation of electron-hole pair is very hard in these materials. This big "rigidity" of excitons is in favour of large hyperpolarizabilities as well as large laser damage threshold[36].

ACKNOWLEDGEMENTS

The authors would like to thank Dr. D. Grec for synthesis of diacetylene monomers as well as Mr. J. Banide, R. Gras and A. Lorin for technical assistance. They are also grateful to Dr. G. L. Baker for supplying the polyacetylene thin film and to Direction des Affaires Industrielles et Internationales (Ministere de PTT) for a partial financial support of this work.

REFERENCES

1. K.C. Rustagi and J. Ducuing, Third-Order Polarizability of Conjugated Organic Molecules, Opt. Commun. 10, 258 (1974).
2. J. Ducuing, in Nonlinear Spectroscopy, N. Bloembergen ed., North Holland Publ. Comp., Amsterdam 1977, p. 276.
3. C. Cojan, G.P. Agrawal and C. Flytzanis, Optical Porperties of One-Dimensional Semiconductor and Conjugated Polymers, Phys. Rev. B15, 909 (1977).
4. G.P. Agrawal, C. Cojan and C. Flytzanis, Nonlinear Optical Properties of One-Dimensional Semiconductors and Conjugated Polymers, Phys. Rev. B17, 776 (1978).
5. J.P. Hermann, D. Ricard and J. Ducuing, Optical Nonlinearities in Conjugated Systems: β-carotene, Appl. Phys. Lett., 23, 178 (1973).
    J.P. Hermann and J. Ducuing, Third Order Polarizabilities of Long-Chain Molecules, J. Appl. Phys., 45, 5100 (1974).
6. C. Sauteret, J.P. Hermann, R. Frey, F. Pradere, R.M. Baughman and R.R. Chance, Phys. Rev. Lett., 36, 956 (1976).
7. For a review of $\chi^{(2)}$ organic molecules, see Nonlinear Optical Properties of Organic Molecules and Crystals, D.S. Chemla and J. Zyss eds., Academic Press, Orlando 1987.
8. S.D. Durbin, S.M. Arakelian, M.M. Cheung and Y.R. Shen, Highly Nonlinear Optical Effects in Liquid crystals, J. Phys. (Paris), Colloque C2, 44, C2-161 (1983).
    I.C. Khoo, Nonlinear Light Scattering by Laser and D.C. Field Induced Molecular Reorientations in Nematic-Liquid-Crystal Films, Phys. Rev. A25, 1040 (1982).
9. P.L. Davies, Polarizabilities of Long Chain Conjugated Molecules, Trans. Faraday Soc., 47, 789 (1952).
10. M. Kuhn, Fortschr. Chem. Org. Nat., 16, 169 (1958).
11. M. Kuhn, Fortschr. Chem. Org. Nat., 17, 404 (1959)
12. Polydiacetylenes Synthesis, Structure and Electronic Properties. D. Bloor and R.R. Chance eds. NATO ASI Series N°102, Martinus Nijhof Publ., Dordrecht (1985).
13. J. Lemoigne, A. Thierry, P.A. Chollet, F. Kajzar and J. Messier, Morphology and Nonlinear Optical Properties of Poly-Dicarbonyl Hexadyine, to be published.
14. F. Kajzar and J. Messier, Third Harmonic Generation in Liquids, Phys. Rev., A32, 2352 (1985).
15. F. Kajzar and J. Messier, Cubic Effects in Polydiacetylene Solutions and Films, in Nonlinear Optical Properties of Organic Molecules and Crystals, Academic Press, Orlando (1987), Vol. 2, p. 51.
16. F. Kajzar and J. Messier, Resonance Enhancement in cubic susceptibility of Langmuir-Blodgett Multilayers of Polydiacetylene, Thin Sol. Films, 132, 11 (1985).
17. The numerical factor in Eq. (12) is erroneous and should read 64 π^4.
18. K. Lochner, H. Bässler, B. Tieke and G. Wegner, Photoconduction in Polydiacetylene Multilayer Structures and Single Crystals. Evidence for Band-to-Band Excitation, Phys. Stat. Sal. (6), 88, 653 (1978).
19. L. Sebastian and G. Weiser, One Dimensional Wide Energy Bands in a Polydiacetylene Revealed by Electroreflectance, Phys. Rev. Lett., 46, 1156 (1981).
20. H. Bäsler, Electrical Transport and Doping of Polydiacetylenes, in Polydiacetylenes. Synthesis, Structure and Electronic Properties, D. Bloor and R.R. Chance eds., NATO ASI Series N°102, Martinus Nijhof Publ., Dordrecht 1985, p. 135.
21. J.F. Ward, Calculation of Nonlinear Optical Susceptibilities Using Diagrammatic Perturbation Theory, Rev. Mod. Phys. 37, 1 (1965).
22. P.W. Lanhof, S.T. Epstein and M. Karplus, Aspects of Time-Dependent Perturbation Theory, Rev. Mod. Phys., 44, 602 (1972).

23. F. Kajzar, L. Rothberg, S. Etemad, P.A. Chollet and D. Grec, Exciton Bleaching in Langmuir-Blodgett Films of Polydiacetylene, this conference.
24. E.W. Hayden and E.J. Mele, Renormalization-Group Studies of the Hubbard-Peierls Hamiltonian for Finite Polyenes, Phys. Rev., B32, 6527 (1985).
25. B.S. Hudson and B.E. Kohler, A Long-Lying Weak Transition in the Polyene α,ω-Diphenyloctatetraene, Chem. Phys. Lett. 14, 299 (1972).
26. B. Hudson and B. Kohler, Linear Polyene Electronic Structure and Spectroscopy, Ann. Rev. Phys. Chem., 25, 437 (1974).
27. B.E. Kohler and D.E. Schikle, Low Lying Singlet States of a Short Polydiacetylene Oligomer, preprint.
28. F. Kajzar, J. Messier and J. Rosilio, Nonlinear Optical Properties of Thin Films of Polysilane, J. Appl. Phys., 60, 3040 (1986).
29. F. Kajzar and J. Messier, Third Harmonic Generation and Two Photon Absorption in a Polydiacetylene Solution, ref. 12, p. 325.
30. P. Prasad, Nonlinear Optical Interactions in Polymer Thin Films, Proceedings of 29th SPIE's Annual Symposium, San Diego, 1986.
31. S. Schmitt-Rink, D.S. Chemla and D.A.B. Miller, Theory of Transient Excitonic Optical Nonlinearities in Semiconductor Quantum-Well Structures, Phys. Rev. B32, 6601 (1985).
32. G.M. Carter, J.V. Hryniewicz, M.K. Thakur, Y.J. Chen and S.E. Meyler, Nonlinear Optical Processes in a Polydiacetylene Measured with Femtosecond Duration Laser Pulses, J. Appl. Phys., 49, 998 (1986).
33. M. Lequime and J.P. Hermann, Reversible Creation of Defects by Light in One Dimensional Conjugated Polymers, Chem. Phys., 26, 431 (1977).
34. P.A. Chollet, F. Kajzar and J. Messier, Electric Field Induced Second Harmonic Generation and Polarization Effects in Polydiacetylene Films, Ref. 12, p. 317.
35. P.A. Chollet, F. Kajzar and J. Messier, Electric Field Induced Optical Harmonic Generation and Polarization Effects in Polydiacetylene Langmuir- Blodgett Multilayers, Thin Sol. Films, 132, 1 (1985).
36. J. Duming, Microscopic and Macroscopic Nonlinear Optics, in Nonlinear Optics, P.G. Harper and B.S. Wherett eds., Academic Press, London 1977, p. 11.

FEMTOSECOND STUDIES OF DEPHASING AND PHASE CONJUGATION WITH INCOHERENT

LIGHT

Takayoshi Kobayashi, Toshiaki Hattori, Akira Terasaki,
and Kenji Kurokawa

Department of Physics, University of Tokyo
Hongo, Tokyo 113, Japan

INTRODUCTION

The fast response of electroactive polymers as nonlinear optical materials is one of the main features which are attracting attention of increasing number of researchers. Especially, the large optical nonlinearity of organic polymeric systems containing conjugated π-electron structures is expected to have response times in femtosecond regime because of their purely electronic origins. Some of them are regarded as attractive candidates for ultrafast optical processing devices, and a switching time of 0.1 ps is suggested for an optical switch utilizing a polydiacetylene as a nonlinear material.[1] Studies of the fast response of optical nonlinearity are most simply performed by transient four-wave mixing, and optical nonlinearities of several organic materials have been studied by transient four-wave mixing using short optical pulses.[2-6] In this article, we describe femtosecond and picosecond measurements of relaxation times of optical nonlinearity by three types of four-wave mixing using temporally incoherent light instead of short pulses.

The dynamical properties of matter has been studied by increasing number of scientists, and information with higher time resolution is being obtained by the development of picosecond and femtosecond spectroscopies. Since picosecond light pulses were first emitted from passively mode-locked ruby laser in 1965,[7] continuous efforts to get shorter pulses have been made, and recently optical pulses as short as 8 fs were obtained[8] by the method of pulse compression of the output from a group-velocity-dispersion-compensated colliding-pulse mode-locked laser. Time-resolved coherent and conventional spectroscopies have been applied to several systems using ultrashort light pulses with pulse width of a few tens to a hundred femtoseconds. However, there are several difficulties in the study of the ultrafast phenomena using such short pulses: (i) Laser systems for the generation of ultrashort pulses are necessarily very expensive and complicated. (ii) The wavelengths of femtosecond laser pulses are limited in the region around 615-625 nm because of the lack of appropriate combination of saturable absorber and gain medium, and the tunability of each laser is generally poor. (iii) It is difficult to avoid broadening of an ultrashort pulse due to linear and/or nonlinear dispersion in optical systems because of its broad power spectrum.

Recently a new spectroscopic technique with incoherent light utilizing coherent transient optical effects has been presented and verified experimentally.[9-13] Since ordinary electronic devices do not have subpicosecond or femtosecond time resolution, optical experiments using that time region are usually performed using nonlinear optical phenomena. In these methods, the signal light generated or modulated by nonlinear optical effects is detected for the measurement of response of matter, using the correlation between excitation and probe light beams. In typical experimental systems, an optical pulse is split into two beams, and they meet again in a sample after passing through variable and fixed optical delay lines, and the intensity of signal or probe light is measured as a function of the delay time. Generally the signal intensity is expressed by the integration of functions of the field amplitude (or intensity) and the response function of the matter.

In the studies of the dynamics taking place in matter, therefore, the time resolution is expected to be determined not by the pulse duration of the light but by the correlation time. According to this principle, extremely high time resolution may be easily obtained by using light having a short enough correlation time, or a broad enough spectral width, for the time region to be measured. The applicability of this principle to short-time measurement has been verified for the dephasing time measurement by degenerate four-wave mixing (DFWM) spectroscopy.[9-14]

In the first part of this article, we describe the study of dephasing in a polydiacetylene film measured by spatial-parametric-mixing type DFWM.[15] Dephasing times in a polydiacetylene (poly-3BCMU) film were resolved for the first time at two wavelengths by DFWM using incoherent light. The measured dephasing times, 30 fs at 648 nm and 130 fs at 582 nm, correspond to excitons in polymer chains with different conjugation lengths.

Though transient DFWM spectroscopy, both using coherent short pulses[16] and incoherent cw light sources as mentioned above, is powerful for the study of dynamic properties of matter, it is not applicable to optically forbidden transition and the range of the available wavelength is limited. Dephasing of Raman active vibrational modes in molecules can be investigated by so-called transient coherent Raman spectroscopy such as CARS (coherent anti-Stokes Raman scattering) and CSRS (coherent Stokes Raman scattering),[17-19] where a pair of picosecond pulses excites a vibrational system coherently and a second pulse of the higher frequency probes the coherence of the system after a certain delay time. The information about the dephasing dynamics of the system can be obtained by the dependence of the coherent Raman scattering intensity on the delay time.

We studied theoretically a possible application of the principle that the correlation time determines the resolution time, to transient coherent Raman spectroscopy. Theoretical derivation of the delay-time dependence of the coherent Raman intensity, and the experimental demonstration of the measurement of the dephasing of the 2915-cm^{-1} mode in dimethylsulfoxide are presented in the second part.[20]

In the last part of this article, we describe the time-resolved study of the relaxation times of the third-order susceptibility for the optical Kerr effect in CS_2 and nitrobenzene performed by phase-conjugation-type DFWM using 5-ns pulses from a broad-band laser. In the case of CS_2 we observed both subpicosecond (about 0.2 ps) and picosecond (about 2 ps) relaxation times for the susceptibility-tensor element of $\chi^{(3)}_{xyyx}$ and also for $\chi^{(3)}_{xxxx}$ both at 553 nm and at 623 nm. The observed relaxation times agree with those obtained utilizing subpicosecond pulses, and the shorter and longer time constants are attributed to the libration and reorientation relaxation, respectively. The relaxation times of nitrobenzene could not be determined because the libration relaxation time is shorter than the resolution time and the reorientation relaxation time is longer than that of CS_2,[21] and it is difficult to determine it precisely due to limited measured delay time span and signal to noise ratio.

ELECTRONIC DEPHASING IN A POLYDIACETYLENE FILM MEASURED BY DEGENERATE FOUR-WAVE MIXING

There has been much interest in the dynamical properties of the excited states of polydiacetylenes (PDAs). They have been studied experimentally by time-resolved absorption, reflection and emission spectroscopy.[5,22-26] The excited state lifetime of soluble PDA (3KAU) was found to be 9 ± 3 ps in aqueous solution[22] and that of PTS to be 2 ps in crystalline phase.[5]

Knowledge of the dephasing dynamics of PDAs is of great importance not only to elucidate the properties of excited states and the mechanism of the optical nonlinearity but also for various applications such as optical switching and optical signal processing. DFWM was applied to the dephasing time measurement,[2-4] but dephasing times have not been resolved so far. Dennis et al.[2] observed DFWM from two PDA (2d and 2j) solutions with 180 ps pulses, but they found that the response times were much shorter than their resolution time. Carter et al.[3] observed DFWM from a PDA (PTS) crystal with 6 ps pulses, and they concluded that the response time was again shorter than 6 ps. Rao et al.[4] performed similar measurements on a PDA (poly-4BCMU) film with 500 fs pulses, but they could not resolve the dephasing time either.

In this study, we applied DFWM with incoherent light to the measurement of the dephasing times in a film of a PDA, poly[4,6-decadiyne-1,10-diol bis((n-butoxycarbonyl)-methyl) urethane], which is abbreviated as poly-3BCMU. By detecting signals diffracted in two directions simultaneously, we could resolve a dephasing time as short as 30 fs. We measured dephasing times of the sample at two wavelengths, 648 nm and 582 nm, and found that the dephasing of the exciton in a chain of the polymer with a longer conjugation length (at 648 nm) is four times faster than that in a chain with a shorter conjugation length (at 582 nm). This result explains the difference in the fluorescence efficiencies between the rod-like and coil-like forms of the acetylene-type phase.

Experimental

The experimental apparatus used for the dephasing time measurement by the spatial-parametric-mixing type DFWM is shown in Fig. 1. The incoherent light source was a broad-band dye laser pumped by a N_2 laser. The output of the oscillator was amplified by a dye cell pumped by the same N_2 laser. In order to obtain very broad-band laser light, the oscillator cavity of the dye laser was constructed with a highly-reflecting aluminum mirror and a glass plate as an output mirror, and no tuning element was placed in the cavity. Rhodamine 6G and rhodamine 640 were used as laser dyes. The center wavelengths and the band widths (FWHM) of the power spectra were 582 nm and 7.7 nm for rhodamine 6G, and 648 nm and 8.7 nm for rhodamine 640, respectively.

Dye laser light was linearly polarized by a Glan-Thompson prism. It was then divided into two beams, n_1 and n_2, by a beam splitter, and n_2 was delayed with respect to n_1 by a delay time t_d. The delay time was varied by a translation stage driven by a stepping motor. The polarization planes of the two beams could be rotated independently by the use of half-wave plates. The pulse energies of the 10 ns dye laser were about 3 μJ at the

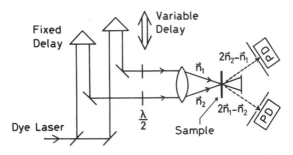

Fig. 1. Experimental setup for spatial-parametric-mixing type degenerate
four-wave mixing measurement. PD and $\lambda/2$ stand for a photodiode
and a half-wave plate (Fresnel rhomb), respectively. The vectors,
n_1 and n_2, are the unit vectors representing the directions of the
two excitation beams.

sample position, and the beam diameters of the focused areas on the sample
were 40 μm.

Degenerate four-wave mixing signals diffracted in two directions, $2n_2$-
n_1 and $2n_1$-n_2, were detected simultaneously by photodiodes, as shown in
Fig. 1, to obtain resolution times shorter than the correlation time of the
incoherent light.[13] Output signals of the photodiodes were processed with
sample-and-hold circuits and an A/D converter, and stored in a microcomputer.

In the experiment a film of poly-3BCMU on a glass plate cast from
chloroform solution was used as a sample. The absorption spectrum of the
film is shown in Fig. 2. The transmittances of the film were 5 % at 582 nm
and 60 % at 648 nm. All the measurements were performed at 26±3°C.

Results and Discussion

Figure 3 shows the data which were obtained with the two excitation
beams under parallel polarization condition. Background signal due to
scattering of the excitation beams was subtracted from the data. No peak

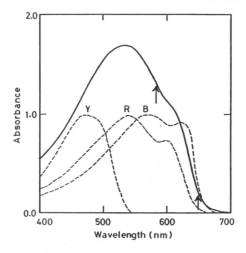

Fig. 2. Absorption spectrum of the film of poly-3BCMU (solid line) with
those of three forms in solution from Ref. 32 (broken lines). B,
R, and Y denote blue, red, and yellow forms, respectively. Two
exciting wavelengths utilized in the present study, 648 nm and
582 nm, are indicated by arrows.

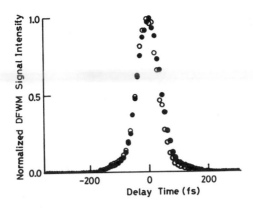

Fig. 3. DFWM signals obtained with the two excitation beams of
polarizations parallel to each other. The wavelength of the
excitation light was 648 nm. Open circles show the signal
intensity diffracted in the direction $2n_1-n_2$, and closed circles
show that in the direction $2n_2-n_1$.

shift or tail was observed. The signal with parallel polarizations was
about thirty times more intense than with perpendicular polarizations.
Therefore, the signal obtained with parallel polarizations can be attributed almost exclusively to diffraction from a thermal grating, which is
generated only when mutual coherence between the two excitation beams
exists.[16]

The contribution of the thermal grating to the DFWM signals is eliminated under perpendicular polarization, and hence we can obtain electronic
DFWM signals,[13] are shown in Fig. 4, after the background scattering intensity being subtracted. The peaks of the signal intensities of the two
directions are shifted from each other, as shown in Fig. 4. Here we define
a peak separation as the distance in the delay time between the two peaks
of the signal intensities of the two directions. The peak separations were
30 and 90 fs at 648 and 582 nm, respectively. The widths (FWHM) of the
signals were 100 fs for 648 nm and 130 fs for 582 nm, which give the correlation time of the incoherent excitation light at each wavelength with a
certain factor according to the signal shape and the definition of correlation time. There is no pronounced asymmetric tails which indicate dephas-

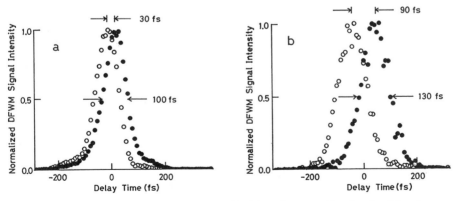

Fig. 4. DFWM signals with the two excitation beams of polarizations
perpendicular to each other. Open circles show the signal
intensity diffracted in the direction $2n_1-n_2$, and closed circles
show that in the direction $2n_2-n_1$. The wavelengths of the
excitation light are a) 648 nm and b) 582 nm.

ing times much longer than the correlation time. The tails seen in the data at 648 nm are due to the spectral shape of the light deviated far from Gaussian, since the field autocorrelation is a Fourier transform of the power spectrum.

When only one-photon resonant DFWM is responsible for the signal, the delay-time (t_d) dependence of the signal intensities is expressed by the following equation:[12]

$$I(t_d) \propto \int_o^\infty dt \int_o^\infty dt' G(t'-t) G(t-t_d) G^*(t'-t_d) \exp[-2(t+t')/T_2] . \tag{1}$$

Here $G(t_d)$ is the autocorrelation function of the incoherent light field:

$$G(t_d) = \langle E(t+t_d) E^*(t) \rangle , \tag{2}$$

where $E(t)$ is the envelope of the electric field of the light. For Eq. (1) to be applicable, the condition that the population lifetime is much longer than the dephasing time must be satisfied. Our preliminary data by a picosecond absorption experiment with second harmonic of a mode-locked Nd:YAG laser (532 nm, 30 ps) show that the population lifetime of excitons in a poly-3BCMU film is of the order of the laser pulse width.[27] When the dephasing time is much longer than the correlation time of the light field, the signal decays exponentially at the rate of $4/T_2$. On the other hand, when the dephasing time is comparable with or even slightly shorter than the correlation time, which will be found to be the case in the present study later, the dephasing time can also be obtained from the peak separation of the two signals diffracted in two directions even though the signal shapes have no prominent tails.[13,16]

In the discussion of the optical nonlinearity of PDAs, however, contribution of the two-photon resonant state cannot be neglected.[28,29] The DFWM signal intensities by two-photon resonance can be expressed as

$$I(t_d) \propto \int_o^\infty dt' \int_o^\infty dt'' \langle E(t-t_d) E^*(t-t_d) E(t-t') E^*(t-t')$$
$$\times E(t-t'') E^*(t-t'') \rangle \exp[-(t'+t'')/T_2^*] . \tag{3}$$

Here T_2^* is the inverse width of the relevant two-photon transition. In this case, the signal will have a tail on the other side of the peak than that of the signal by one-photon resonance,[30] and the decay rate is $2/T_2^*$, since this process is a two-photon version of free induction decay, and therefore, the dephasing by the inhomogeneous broadening is irreversible.

We must notice here, however, that the signal intensity given by Eq. (3) does not vanish at a large t_d. The ratio of the intensity at $t_d = 0$ to the background depends on the statistical property of the incoherent field and becomes three when a Gaussian random process[31] is assumed for a model of the incoherent field, while in the present experiments on poly-3BCMU, no background signal was detected other than scattering due to a linear process. Therefore, at the two wavelengths investigated in the present study, the contribution of the two-photon state to the DFWM signal is negligible, and the observed peak separations are attributed to dephasing of one-photon resonant transitions. Using Eq. (1) and the observed peak separations the dephasing times are calculated to be 30 fs at 648 nm and 130 fs at 582 nm by assuming Gaussian autocorrelation functions. The effect of spectral diffusion on these dephasing times can be excluded since spectral diffusion in polymeric systems such as PDA may take place in subpicoseconds or slower.

This result that the dephasing time at a longer wavelength is shorter than that at a shorter wavelength is contrary to those of previous studies.[13,16] A larger peak separation was reported at a longer wavelength than at a shorter wavelength with cresyl fast violet in cellulose, and was explained in terms of the difference in the intramolecular relaxation rate at the two wavelengths.[13] Dephasing time measurements of three dyes, cresyl violet, Nile blue, and oxazine 720, in polymethylmethacrylate (PMMA) at 15 K at 620 nm were also reported.[16] The peak separation for cresyl violet was 60 fs, whereas for the other two dyes shifts were shorter than 20 fs. The difference in the dephasing times was attributed to that in the excess photon energy of the excitation light from the absorption edge of each sample.

The present sample is a polymeric material and has its own characteristics different from ordinary dyes.[32,33] It is known that poly-3BCMU in solution has three forms, blue (B), red (R), and yellow (Y) forms.[32] The blue form is considered to have a polymer main chain with rod-like conformation, while R and Y forms have main chains with coil-like conformation. These three forms are realized depending on the temperature and the composition of the solvent. The shoulder of the absorbance of the sample in the present study at 620 nm (2.0 eV) corresponds to the B band, and the peak at 530 nm (2.3 eV) corresponds to the R band. They are attributed to the π-π^* exciton transition in each type of the polymer chain. Therefore, our sample is thought to be a mixture of a coil-like conformation (with shorter conjugation lengths of π-electron) and a rod-like conformation (with longer conjugation lengths), or a mixture of the polymer chains with continuously distributed conjugation lengths between these two extreme forms realized in solution.

In the present experiment, the wavelength of 648 nm is on the absorption edge of the rod-like form exciton, while that of 582 nm is on resonance with the exciton in polymer chains with shorter conjugation lengths. Therefore, from the present results on the dephasing times at the two wavelengths, it can be concluded that the exciton dephasing is about four times faster in the rod-like form than in a chain with a shorter conjugation length.

The present result may be explained in two ways. One explanation is that excitons are more mobile and the phases of them are changed more often in longer conjugated chains than shorter ones, where excitons do not move over long distances. The other explanation is that exciton levels lie more closely in longer conjugated chains than in shorter conjugated chains, and therefore, dephasing due to multilevel excitation is faster.[16] We cannot determine which is the case from the present data only. An extended study at other wavelengths and temperatures is in progress.

It has been reported that the fluorescence intensities are suppressed when the solution of poly-3BCMU is converted from a yellow (coil-like) form to a blue (rod-like) form, and when the solution of PDA (poly-4BCMU) is converted from the coil form to the rod form.[32] It is also reported that only partially polymerized crystal of PDA (PTS) emits fluorescence.[34] Our results are consistent with the explanation[32,34] by which the changes in the fluorescence quantum efficiencies were attributed to the increase in the nonradiative decay rates with exciton delocalization.

VIBRATIONAL DEPHASING TIME MEASUREMENT BY COHERENT STOKES RAMAN SCATTERING

The method for the observation of picosecond to femtosecond electronic dephasing dynamics discussed in the previous section is extended to the vibrational dephasing time measurement. For the latter measurement a type of nondegenerate four-wave mixing with both coherent and incoherent light is utilized. The first theoretical study is presented and experimental

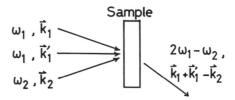

Fig. 5. Schematic of CSRS experiment for vibrational dephasing measurement.

description follows. It is based on a transient coherent Raman process
with three beams (see Fig. 5), two of which are incoherent light from a
single broad-band laser, and a delay time between the two is variable. A
third beam has a higher frequency than the incoherent light by a vibrational energy in a molecule of interest, and is coherent in the delay-time
range of the measurement. The delay-time dependence of coherent Stokes
Raman scattering (CSRS) intensity offers the information about the coherence dynamics of the vibrational transition with a resolution time limited
by the correlation time of the incoherent light. When the frequency of the
third beam is lower than the others, coherent anti-Stokes Raman scattering
(CARS) takes place. For the theoretical calculation, a three-level model
of molecular system with homogeneous broadening is used, and the delay-time
dependence of CSRS intensity was calculated. Dephasing dynamics of the
2915-cm^{-1} mode in dimethylsulfoxide was observed experimentally by the new
method using nanosecond laser pulses, and the result was found to agree
well with that obtained with picosecond pulses.

Description of a Model and Time Dependence of Signal Intensity

Until now all reported studies of transient spectroscopy with incoherent light, including photon echo and pump-probe spectroscopy, were concerned with broad-band (about 100 cm^{-1}) with only one center frequency
which is resonant with a two-level system. However, there exist various
coherent transient phenomena where light beams of two or more different
frequencies are concerned, and the time resolution of the transient coherent spectroscopies using these phenomena is also expected to be determined
not by the duration of light pulses used but by the correlation time of the
radiation field.

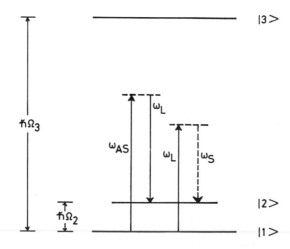

Fig. 6. Energy diagram of the model system for vibrational dephasing
experiment.

In this section, theoretical expectation values of the coherent Raman signal intensity using incoherent light will be presented with a simple model. Calculation of the time-dependent CSRS intensity will be presented for the correspondence with the experimental study described in the following section, although CARS is substantially the same.

A simple model of a three-level system (see Fig. 6) is usually taken for theoretical considerations of coherent Raman phenomena.[35,36] Two vibrational levels |1> and |2> belong to the ground electronic state, whereas level |3> belongs to an electronically excited state. Levels |1> and |3> and levels |2> and |3> are connected with each other by electronic transition dipoles. The energy difference between |3> and |1> is $\hbar\Omega_3$, and that between |2> and |1> is $\hbar\Omega_2$. In ordinary coherent Raman experiments, this system is placed in radiation field which consists of light beams of two frequencies, the difference between which is resonant with the transition between |1> and |2>. The higher frequency is denoted by ω_{AS} and the lower by ω_L.

CSRS is a third-order effect, and the intensity is proportional to the light intensity of frequency ω_{AS} and to the squared light intensity of frequency ω_L. In usual CSRS (or CARS) experiments two beams are used, but for the purpose of time-resolved measurement, a triple-beam (BOXCARS) configuration was applied,[37] where two beams of frequency ω_L are used, and three waves are mixed to generate a wave of frequency $2\omega_L - \omega_{AS}$.

The electric field is given as

$$E(\mathbf{r},t) = E_{AS}(t)\exp[i(\mathbf{k}_{AS}\mathbf{r} - \omega_{AS}t)] + E_{L1}(t)\exp[i(\mathbf{k}_{L1}\mathbf{r} - \omega_L t)]$$

$$+ E_{L2}(t)\exp[i(\mathbf{k}_{L2}\mathbf{r} - \omega_L t)] + \text{c.c.} \quad , \tag{4}$$

where c.c. stands for complex conjugates of the preceding terms and $E_{AS}(t)$, $E_{L1}(t)$, and $E_{L2}(t)$ are functions of t slowly varying compared to the optical frequencies. In BOXCARS experiments, light of frequency $\omega_S = 2\omega_L - \omega_{AS}$ and with wave vector $\mathbf{k}_S = \mathbf{k}_{L1} + \mathbf{k}_{L2} - \mathbf{k}_{AS}$ is detected. This can be easily performed by separating spatially the signal from the other light beams because of the directionality of the signal and the laser beams.

The polarization with frequency ω_S and wave vector $\mathbf{k}_S$ is derived by a perturbational method. The following conditions are assumed; (i) Light frequencies are tuned exactly to the vibrational energy. (ii) The light of frequencies ω_{AS} and ω_L are off-resonance with electronic transitions. (iii) The broadening of the relevant energy level of the molecular system is homogeneous. Under these conditions, a third-order polarization $P^{(3)}(\mathbf{k}_S, \omega_S)$ is given by

$$P^{(3)}(\mathbf{k}_S, \omega_S) = C \cdot \exp[i(\mathbf{k}_S\mathbf{r} - \omega_S t)] \int_0^\infty dt' \, \exp[-(t-t')/T_2]$$

$$\times [E_{L1}(t)E_{L2}(t') + E_{L2}(t)E_{L1}(t')]E_{AS}^*(t') , \tag{5}$$

where T_2 is the dephasing time of the vibrational transition, and C is a time-independent proportionality coefficient. This expression has two terms, in which E_{L1} and E_{L2} are exchanged with each other.

In the present CSRS experiments using incoherent light, two incoherent light beams of central frequency ω_L are obtained by splitting a beam from a broad-band dye laser. One of the two beams is delayed to the other with a delay time t_d, and the light of frequency ω_{AS} is assumed to be coherent in the time scale of observation.

When one assumes that the correlation time of the incoherent light is

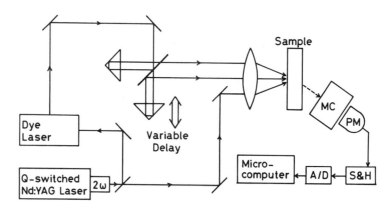

Fig. 7. Experimental setup for CSRS measurement using incoherent and
coherent light. MC is the monochromator; PM, the photomultiplier.

much shorter than the dephasing time, and that the stochastic property of
the incoherent light is expressed in terms of a Gaussian random process,
one can derive a simple expression for the signal intensity as

$$I(t_d) = 1 + G(t_d) + (2\tau_c/T_2)\exp(-2|t_d|/T_2) \quad . \tag{6}$$

Here $G(t)$ is the autocorrelation function of the incoherent light field
amplitude normalized to unity at its peak, and the correlation time τ_c is
defined by

$$\tau_c = \int_0^\infty G(t)dt \tag{7}$$

In the above expression, the fraction of the third term in the signal
intensity is approximately proportional to the ratio of the correlation
time to the dephasing time. Therefore, for the dephasing time to be determined with high precision, incoherent light with an adequate correlation
time must be used, since low intensity signal cannot be distinguished from
the background due to inevitable noises from various sources.

Experiment and Results

CSRS signal by the symmetric CH-stretching vibration of dimethylsulfoxide (DMSO) with a wavenumber 2915 cm^{-1} was measured by the apparatus
shown in Fig. 7. Coherent light source was the second harmonic (532 nm) of
a Q-switched Nd:YAG laser operated at 8 Hz. Since the correlation time of
this light source is estimated to be about 30 ps, it can be regarded safely
as coherent for the time period shorter than 10 ps. The main part (about
90% intensity) of this beam was split off and used to pump a broad-band dye
laser. By changing the laser dye concentration, the oscillation wavelength
was tuned to 630 nm which is resonant with the Raman mode of DMSO. The
spectral width (FWHM) of the laser light was 7 nm, which corresponds to a
correlation time of about 100 fs.

The dye laser beam was split into two with nearly equal intensity
using a beam splitter. Both of them were made parallel again by the same
beam splitter after passing through delay lines. The length of one of the
delay lines was varied by a stepping motor. The three beams were focused
in DMSO in a 10-mm sample cell by a lens of 20-cm focal length. To eliminate the scattered laser light, an iris with 5-mm diameter, a 10-cm monochromator, and color filters were placed between the sample and the photomultiplier.

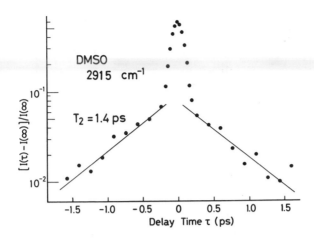

Fig. 8. The delay-time dependence of the CSRS intensity normalized by the
background intensity. The background intensity has been
subtracted.

In Fig. 8, the CSRS signal intensity is plotted as a function of the
delay time in a semilogarithmic scale after the background is subtracted.
From the slope of the trailing part, the vibrational dephasing time T_2 is
found to be 1.4 ps in agreement with the value obtained with picosecond
pulses.[18] The intensity ratio between the peak and the tail extrapolated
to $t_d=0$ is about seven, which is also in good agreement with the theoretically expected ratio for the values of T_2 (1.4 ps) and the correlation time
(100 fs) determined above.

Since CSRS or CARS does not include a rephasing process, which takes
place in photon echoes, the obtained value of the dephasing time of three-
level systems has some ambiguities when the system is inhomogeneously
broadened.[38,39] Raman echo[35] can provide the information about the dephasing dynamics without the ambiguities.[36] However, it is a difficult experiment since it is a higher-order process. Raman echo experiments have been
carried out in solids,[40] gases,[41,42] and liquid nitrogen,[43] but not yet in
liquids at room temperature, and the first experiment may possibly be
performed with the use of incoherent light applying a similar method to the
present study.

MOLECULAR DYNAMICS OF KERR LIQUIDS STUDIED BY OPTICAL PHASE CONJUGATION

We extend the incoherent method to another type of the degenerate
four-wave mixing, i.e., the phase conjugation in a Kerr medium. Liquid CS_2
is one of the most familiar and simplest optical Kerr media with a large
nonlinear refractive index n_2.[44] In 1975, a direct time-resolved measurement using a picosecond laser revealed that the relaxation time of optically induced birefringence in CS_2 is 2.1 ps.[45] The recent development of the
generation technique of subpicosecond and femtosecond optical pulses made
easier the relaxation-time measurement of the optical Kerr effect with
higher time resolution. Using a time-resolved interferometric technique
with 70-fs pulses at 619.5 nm, Halbout and Tang observed both a 2.00-ps
response and a 0.33-ps response.[46] Greene and Farrow measured two relaxation times, 2.16±0.1 ps and 0.24±0.02 ps, utilizing a biased Kerr shutter
with 150-fs pulses at 620 nm.[47] Using a usual Kerr shutter with 200-fs
pulses, Etchepare et al. observed a response of 1.4±0.1 ps and a response
of 0.20±0.05 ps. Pulses with two wavelengths were used in their measurement: 615 nm for the pump pulse and 650 nm for the probe pulse.[48] In all

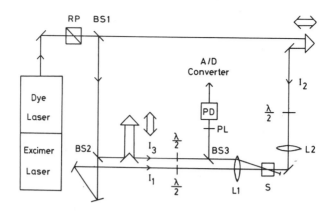

Fig. 9. Experimental setup for the phase-conjugation type degenerate four-wave mixing using incoherent light. BS1,2 is the beam splitter; BS3, the glass plate; L1,2, the lens; PD, the photodiode; PL, the polarizer; RP, the Rochon prism; S, the sample; $\lambda/2$, the half-wave retarder.

the three measurements subpicosecond (0.2–0.3 ps) and picosecond (1.4–2.2 ps) relaxations with almost the same corresponding time constants were observed. Also in the frequency domain, Trebino et al. measured the relaxation times of four nonzero third-order-susceptibility elements for the optical Kerr effect in CS_2 using a tunable-laser-induced-grating technique with an excitation wavelength of 575 nm and a probe wavelength of 570 nm. They obtained, for example, two relaxation times of 2.49±0.5 ps and 0.21±0.04 ps for $\chi^{(3)}_{xxxx}$.[49]

The picosecond relaxation in CS_2 is considered to be associated with orientational randomization by rotational diffusion of molecules. On the other hand, the mechanism of the subpicosecond response is not well explained. Several mechanisms have been proposed as its possible physical origin: the damped librational motion, the time-dependent behavior resulting from collisions, and the short-time behavior associated with the rotational diffusion of molecules possesing a large anisotropic polarizability.[46,47,49,50]

Liquid nitrobenzene is also one of the well-known optical Kerr media. Tang and Halbout observed a subpicosecond relaxation time (about 150 fs) with an interferometric technique.[46] Etchepare et al. observed subpicosecond (shorter than 0.2 ps) and picosecond (4 ps and 42 ps) relaxation times by the method of the Kerr shutter.[48]

In this article we report for the first time the time-resolved optical phase conjugation in Kerr liquids, CS_2 and nitrobenzene, using incoherent light. The subpicosecond (about 0.2 ps) and the picosecond (about 2 ps) relaxation times were observed in CS_2. The details of the experimental procedure and results obtained by the time-resolved phase conjugation in CS_2 and nitrobenzene are presented in the following.

Experimental

The experimental setup used for the measurements of the phase-conjugate signal in both Kerr liquids is shown in Fig. 9. The incoherent light source was similar to that used in the experiment on PDA except that the excitation source was an excimer laser. Rhodamine 560 (Rh 560) in methanol or rhodamine 640 (Rh 640) in ethanol was used as the laser dye both in the oscillator and the amplifier. The center wavelength was 553 nm

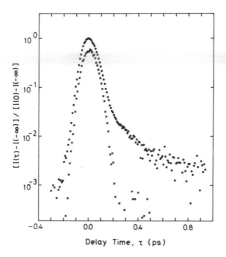

Fig. 10. Normalized signal intensity as a function of delay time τ of I_1 to
I_3. (•) CS_2 in type I polarization configuration, where, I_3 is
vertically polarized and both I_1 and I_2 are horizontally
polarized. (x) reference sample (DEANS in CS_2) in type II
polarization configuration, where I_1, I_2, and I_3 are all
vertically polarized. The wavelength is 553 nm. Plot of the
normalized signal intensity in reference sample is slightly
shifted downward.

and the width (FWHM) of the spectrum was 5.5 nm for Rh 560, and these
values for Rh 640 were 623 nm and 5.0 nm.

The dye laser beam polarized linearly by a Rochon prism was split into
three beams I_1, I_2, and I_3. After passing through delay lines beams I_3 and
I_2 were focussed in the sample. Lengths of the delay lines were varied
with stepping motors. In all the measurements beam I_2 was delayed or
advanced by 0.2 ns to the other two beams, I_1 and I_3, and had no correlation with them because the correlation time of the dye laser light was in
subpicosecond region. The signal intensity was measured as a function of
the delay time (τ) of beam I_1 with respect to beam I_3. Beam I_1 was made
parallel with beam I_3 and was focussed in the sample. Beams I_1 and I_2 were
exactly counterpropagating to each other. The duration of the pulses was
5 ns. The peak intensities of beams I_1, I_2, and I_3 in the focal plane when
Rh 560 was used as the laser dye were about 200, 400, and 300 MW/cm^2,
respectively. These values for Rh 640 were about 70, 200, and 100 MW/cm^2.
Polarization planes of the three beams can be independently changed with
three half-wave retarders, each of which is composed of coupled Fresnel
rhombs. Measurements were performed in two types of polarization configurations. The two type are denoted as follows;

Type I: I_1 and I_2, are horizontally polarized and I_3 is
vertically polarized,
Type II: I_1, I_2, and I_3 are all vertically polarized.

The sample was in a 1-cm cell. The angel between beams I_1 and I_3 was 4° in
the sample cell. The phase-conjugate signal was split by a glass plate BS3
and detected with a photodiode after passing through an analyser PL.

Results and Discussion

The semilogarithmic intensity of the normalized phase-conjugate signal
in CS_2 obtained with 553 nm laser is shown in Fig. 10. Beam I_2 was delayed

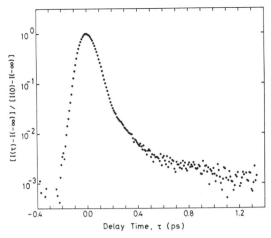

Fig. 11. Normalized signal intensity as a function of delay time in CS_2 at 553 nm. Polarization configuration is type I.

by about 0.2 ns with respect to the other two beams. The polarization configuration was type I, that is, I_1 and I_2 were horizontally polarized and I_3 was vertically polarized. Figure 10 also shows a semilogarithmic plot of the normalized signal with type II polarization configuration in a reference sample, which is a solution of 4-diethylamino-4'-nitrostilbene (DEANS) in CS_2 with a concentration of 1×10^{-3} mole/l. The absorbance of the reference sample was 0.08 at 553 nm. Since the signal intensity in the reference sample is mainly due to the thermal grating, and it is more than two orders of magnitude larger than that in neat CS_2, the effect of the signal due to the nonlinearity of CS_2 itself is negligible. As shown in Fig. 10 the signal obtained for the reference sample is symmetric with respect to the decay time and its FWHM is 0.12 ps. Since I_2 has no corre- lation with the other two, the observed signal in the reference sample is considered to be due to a thermal grating produced by the interference between beams I_1 and I_3. The observed signal, therefore, gives a correla- tion profile $|G(\tau)|^2$ of the incident incoherent light, where $G(\tau)$ is the autocorrelation of the incoherent field.

On the other hand, the signal in CS_2 is clearly asymmetric and con- sists of three components with different decay times. The signal shape was independent of the scan direction of the delay time. The time dependence of the signal in the case when the beam I_2 was advanced by about 0.2 ns to the other two was the same as in Fig. 10. In Fig. 10 the fastest decay component among the three is very intense and its time dependence coincides with the reference signal, that is, with the correlation profile $|G(\tau)|^2$. In order to determine the longest decay time, measurement in a wider delay- time range was performed, the result being shown in Fig. 11. The time constant τ_1 was obtained as $0.9^{+0.2}_{-0.1}$ ps. By subtracting the slowest compo- nent, the time constant τ_s of the middle component was obtained as 88+8 fs. Twice these obtained values, that is, $2\tau_1 = 1.8^{+0.4}_{-0.2}$ ps and $2\tau_s = 0.18 + 0.02$ ps, agree within experimental error with the long and short time constants, respectively, obtained by ultrashort pulses or by the frequency-domain measurements as mentioned above. Signal intensities were also measured with type II polarization configuration at this wavelength, 553 nm, and the two time constants obtained with the data agree with the result of type I polarization configuration within experimental error.

Figure 12 shows a semilogarithmic plot of the signal intensity in CS_2 of type I polarization configuration at 623 nm. The time constants τ_1 and

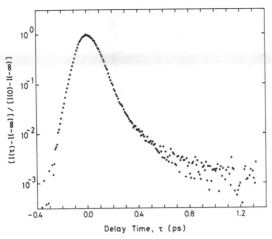

Fig. 12. Normalized signal intensity as a function of delay time in CS_2 at 623 nm. Polarization configuration is type I.

τ_s were obtained as $0.8^{+0.4}_{-0.2}$ ps and 100 ± 15 fs, respectively. These value agree within experimental errors with those at 553 nm. Two time constants τ_l and τ_s of type II polarization configuration were also obtained at 623 nm and agree within experimental errors with respective values of the type I polarization configuration.

According to the theory based on the Born-Oppenheimer approximation,[51] the third-order nonlinear polarization $P^{(3)}$ in optically transparent media has the form:

$$P_i^{(3)}(t) = \sigma_{ijkl}E_j(t)E_k(t)E_l(t)$$
$$+ E_j(t)\int_{-\infty}^{t}dt'd_{ijkl}(t-t')E_k(t')E_l(t') , \quad i=x,y,z , \tag{8}$$

where σ_{ijkl} is a constant tensor which describes the electronic nonlinearity, and $d_{ijkl}(t)$ represents the nuclear contribution to the third-order polarization. $E_i(t)$ denotes i (=x,y,z) component of the total electric field in the medium. For the following discussion, the two contributions are put together for convenience as follows:

$$P_i^{(3)}(t) = E_j(t)\int_{-\infty}^{t}dt'r_{ijkl}(t-t')E_k(t')E_l(t') \tag{9}$$

with

$$r_{ijkl}(t) = \sigma_{ijkl}\delta(t)+d_{ijkl}(t) . \tag{10}$$

The total electric field $E_i(t)$ can be written in the present case as

$$E_i(t,\mathbf{r}) = E_{(1)i}(t)\exp[i(\omega t-\mathbf{k}_1\mathbf{r})]+E_{(2)i}(t)\exp[i(\omega t-\mathbf{k}_2\mathbf{r})]$$
$$+ E_{(3)i}(t)\exp[i(\omega t-\mathbf{k}_3\mathbf{r})]+c.c. , \tag{11}$$

with

$$\mathbf{k}_1+\mathbf{k}_2=0 , \tag{12}$$

where $E_{(\alpha)i}(t)$ is the time-varying amplitude of the electric field corresponding to beam I and c.c. stands for complex conjugates of the preceding terms.

To derive an expression for the delay-time dependence of the intensity of the phase-conjugate signal, we assume that the electric field of the incoherent light can be expressed in terms of a stationary complex Gaussian random process,[31] and the autocorrelation function of the field decays exponentially:

$$G(\tau) = \exp(-|\tau|/\tau_c) . \tag{13}$$

The incoherent light utilized in experiments may not have the characteristic expressed in Eq. (13), but the signal intensity calculated under the assumption is expected to offer a good approximate value in the delay-time region longer than τ_c. Furthermore, we neglect the two-photon resonance contribution to the nonlinear susceptibility, and the relaxation of r_{ijkl} is assumed to be exponential with a decay-time constant T,

$$r_{ijkl}(t-t') = r_{ijkl}(0)\exp[-(t-t')/T] , \tag{14}$$

and the decay-time constant T is assumed to be much larger than the correlation time τ_c.

Using these relations we can obtain the delay-time dependence of the signal intensity as

$$I_s(\tau) \propto T^2\exp(-2|\tau|/\tau_c)+B+T\tau_c \tag{15}$$

with

$$B = \begin{cases} \dfrac{\tau_c^2}{2} \exp(-2|\tau|/\tau_c) , & \tau \leq 0 \\[3mm] \dfrac{3\tau_c^2}{2} \exp(-2\tau/T)-\tau_c(\tau_c+\tau)\exp(-2\tau/\tau_c) , & \tau > 0 . \end{cases} \tag{16}$$

The first term of Eq. (15) form a spike at $\tau=0$ in $I_s(\tau)$. We denote this term as a 'spike' term. The second term of Eq. (15) decays with a time constant of T/2 for long delay time $\tau > \tau_c$. Therefore, this term provides direct information on the relaxation time which is much shorter than the pulse duration. We denote this term as a 'signal' term. Since this term has a sharp rise near $\tau \sim \tau_c$, the time resolution is determined by the correlation time τ_c. The third term in Eq. (15) appears as a background which is independent of the delay time. We denote this term as a 'background' term. From Eqs. (15) and (16) relative contributions of the 'spike' term, the 'signal' term, and the 'background' term estimated at $\tau=0$ are T^2, $\tau_c^2/2$, and $T\tau_c$, respectively. Therefore, the first component in the observed signal is considered to be due to the 'spike' term. The effect described by the 'spike' term is the same one known as the coherence effect involved in the time-resolved optical Kerr effect.[52]

The middle and the slowest decay components in the observed signals shown in Figs. 10, 11, and 12 are considered to be due to the 'signal' term. In the delay-time range where the delay time τ is much longer than the correlation time τ_c the 'signal' term decays with time constant T/2. Thus, from the measured decay-time constants $\tau_1=0.9^{+0.2}_{-0.1}$ ps and $\tau_s=88\pm8$ fs, relaxation times of the third-order nonlinear polarization in CS_2 are obtained to be $1.8^{+0.4}_{-0.2}$ ps and 0.18 ± 0.02 ps. As previously mentioned, these values agree well with the two time constants determined by the measurements using ultrashort optical pulses.

If we assume the third-order nonlinear polarization is purely elec-

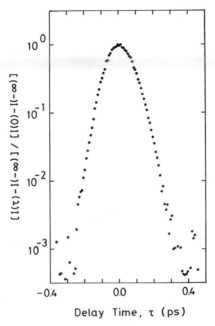

Fig. 13. Normalized signal intensity as a function of delay time in nitrobenzene at 553 nm. The polarization configuration is type I.

tronic, that is, $r_{xyyx}(t)=\sigma_{xyyx}\delta(t)$, then $I_s(-\infty)$ becomes a half of $I_s(0)$. In the present measurements in the type I polarization configuration $I_s(-\infty)$ was about 3% of $I_s(0)$. Therefore, the nuclear contribution is considered to play a major role in r_{xyyx}. This result coincides with the previous study using ultrashort pulses.[47,53]

Figure 13 shows semilogarithmic plots of the signal in nitrobenzene in the type I polarization configurations at 553 nm, with the beam I_2 being delayed by about 0.2 ns with respect to the others. Signal was also observed in type II polarization configuration, and the both results coincide with each other and are symmetric and composed of only one component, and no asymmetric component could be observed. The signal intensity of type II was about thirty times larger than that of type I. Therefore, the type II signal is attributed mainly to a thermal grating and gives a correlation profile $|G(\tau)|^2$ of the incident incoherent light. Thus, the observed signal of type I is due to the 'spike' term mentioned above. It is considered that because of the large ratio of $T^2:\tau_c^2$ due to a large rotational relaxation time in nitrobenzene [4 ps and 42 ps (Ref. 48)] than that in CS_2. and because of a shorter relaxation time[46,48] than the correlation time, no asymmetric component corresponding to the 'signal' term could be observed in nitrobenzene.

In summary, we made time-resolved study of subpicosecond-picosecond molecular dynamics in CS_2 and nitrobenzene by phase-conjugation type degenerate four-wave mixing using incoherent light. The observed signals with CS_2 both at 553 nm and at 623 nm consist of three decay components, two of which are characterized by exponential decays. The fastest decaying component of the three coincides with the correlation profile $|G(\tau)|^2$ of the incident incoherent light. This component is considered as due to the same one as the coherence effect in the time-resolved optical Kerr effect. The middle and the slowest decay components provide direct information on the relaxation times in CS_2. The two relaxation times obtained from these components agree with the two time constants measured by the use of ultrashort optical pulses. In the case of nitrobenzene the observed signal is symmetric and consists of one component corresponding to the fastest decaying component of the signal observed in CS_2. It is considered that because of a longer rotational relaxation time than that in CS_2, and because of a shorter relaxation time than the correlation time of the incoherent light, no asymmetric component which provides direct information on the relaxation time could be observed in nitrobenzene.

ACKNOWLEDGEMENTS

We wish to thank Professor T. Kotaka for providing us with poly-3BCMU, and Professor E. Hanamura for helpful discussion. This work is supported partly by a Grant-in-Aid for Special Distinguished Research (56222005) from the Ministry of Education, Science and Culture of Japan.

REFERENCES

1. P. W. Smith, Bell. Syst. Tech. J. 61, 1975 (1982).
2. W. M. Dennis, W. Blau, and D. J. Bradley, Appl. Phys. Lett. 47, 200 (1985).
3. G. M. Carter, M. K. Thakur, Y. J. Chen, and J. V. Hryniewicz, Appl. Phys. Lett. 47, 457 (1985).
4. D. N. Rao, P. Chopra, S. K. Ghoshal, J. Swiatkiewicz, and P. N. Prasad, J. Chem. Phys. 84, 7049 (1986).
5. G. M. Carter, J. V. Hryniewicz, M. K. Thakur, Y. J. Chen, and S. E. Meyler, Appl. Phys. Lett. 49, 998 (1986).
6. D. N. Rao, J. Swiatkiewicz, P. Chopra, S. K. Ghoshal, and P. N. Prasad, Appl. Phys. Lett. 48, 1187 (1986).
7. H. W. Mocker and R. J. Collins, Appl. Phys. Lett. 7, 270 (1965).
8. W. H. Knox, R. L. Fork, M. C. Downer, R. H. Stolen, and C. V. Shank, Appl. Phys. Lett. 46, 1120 (1985).
9. N. Morita and T. Yajima, Phys. Rev. A 30, 2525 (1984).
10. S. Asaka, H. Nakatsuka, N. Fujiwara, and M. Matsuoka, Phys. Rev. A 29, 2286 (1984).
11. R. Beach and S. R. Hartmann, Phys. Rev. Lett. 53, 663 (1984).
12. H. Nakatsuka, M. Tomita, M. Fujiwara, and S. Asaka, Opt. Commun. 52, 150 (1984).
13. M. Fujiwara, R. Kuroda and H. Nakatsuka, J. Opt. Soc. Am. B 2, 1634 (1985).
14. S. R. Meech, A. J. Hoff, and D. A. Wiersma, Chem. Phys. Lett. 121, 287 (1985).
15. T. Hattori and T. Kobayashi, Chem. Phys. Lett. 133, 230 (1987).
16. A. M. Weiner, S. De Silvestri, and E. P. Ippen, J. Opt. Soc. Am. B 2, 654 (1985).
17. A. Laubereau and W. Kaiser, Rev. Mod. Phys. 50, 607 (1978).
18. S. M. George, H. Auwester, and C. B. Harris, J. Chem. Phys. 73, 5573 (1980).
19. S. M. George, A. L. Harris, M. Berg, and C. B. Harris, J. Chem. Phys. 80, 83 (1984).
20. T. Hattori, A. Terasaki, and T. Kobayashi, Phys. Rev. A 35, 715 (1987).
21. K. Kurokawa, T. Hattori, and T. Kobayashi, submitted to Phys. Rev. A.
22. S. Koshihara, T. Kobayashi, H. Uchiki, T. Kotaka, and H. Ohnuma, Chem. Phys. Lett. 114, 446 (1985).
23. T. Kobayashi, J. Iwai, and M. Yoshizawa, Chem. Phys. Lett. 112, 360 (1984).
24. T. Kobayashi, H. Ikeda, and S. Tsuneyuki, Chem. Phys. Lett. 116, 515 (1985).
25. J. Orenstein, S. Etemad, and G. L. Baker, J. Phys. C 17, L297 (1984).

26. L. Robins, J. Orenstein, and R. Superfine, Phys. Rev. Lett. _56_, 1850 (1986).
27. K. Ichimura, T Kobayashi and T. Kotaka, to be published.
28. Y. Tokura and T. Koda, J. Chem. Phys. _85_, 99 (1986).
29. R. R. Chance, M. L. Shand, C. Hogg and R. Silbey, Phys. Rev. B _22_, 3540 (1980).
30. J.- C. Diels and I. C. McMichael, J. Opt. Soc. Am. B _3_, 535 (1986).
31. J. W. Goodman, Statistical Optics (Wiley, New York, 1985).
32. T. Kanetake, Y. Tokura, T. Koda, T. Kotaka, and H. Ohnuma, J. Phys. Soc. Jpn. _54_, 4014 (1985).
33. R. R. Chance, G. N. Patel, and J. D. Witt, J. Chem. Phys. _71_, 206 (1979).
34. H. Sixl and R. Warta, Chem. Phys. Lett. _116_, 307 (1985).
35. S. R. Hartmann, IEEE. J. Quantum Electron. _QE-4_, 802 (1968).
36. R. F. Loring and S. Mukamel, J. Chem. Phys. _83_, 2116 (1985).
37. A. C. Eckbreth, Appl. Phys. Lett. _32_, 421 (1978).
38. W. Zinth, H.- J. Polland, A. Laubereau, and W. Kaiser, Appl. Phys. B _26_, 77 (1981).
39. S. M. George and C. B. Harris, Phys. Rev. A _28_, 863 (1983).
40. P. Hu, S. Geschwind, and T. M. Jedju, Phys. Rev. Lett. _37_, 1357 (1976).
41. K. P. Leung, T. W. Mossberg, and S. R. Hartmann, Phys. Rev. A _25_, 3097 (1982).
42. V. Brückner, E. A. J. M. Bente, J. Langelaar, and D. Bebelaar, Opt. Commun. _51_, 49 (1984).
43. J. D. W. van Voorst, D. Brandt, and B. L. van Hensbergen, in Technical Digest of Topical Meeting on Ultrafast Phenomena, (1986).
44. R. Y. Chiao, E. Garmire, and C. H. Townes, Phys. Rev. Lett. _13_, 479 (1964).
45. E. P. Ippen and C. V. Shank, Appl. Phys. Lett. _26_, 92 (1975).
46. J. M. Halbout and C. L. Tang, Appl. Phys. Lett. _40_, 765 (1982); C. L. Tang and J. M. Halbout, in Picosecond Phenomena III edited by K. B. Eisenthal. R. W. Hochstrasser, W. Kaiser, and A. Laubereau (Springer, Berlin, 1982), p.212.
47. B. I. Greene and R. C. Farrow, J. Chem. Phys. _77_, 4779 (1982); B. I. Greene and R. C. Farrow, in Picosecond Phenomena III edited by K. B. Eisenthal. R. W. Hochstrasser, W. Kaiser, and A. Laubereau (Springer, Berlin, 1982), p.209.
48. J. Etchepare, G. Grillon, R. Astier, J. L. Martin, C. Bruneau, and A. Antonetti, in Picosecond Phenomena III edited by K. B. Eisenthal. R. W. Hochstrasser, W. Kaiser, and A. Laubereau (Springer, Berlin, 1982), p.217; J. Etchepare, G. Grillon, and A. Antonetti, Chem. Phys. Lett. _107_, 489 (1984).
49. R. Trebino, C. E. Barker, and A. E. Siegman, IEEE J. Quantum Electron. _QE-22_, 1413 (1986).
50. M. Golombok and G. A. Kenney-Wallace, in Ultrafast Phenomena IV edited by D. H. Auston and K. B. Eisenthal (Springer, Berlin, 1984), p.383.
51. R. W. Hellwarth, Prog. Quantum Electron. _5_, 1 (1977).
52. J. L. Oudar, IEEE J. Quantum Electron. _QE-19_, 713 (1983).
53. J. Etchepare, G. Grillon, I. Thomazeau, A. Migus, and A. Antonetti, J. Opt. Soc Am. B _2_, 649 (1985).

STUDIES OF MONOMER AND POLYMER MONOLAYERS USING OPTICAL SECOND AND THIRD HARMONIC GENERATION

Garry Berkovic and Y. R. Shen

Department of Physics, University of California
Center for Advanced Materials, Lawrence Berkeley Lab.
Berkeley, California 94720

Optical second harmonic generation (SHG) is a highly surface sensitive technique for studying ultrathin molecular layers. This technique has been used to study monolayers of various organic monomers -- vinyl stearate, octadecyl methacrylate and some diacetylene derivatives -- and their corresponding polymers spread at air/water interfaces. Different SHG signals are obtained from a pure water surface, and from water covered with monomer and polymer monolayers. During polymerization of monomer monolayers the SHG signal intensity is observed to change continuously from the monomer monolayer value to the polymer monolayer value. Monolayer polymerization of some amphiphilic diacetylenes was also studied by third harmonic generation (THG), which unlike SHG is not a surface sensitive probe. However, the extremely high third order nonlinearity of polydiacetylenes enables THG from a single monolayer to be observable above the water background signal. In addition, evaluation of surface molecular orientations and molecular nonlinear optical coefficients from monolayer SHG and THG is also discussed.

INTRODUCTION

Monolayers of polymerizable material have been the subject of numerous studies in many fields of basic and applied chemistry research. Polymerizable monolayers have been employed as ultrathin coatings in device technology in microlithography[1] and microelectronics,[2] and show potential for application in nonlinear optical devices.[3] More basic research has involved studying monolayer polymerization as two dimensional systems in which molecular separations and orientations may be varied by the experimentalist in order to study the basic factors affecting reactivity and kinetics.[4] In biological chemistry polymerizable monolayers have been employed to fabricate synthetic lipids[5] and vessicles.[6]

Unfortunately, characterization and analysis of monolayers and ultrathin films is made difficult by two interrelated limitations -- the small number of molecules in a monolayer, and the fact that a monolayer must invariably be supported on some much thicker substrate which will often interfere with or mask the adsorbate analysis. Removal of the monolayer film into solution may enable more convenient

157

analysis, but only at the price of removing the monolayer from its environment of interest. Although some very recent advances have enabled a small number of in situ studies of monolayers using specially designed and adapted instrumentation for infrared[7] and X-ray diffraction analysis,[8] there clearly is a need for other special experimental techniques for the study of monolayers.

One technique which provides in situ, nondestructive, surface sensitive analysis for monolayers is second order nonlinear optics. Using the processes of second harmonic generation (SHG) and sum frequency generation (SFG) a variety of studies of molecular monolayers have been performed. Using merely the SHG of a fixed frequency laser input it is possible to observe the surface concentration,[9] orientation[9,10] and two dimensional phase transitions[11] of molecular monolayers on various substrates. Using tunable frequency input it is possible to measure the electronic[12] and vibrational[13] absorption spectrum of the adsorbed monolayer.

Second harmonic generation in a medium arises from the second order polarization $\vec{P}^{(2)}$ induced in a medium by an applied electric field $\vec{E}$, given by

$$\vec{P}^{(2)} = \overset{\leftrightarrow}{\chi}^{(2)} : \vec{E}\vec{E} \tag{1}$$

where $\overset{\leftrightarrow}{\chi}^{(2)}$ is the second order nonlinear susceptibility. It then follows that a laser field at frequency ω (i.e. $E = E_0 e^{i\omega t}$) will generate a second order polarization oscillating at frequency 2ω. This polarization is observable as photons generated at frequency 2ω, whose intensity is proportional to $|P(2\omega)|^2$. Sum frequency generation is the analogous process using two input laser fields of different frequencies.

The reason that SHG is a surface sensitive probe is that it follows from Eq. (1) that $\chi^{(2)} = 0$ in a medium possessing a center of symmetry. However, it is well established[14] that this is not strictly correct in a centrosymmetric medium, which usually has a weak second order polarization resulting from a weak quadrupole contribution that has been neglected in Eq. (1).[14] Nevertheless SHG is sufficiently repressed in a centrosymmetric or isotropic bulk medium to generally enable easy observation of SHG from a non-centrosymmetric monolayer adsorbed on it.

The second harmonic signal, $I(2\omega)$, generated in reflection from a monolayer covered water surface is given by[9]

$$I(2\omega) = \frac{32\pi^3\omega^2}{c^3\varepsilon(\omega)\varepsilon^{1/2}(2\omega)} |\vec{e}_{2\omega}\overset{\leftrightarrow}{\chi}_S^{(2)} : \vec{e}_\omega\vec{e}_\omega|^2 I^2(\omega), \tag{2}$$

where $\vec{e}_\Omega = L_\Omega \hat{e}_\Omega$ with $\hat{e}_\Omega$ denoting the unit polarization vector of the field at frequency Ω, and L_Ω the Fresnel factor for the field; $\varepsilon(\omega)$ is the substrate dielectric constant; $I(\omega)$ is the laser intensity and $\chi_S^{(2)}$ is the (complex) surface second order susceptibility tensor.

The contributions to $\overset{\leftrightarrow}{\chi}_S^{(2)}$ may be represented by

$$\overset{\leftrightarrow}{\chi}_S^{(2)} = \overset{\leftrightarrow}{\chi}_W^{(2)} + \overset{\leftrightarrow}{\chi}_m^{(2)}, \tag{3}$$

where $\overset{\leftrightarrow}{\chi}_W^{(2)}$ and $\overset{\leftrightarrow}{\chi}_m^{(2)}$ are the susceptibilities of the substrate (water) and the adsorbate monolayer, respectively. Any interaction between adsorbate and substrate may be taken into account[14] by including an additional term $\overset{\leftrightarrow}{\chi}_I^{(2)}$ in $\overset{\leftrightarrow}{\chi}_S^{(2)}$. As discussed above, due to its isotropy,

$\overset{\leftrightarrow}{\chi}_w^{(2)}$ is not expected to be strong, and thus the monolayer susceptibility is readily observable.

The next higher order nonlinear optical process is third harmonic generation (THG) caused by a third order polarization, $\vec{P}^{(3)}$, given by

$$\vec{P}^{(3)} = \overset{\leftrightarrow}{\chi}^{(3)} : \vec{E}\vec{E}\vec{E}. \tag{4}$$

However, THG is electric dipole allowed in all media, and so is not surface specific like SHG. The third harmonic signal, $I(3\omega)$, reflected off a monolayer covered water surface is

$$I(3\omega) = \frac{576\pi^4\omega^2}{c^4\varepsilon(\omega)\varepsilon(3\omega)} \ |\vec{e}_{3\omega}\overset{\leftrightarrow}{\chi}_s^{(3)} : \vec{e}_\omega\vec{e}_\omega\vec{e}_\omega|^2 I^3(\omega), \tag{5}$$

where $\overset{\leftrightarrow}{\chi}_s^{(3)}$ is the effective surface third order susceptibility given by:[15]

$$\overset{\leftrightarrow}{\chi}_s^{(3)} = \overset{\leftrightarrow}{\chi}_m^{(3)} + i\overset{\leftrightarrow}{\chi}_w^{(3)}\ell_{eff} \tag{6}$$

with

$$\ell_{eff}^{-1} = (3\omega/c)(n_\omega\cos\theta_\omega + n_{3\omega}\cos\theta_{3\omega}). \tag{7}$$

Here, $\overset{\leftrightarrow}{\chi}_m^{(3)}$ is the third order susceptibility of the monolayer and $\overset{\leftrightarrow}{\chi}_w^{(3)}$ is the susceptibility of a unit volume of water. The effective interaction depth, ℓ_{eff}, is defined by the refractive index (n_Ω) and propagation direction (θ_Ω) of the fundamental and third harmonic beams in the water. For 1.06 μm light incident on water at an angle of 60°, we calculate $\ell_{eff} = 280$ Å. Thus, in order to observe THG from an adsorbate monolayer, its third order susceptibility must be at least comparable to that of a 280 Å thickness of water. Since monolayer thicknesses are of the order of 10 Å, THG from adsorbate monolayers should only be observable from adsorbates whose molecular third order polarizability is very much greater than that of water. We also note in Eq. (6) that in the effective surface susceptibility $\overset{\leftrightarrow}{\chi}_s^{(3)}$ there is a phase factor i associated with the bulk term $\overset{\leftrightarrow}{\chi}_w^{(3)}$.

In this paper we describe some of our experiments using nonlinear optics to probe monomer and polymer monolayers spread at an air/water interface. The materials studied are all spreadable on water by virtue of being amphiphilic, i.e. they contain both hydrophilic and hydrophobic parts. We have used SHG to study the polymerization of two long chain monomers -- vinyl stearate (VS) and octadecyl methacrylate (ODMA) -- shown in Fig. 1. Although these materials do not have large second order nonlinearities it is shown that the SHG signal can still be used to follow the extent and kinetics of polymerization undergone by the monomer, without disturbance or destruction of the monolayer film. The orientation of molecules at the air/water interface could also be deduced from the polarization of the SHG signal.

We have also used SHG to probe the polymerization of some amphiphilic diacetylenes (see Fig. 2). It was found that the SHG signals from these monomers and their corresponding polymers did not differ sufficiently to allow SHG to monitor the extent and rate of the polymerization reaction. However, we have also been able to use THG to monitor the polymerization of a diacetylene monolayer in spite of the fact that THG is not surface-specific. This is because of the very high third-order nonlinearity of polydiacetylenes. We have also used THG to monitor the very interesting phase transition of one diacetylene polymer, poly-4-BCMU (see Fig. 2), from its "yellow" form to its "red" form when its monolayers on water are compressed above a certain density.[16]

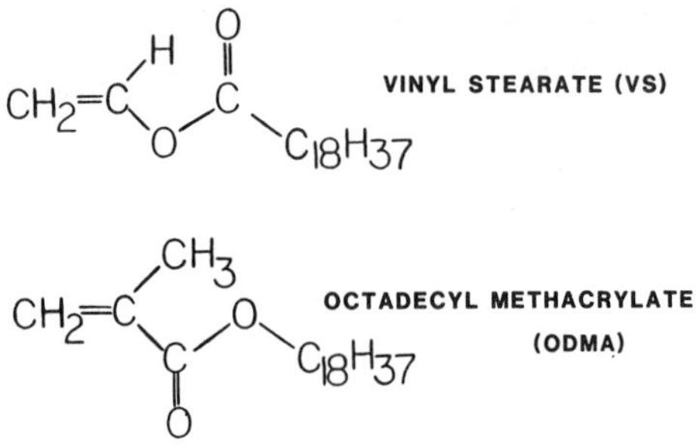

VINYL STEARATE (VS)

$CH_2=C$ ... $C_{18}H_{37}$

OCTADECYL METHACRYLATE (ODMA)

Fig. 1.

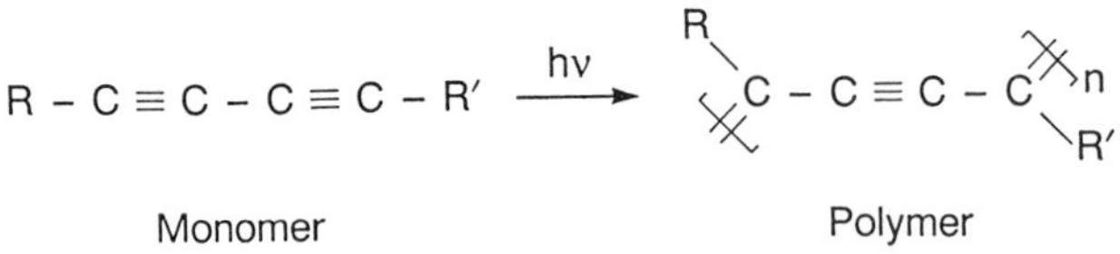

$$R - C \equiv C - C \equiv C - R' \xrightarrow{h\nu} \left(\begin{array}{c} R \\ C - C \equiv C - C \\ R' \end{array} \right)_n$$

Monomer   Polymer

MONOMER	POLYMER	SUBSTITUENTS
12-8 DA	Poly-12-8 DA	$R = C_{12}H_{25}$ $R' = C_8H_{16}COOH$
9-9 DA	Poly-9-9 DA	$R = C_9H_{19}$ $R' = C_9H_{18}COOH$
4-BCMU	Poly-4-BCMU	$R,R' = (CH_2)_4OOC-$ $NHCH_2COOC_4H_9$

Fig. 2. Diacetylene monomers and polymers used in this study.

Finally, we show how monolayer SHG and THG data may be analyzed to evaluate molecular nonlinear optical constants. This represents a convenient alternative method[17] for evaluation of molecular nonlinearities.

EXPERIMENTAL

Our experimental set us is depicted in Fig. 3. Laser light is focused onto the surface of a Langmuir trough onto which a monolayer of the material under study has been spread by pipetting an appropriate volume of a solution in a volatile solvent. Before reaching the water surface the laser light passes through a narrow bandpass color filter which transmits only the fundamental frequency, and thus blocks any pregenerated harmonic radiation. Light reflected from the water

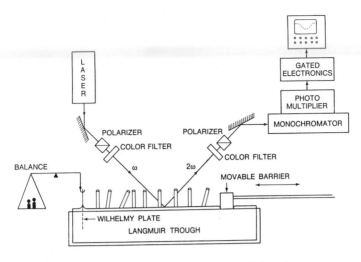

Fig. 3. Schematic of experimental set up.

surface passes through a second color filter, which blocks the
fundamental, but transmits the desired harmonic frequency (2ω or 3ω).
Harmonically generated photons are detected through a monochromator and
a gated photon counting system, and averaged over at least 3×10^4
laser pulses. These signals were verified as being from true harmonic
generation by virtue of their very sharply peaked, background free
responses detected when the output monochromator is scanned through the
harmonic wavelength, and the correct dependence of the signal intensity
on input laser power.

The laser employed was a 500 Hz Q-switched Nd:YAG laser, where each
Q-switched envelope contained approximately 10 mode-locked pulses of 60
ps duration. The laser light was incident on the water surface at 60°
from the normal, and was focused onto the surface using lenses of 15-30
cm focal length. For THG measurements the laser output at 1.06 μm was
employed; while for SHG both this wavelength and the frequency doubled
output (532 nm) were employed. Pulse energies at these wavelengths
were 0.7 mJ (1.06 μm) and 0.3 mJ (532 nm) per Q-switched pulse.

RESULTS AND DISCUSSION

SHG Studies of Polymerization of Vinyl Stearate and Octadecyl
Methacrylate Monolayers

The SHG signal generated from a pure water surface has been
compared with that of a water surface covered with a monomer monolayer
of vinyl stearate (or octadecyl methacrylate), and with water covered
with a monolayer of a commercially available bulk polymerized sample of
the corresponding polymer.[18] As shown in Table I for both VS and ODMA
the SHG signal intensity shows clear differences between pure water,
monomer covered water and polymer covered water. Furthermore, after
monomer monolayers were irradiated for 2 hrs with a weak UV lamp under
nitrogen atmosphere the SHG signals became very similar to those of the
authentic polymer monolayers, indicating that the monomer monolayer had
undergone almost complete UV initiated polymerization (no change was
induced in the absence of UV radiation by either the probe laser or the
ambient thermal conditions).

The observation that, in our case, SHG from monomers is larger than

Table I. Relative intensities and polarization of second harmonic generation from a water surface covered with various monolayers.

System	Relative SHG Intensity[a]	Polarization Ratio[b]
Water only	100	2
Water + VS monolayer (27 Å^2/molecule)	260	1.5
Water + poly VS monolayer[c] (27 Å^2/monomeric unit)	170	0.5
Water + VS monolayer after UV irradiation	180	
Water + ODMA monolayer (26 Å^2/molecule)	370	
Water + poly ODMA monolayer[c] (26 Å^2/monomeric unit)	220	0.5
Water + ODMA monolayer after UV irradiation	250	

[a]The total output SHG signal generated using an input 532 nm laser field polarized at 45° to the plane of incidence.
[b]The ratio of s-polarized to p-polarized SHG output.
[c]Bulk polymerized polymer spread on water.

that of their corresponding polymers can be understood as follows: Second order optical nonlinearity of molecules arises mainly from chemical bonds in which the electron distributions are more readily distorted by optical excitation.[19] In VS and ODMA the π electrons in the double bonds are likely to dominate the nonlinearity. Since the polymerization process breaks a carbon-carbon double bond, the optical nonlinearity decreases. It thus may also be expected that SHG from bulk and surface polymerized samples of the same polymer should be essentially the same, although their molecular weight distributions may be quite different.[20]

In order to follow the kinetics of polymerization we also made SHG measurements during the UV irradiation.[18] As shown in Fig. 4 the SHG intensity decreases continuously during the reaction. Unfortunately, due to the low optical nonlinearity of these materials and the relatively small changes in signal intensity during polymerization our measurement was not accurate enough to equivocably distinguish between first and second order kinetics for the polymerization.

SHG and THG Studies of Diacetylene Monolayer Polymerization

In a similar way we have used SHG to study polymerization of the two monomeric diacetylenes 12-8 DA and 9-9 DA (see Fig. 2).[21] Once again, the SHG signal changes noticeably when water is covered with a monomeric monolayer. However, in this case, not much change occurred in the SHG signal when the monomer monolayer was polymerized by UV irradiation. This behavior was observed for both 12-8 DA and 9-9 DA, using different fundamental laser wavelengths (1.06 μm or 532 nm, as well as 640 nm and 590 nm generated from a dye laser). Results from a typical experiment are shown in Table II.

The fact that the SHG signal is essentially unchanged after

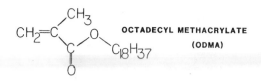

OCTADECYL METHACRYLATE
(ODMA)

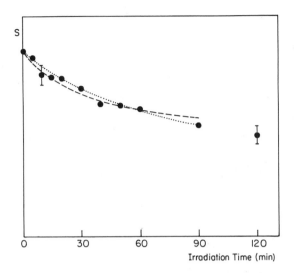

Fig. 4. The relative SHG intensity (S) is plotted against irradiation time for UV polymerization of ODMA. The data (●) can be fitted satisfactorily by both first order (•••) and second order (---) kinetics (see the text). All experimental data points have the same uncertainty, although error bars for most points have been omitted for clarity.

Table II. Relative second harmonic generation intensities from water and water covered with monomeric and UV polymerized 12-8 DA monolayers.[a]

Sample	s-polarized SHG output	p-polarized SHG output
Water only[b]	100	60
Water + 12-8 DA monolayer (25 $Å^2$/molecule)	180	200
Water + 12-8 DA monolayer after UV irradiation	160	180

[a] Measurements were performed using a 1.06 μm input laser field polarized at 45° to the plane of incidence. Relative SHG intensities have an uncertainty of ± 5%.
[b] The water SHG signal is set on a relative scale at 100 units in the s-polarized output.

polymerization indicates that the SHG arises from a part of the diacetylene molecule unchanged by polymerization. It is reasonable to

163

attribute the signal to the carboxylic acid end group. The magnitude of the signal relative to water and the stronger p-polarized component are in agreement with studies of other molecules where the SHG signal arises mainly from a carboxylic acid end group,[11] which has larger nonlinearity than a hydrocarbon chain.[17]

The lack of SHG from the diacetylenic core of both monomers and polymers is understood by the fact that this part of the molecule is essentially centrosymmetric -- its symmetry is only broken by the carboxylic acid group which is at least 8 carbon atoms away. Significant SHG from diacetylenes has only been reported[22] when two different, highly asymmetric groups are substituted immediately after the diacetylene bond.

Due to the very large third order nonlinearity of polydiacetylene we also examined the possibility of using THG to study the polymerization of diacetylene monomers.[21] Since THG is an electric dipole allowed process in all media, the water substrate had, as expected, a large THG signal. Addition of a monomer monolayer of 12-8 DA or 9-9 DA did not cause a noticeable change in THG. However, when these monolayers were polymerized the THG signal increased significantly, but exhibited large fluctuations both from experiment to experiment and from within one experiment. It generally varied between 3-10 times that of pure water. This variation apparently comes from nonuniformity of the polymer density in the monolayer film.[23] As the higher density regions float in and out of the laser beam the THG signal should fluctuate strongly (such fluctuations were not evident in the SHG studies because of the much smaller magnitudes and differences in the SHG signals from water, monomer and polymer).

THG Study of 2-Dimensional Conformational Phase Transition in Poly-4-BCMU

Poly-4-BCMU is a diacetylene polymer soluble in some organic solvents which may then be directly spread to give a monolayer on water.[16] In bulk phases this material exhibits two structural conformers -- the "red" and the "yellow" forms, in which there is stronger intramolecular hydrogen bonding in the former.[24] It was recently shown that when monolayers of this material are spread on water the yellow form is obtained at low surface densities, while if the monolayer is compressed to higher densities the red form is obtained.[16] We have studied this transformation using THG.[21] The good surface spreading characteristics of this material enabled stable, reproducible THG signals, unlike the in situ polymerized monolayers of 12-8 DA and 9-9 DA. Figure 5 shows the THG signal from water covered with poly-4-BCMU at various surface densities of the monolayer. It is seen that at low coverages ($\geq$ 120 A^2/repeating unit) the THG signal increases only slightly from the pure water value. However, at about 100 A^2/repeating unit the THG signal increases abruptly in the s-polarized output (the p-polarized output remains very weak). This surface density is in excellent agreement with the density at which the red form has been reported to appear.[16] Furthermore, the much higher THG signal observed when the red form appears also agrees very well with results from four wave mixing experiments in bulk phases of poly-4-BCMU which also showed[25,26] that the red form has a much higher third order nonlinearity than the yellow form.

An additional feature evident in Fig. 5 is that at higher surface densities (< 50 A^2/repeating unit) the polarization of the THG output changes significantly. This can be attributed to a change in the orientation of the polymer chains at the water surface when the monolayer is compressed to such high densities (see below).

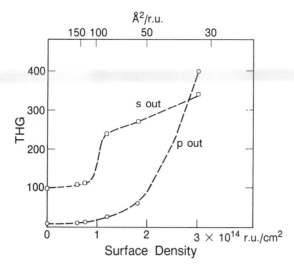

Fig. 5. Third harmonic generation from water covered with a monolayer of poly-4-BCMU at various surface densities [expressed as either number of polymer repeating units (r.u.) per cm^2, or Å^2/r.u.]. The input laser field was polarized at 45° to the plane of incidence; THG was measured for both s- and p-polarized outputs. The s-polarized pure water (zero surface density of polymer) THG signal is set at 100 arbitrary units. The broken curve joining the data points is for guidance only.

Evaluation of Molecular Orientations and Molecular Nonlinearities from Monolayer SHG and THG Measurements

In many cases, both molecular orientations and molecular nonlinearities may be calculated from monolayer SHG and THG measurements such as those described above. When measurements are made relative to the "internal standard" provided by the pure water substrate whose nonlinear coefficients are known, the absolute values of surface susceptibility components are thus determined. Subtraction of the substrate contribution from the total surface susceptibility can be made if the relative phases of the substrate and total surface susceptibilities are known. A simple method for determining these relative phases has been described.[9,27]

Although there are in principle 18 independent components of the second order surface susceptibility tensor $\overset{\leftrightarrow}{\chi}_s^{(2)}$, the situation is greatly simplified in the case of a surface isotropic (on the scale of the laser spot size) about its normal. This condition, which should generally hold for monolayers, reduces the number of independent nonvanishing components of $\overset{\leftrightarrow}{\chi}_s^{(2)}$ (and $\overset{\leftrightarrow}{\chi}_m^{(2)}$) to three.[14] A further simplification occurs if the polarizability tensor of the molecule forming the monolayer is dominated by a single component $\alpha_{\xi\xi\xi}^{(2)}$ along some molecular axis ξ. In such cases the only nonvanishing components of $\overset{\leftrightarrow}{\chi}_m^{(2)}$ are $\chi_{m,zzz}^{(2)}$ and $\chi_{m,zii}^{(2)} = \chi_{m,izi}^{(2)}$ (i = x,y), which are simply related to $\alpha_{\xi\xi\xi}^{(2)}$ and θ, the average orientation of ξ to the surface normal.[9,17] Equation (2) demonstrates that, depending on the input (fundamental) and output (harmonic) polarizations employed, each SHG measurement yields a certain linear combination of the nonvanishing components of $\overset{\leftrightarrow}{\chi}_s^{(2)}$ (and hence $\overset{\leftrightarrow}{\chi}_m^{(2)}$). Thus in this special case just two independent SHG measurements (such as the s- and p-polarized SHG outputs following 45° polarized input) are sufficient to evaluate $\chi_{m,zzz}$ and $\chi_{m,zii}$, which then yield $\alpha^{(2)}_{\xi\xi\xi}$ and θ.

Thus in the case of poly VS and poly ODMA, where the nonlinearity is expected to be dominated along the C=O bond, such analysis of the SHG signal intensity and polarization ratio (see Table I) has been performed.[18] After showing that the monolayer SHG signal is in phase with that of water, it was deduced that the C=O bond is essentially perpendicular to the water surface and $\alpha^{(2)}_{\xi\xi\xi} \sim 3 \times 10^{-32}$ esu (poly VS) and $\alpha^{(2)}_{\xi\xi\xi} \sim 5 \times 10^{-32}$ esu (poly ODMA). The conclusion that the C=O bond is directed perpendicular to the water surface is in agreement with both theoretical predictions[28] and infrared analysis of monolayers which had been transferred onto various substrates.[29-31]

The SHG data of VS and ODMA monomers cannot unfortunately be analyzed by this model. The nonlinearity of these molecules is expected to arise from both their C=C and C=O bonds which are not collinear, and thus it is not reasonable to represent their nonlinearity by only one tensor component.

In an analogous way, the THG data of poly-4-BCMU can be analyzed[21] to find the dominant tensor component of the molecular polarizability $\alpha^{(3)}_{\xi\xi\xi\xi}$ along the polydiacetylene chain, and the orientation of the chain segments relative to the water surface. We found $|\alpha^{(3)}_{\xi\xi\xi\xi}| \sim 9 \times 10^{-34}$ esu for the yellow form and $|\alpha^{(3)}_{\xi\xi\xi\xi}| \sim 6 \times 10^{-33}$ esu for the red form. This value extrapolates to a value of $\chi^{(3)}_{xxxx} \sim 3 \times 10^{-11}$ esu for bulk poly-4-BCMU in its red form, which agrees well with other measurements of the third order nonlinearity of polydiacetylenes at 1.06 μm.[32-34]

Analysis of the change in THG polarization as the monolayer is compressed to high density (see Fig. 5) yielded[21] the result that the polymer chain segments lie quite flat ($\sim 70°$ to the normal) at low surface densities, but lie at $\sim 40°$ to the normal when at a surface density of 30 Å^2/repeating unit.

CONCLUSION

We have shown that monolayer SHG and THG measurements are a sensitive probe for monolayer polymerization reactions, phase transitions and molecular surface orientation, and can be used to evaluate molecular nonlinear optical coefficients. SHG measurements can be applied to monolayers of molecules of quite low nonlinearity (such as vinyl stearate and octadecyl methacrylate monomers and polymers). However, THG measurements on monolayers are only applicable to materials of very high third order nonlinearity like polydiacetylenes.

ACKNOWLEDGEMENTS

The authors have enjoyed collaboration with Th. Rasing, P. Guyot-Sionnest, R. Superfine (Berkeley) and P. N. Prasad (SUNY, Buffalo) on some of the projects described in this chapter. G.B. acknowledges the financial support of a Chaim Weizmann Postdoctoral Fellowship. This work was supported by the Director, Office of Energy Research, Office of Basic Energy Sciences, Materials Sciences Division of the U.S. Department of Energy under Contract No. DE-AC03-76SF00098.

REFERENCES

1. A. Barraud, C. Rosilio, and R. Ruaudel-Teixier, Thin Solid Films 68, 91 (1980).
2. G. L. Larkins, E. D. Thompson, M. J. Deen, C. W. Burkhart, and J. B. Lando, IEEE Trans. Magn. 19, 980 (1983).
3. J. Zyss, J. Mol. Elect. 1, 25 (1985).
4. M. Puterman, T. Fort, Jr., and J. B. Lando, J. Colloid Int. Sci. 47, 705 (1974); S. A. Letts, T. Fort, Jr., and J. B. Lando, J. Colloid Inst. Sci. 56, 64 (1976).
5. R. Elbert, A. Laschewsky, and H. Ringsdorf, J. Am. Chem. Soc. 107, 4134 (1985).
6. P. Yager and P. E. Schoen, Mol. Cryst. Liq. Cryst. 106, 371 (1984).
7. J. F. Rabolt, F. C. Burns, N. E. Schlotter, and J. D. Swalen, J. Chem. Phys. 78, 946 (1983).
8. M. Seul, P. Eisenberger, and H. M. McConnell, Proc. Natl. Acad. Sci. USA 80, 5795 (1983).
9. T. F. Heinz, H. W. K. Tom, and Y. R. Shen, Phys. Rev. A 28, 1883 (1983).
10. Th. Rasing, Y. R. Shen, M. W. Kim, P. Valint, Jr., and J. Bock, Phys. Rev. A 31, 537 (1985).
11. Th. Rasing, Y. R. Shen, M. W. Kim, and S. Grubb, Phys. Rev. Lett. 55, 2903 (1985).
12. T. F. Heinz, C. K. Chen, D. Ricard, and Y. R. Shen, Phys. Rev. Lett. 48, 478 (1982).
13. J. H. Hunt, P. Guyot-Sionnest, and Y. R. Shen, Chem. Phys. Lett. 133, 189 (1987); X. D. Zhu, H. Suhr, and Y. R. Shen, Phys. Rev. B 35, 3047 (1987).
14. P. Guyot-Sionnest, W. Chen, and Y. R. Shen, Phys. Rev. B 33, 8254 (1986).
15. H. W. K. Tom, T. F. Heinz, and Y. R. Shen, Phys. Rev. Lett. 51, 1983 (1983).
16. J. E. Biegajski, R. Burzynski, D. A. Cadenhead, and P. N. Prasad, Macromolecules 19, 2457 (1986).
17. Th. Rasing, G. Berkovic, Y. R. Shen, S. G. Grubb, and M. W. Kim, Chem. Phys. Lett. 130, 1 (1986).
18. G. Berkovic, Th. Rasing, and Y. R. Shen, J. Chem. Phys. 85, 7374 (1986).
19. D. J. Williams, Angew. Chem. Int. Ed. 23, 690 (1984).
20. K. C. O'Brien, J. Long, and J. B. Lando, Langmuir 1, 514 (1985).
21. G. Berkovic, R. Superfine, P. Guyot-Sionnest, Y. R. Shen, and P. N. Prasad, submitted for publication in J. Opt. Soc. Amer. B.
22. A. F. Garito, C. C. Teng, K. Y. Wong, and O. Zammani'Khamiri, Mol. Cryst. Liq. Cryst. 106, 219 (1984).
23. D. Day and J. B. Lando, Macromolecules 13, 1478 (1980).
24. R. R. Chance, G. N. Patel, and J. D. Witt, J. Chem. Phys. 71, 206 (1979); K. C. Lim and A. J. Heeger, J. Chem. Phys. 82, 522 (1985).
25. R. R. Chance, M. L. Shand, C. Hogg, and R. Sibley, Phys. Rev. B 22, 3540 (1980).
26. D. N. Rao, P. Chopra, S. K. Ghoshal, J. Swiatkiewicz, and P. N. Prasad, J. Chem. Phys. 84, 7049 (1986).
27. K. Kemnitz, K. Bhattacharyya, J. M. Hicks, G. R. Pinto, K. B. Eisenthal, and T. F. Heinz, Chem. Phys. Lett. 131, 285 (1986).
28. D. Naegele and H. Ringsdorf, J. Polym. Sci. Polym. Chem. Ed. 15, 2821 (1977).
29. V. Enkelmann and J. B. Lando, J. Polym. Sci. Polym. Chem. Ed. 15, 1843 (1977).
30. K. Fukuda and T. Shiozawa, Thin Solid Films 68, 55 (1980).
31. S. J. Mumby, J. D. Swalen, and J. F. Rabolt, Macromolecules 19, 1054 (1986).

32. F. Kajzar, J. Messier, J. Zyss, and I. Ledoux, Opt. Commun. $\underline{45}$, 133 (1983).
33. F. Kajzar and J. Messier, Thin Solid Films $\underline{132}$, 11 (1985).
34. G. M. Carter, Y. J. Chen, and S. K. Tripathy, Appl. Phys. Lett. $\underline{43}$, 891 (1983).

DEVELOPMENT OF POLYMERIC NONLINEAR OPTICAL MATERIALS

R.N. DeMartino, E.W. Choe, G. Khanarian, D. Haas,
T. Leslie, G. Nelson, J. Stamatoff, D. Stuetz,
C.C. Teng, and H. Yoon

Hoechst Celanese Corporation
R.L. Mitchell Technical Center
86 Morris Avenue
Summit, NJ 07901

INTRODUCTION

New requirements for communications, computing, medical, and dedicated military systems stress the need for increasing speed, information content, and the controlled delivery of optical energy in specific forms. It is obvious that to meet these requirements, lightwave technology must be developed. Laser developments have satisfied many of the source requirements for these needs. Nonlinear optical (NLO) devices are desirable to process light derived from lasers so that systems capable of rapid, high information content communications or computing may be developed. NLO devices are needed for controlled delivery of energy at appropriate wavelengths. Currently available NLO inorganic crystalline materials are not adequate for the desired device characteristics. The intrinsic properties of organic NLO materials offer significant advantages in the magnitude of NLO effects over a broad range of wavelengths and the speed of NLO processes. Although organic crystalline materials eliminate these two major inadequacies of inorganic crystalline materials, fabrication requirements for NLO devices make single crystal technology unacceptable for most device designs.

Celanese is engaged in a program to develop NLO polymers. Such materials possess not only the enhanced magnitude and speed of organic materials but also the very attractive advantages of polymer processing. These materials are projected to be wholly adequate for NLO device development. Further, these materials are the only potential materials which are projected to meet device requirements for the entire range of communications, computing, medical and dedicated military systems. In this paper, these design concepts will be highlighted in an attempt to define an industrial approach to this material problem.

MOLECULAR DESIGN OF NLO MOLECULES

Nonlinear optics is primarily concerned with response of a dielectric material to a strong electromagnetic field. The material response is formally represented by a dielectric constitutive equation of the following form (1):

$$P_i = P_{o,i} + \chi_{ij}^{(1)}E_j + \chi_{ijk}^{(2)}E_jE_k + \chi_{ijkl}^{(3)}E_jE_kE_l + \ldots \qquad \text{I}$$

where P_i is the i-th component of the polarization, $P_{o,i}$ the permanent polarization, $\chi^{(1)}$ the linear electronic susceptibility, and $\chi^{(2)}$ and $\chi^{(3)}$ the second and third order susceptibilities. In Equation I, summation over repeated indices is assumed. In this paper, we are primarily concerned with second order NLO materials, i.e., materials exhibiting a large $\chi^{(2)}$. Applications of second order materials may be a linear electrooptic modulator, such as a directional coupler in optical communication, a frequency doubler using second harmonic generation, or a parametric oscillator.

The macroscopic NLO behavior of organic NLO materials originates from the polarization response of molecular electrons. Similar to Equation I, the molecular polarization behavior can formally be expressed as

$$\mu_i = \mu_{o,i} + a_{ij}E_j + \beta_{ijk}E_jE_k + \gamma_{ijkl}E_jE_kE_l + \ldots \qquad \text{II}$$

where μ is the dipole moment of the molecule in electrical fields, $\mu_{o,i}$, the permanent dipole moment, a, the polarizability, β and γ the second and third order hyperpolarizabilities, respectively. When molecules are immobile in the applied electrical or optical field (which is the case with most solid samples), $\chi^{(2)}$ results directly from the molecular β. Optimization of $\chi^{(2)}$ of NLO polymers, therefore, can begin from optimizing the molecular structure for larger β.

Because of their odd ordered tensorial character, both $\chi^{(2)}$ and β automatically vanish in a centrosymmetric material system. It then follows that for a material to display second order NLO behavior, it not only must contain noncentrosymmetric NLO moieties, but also these must be spatially arranged in a noncentrosymmetric manner. Thus, the development of NLO polymeric materials requires: (1) the design of noncentric NLO organic moieties with large molecular susceptibilities (e.g. β for $\chi^{(2)}$), (2) the incorporation of these moieties into tractable, optically clear polymers, and (3) the fabrication of polymers to control orientation and symmetry of the final material (e.g. a noncentric oriented structure is required for $\chi^{(2)}$ materials). Therefore, design of noncentric high β molecules is an essential but not wholly sufficient step in the development of NLO polymers.

An absolute measurement technique for β is DC second harmonic generation (DCSHG). The DCSHG technique[2], however, requires a complicated set of instruments and relatively high concentrations of NLO molecules in solution. In view of the large number of samples that were generated and a need to rapidly feed back the information to the synthetic chemists, an alternative experimental technique called "solvatochromism" was used for this study[3]. The method is based on the experimental observation that the light absorption frequency of dipolar molecules is dependent on the solvent polarity and the solvent dependent frequency shift originates from the same molecular electronic factors as the molecular second hyperpolarizability. An example of this solvent induced frequency shift is shown in Figure 1. Furthermore, the solvatochromism experiments are relatively easy laboratory experiments and data can be quickly analyzed and readily fed back for the next research iteration. This technique permits the generation of an extensive data base of β for various organic molecules (many of which show limited solubility). The data base is required to delineate general structural design criteria for increasing β .

Solvatochromic Determination of β

β , for a linear two state molecule, is given by:

$$\beta = \frac{3}{2h^2} \left| \mu_{ge}^2 \right| \left(\mu_e - \mu_g \right) \frac{\omega_0^2}{\left(\omega_0^2 - 4\omega^2 \right) \left(\omega_0^2 - \omega^2 \right)} \qquad \text{III}$$

where μ_{ge} = transition dipole moment

μ_e = dipole moment in the excited state

μ_g = dipole moment in the ground state

ω_0 = absorption angular velocity

ω = angular velocity of excitation light

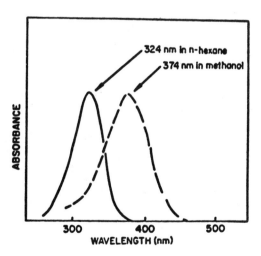

Figure 1.   Solvatochromic Shift of UV/VIS Absorption Band of MNA

The transition dipole moment of the electronic transition is related to the intensity of the UV/VIS absorption band of a molecule. The oscillator strength (f) which is a measure of the number of electrons participating in the optical excitation is given by:

$$f = \left(4.381 \times 10^{-9}\right) \int_{band} \epsilon_\nu \, d\nu \qquad\qquad \text{IV}$$

where ϵ_ν = molar extinction coefficient in $cm^{-1} \, mole^{-1}$

ν = absorption wave number in cm^{-1}

The oscillator strength is related to the transition dipole moment by:

$$f = \left(4.703 \times 10^{29}\right) \nu_m \left| \mu^2_{ge} \right| \qquad\qquad \text{V}$$

where ν_m = peak absorption wave number in cm^{-1}. Thus, one can calculate μ_{ge} from Equations IV and V.

The dipole moment difference, $\mu_e - \mu_g$, can be determined from the solvatochromic shift of the absorption band of a molecule in two solvents of different polarities:

$$\Delta\nu = \left(\mu_g / \ell^3\right) \left(\mu_e - \mu_g\right) \Delta\left(\frac{\epsilon-1}{\epsilon-2}\right) \qquad\qquad \text{VI}$$

where $\Delta\nu$ = the shift of the absorption band

ℓ = molecular length

ϵ = dielectric constant of solvent

Calculations of μ_{ge} from Equation V and $\mu_e - \mu_g$ from Equation VI allows direct substitution of these values into Equation III and thus β can be determined at the frequency, ω, of interest.

This development of solvatochromic methods has permitted the rapid estimation of the second order molecular NLO susceptibility, β, of organic molecules.

This technique, however, relies upon a simple two state model and is only useful for characterizing linear conjugated molecules possessing a large β_{zzz} component. Using solvatochromic methods, various series of molecules have been synthesized and evaluated leading to a determination of the following characteristics of high β molecules:

1. They contain highly delocalized electrons over a long conjugation length. In order to achieve this, the molecule must be:

   a. Long
   b. Straight
   c. Planar
   d. Conjugated (contains no features which electronically block conjugation)

2. Large dipole moment (i.e. asymmetric electronic environment) which is achieved through the use of effective electron donor and acceptor substituents.

By synthesis of series of molecules and exploring many different variations of molecular structure, these design features were readily identified. These features are really quite obvious in terms of constructing molecular units which, in an intense optical field, produce a large nonlinear induced dipole moment. Conjugation, planarity, and length all play a direct role in determining the magnitude of the induced dipole moment. An asymmetric electronic environment is achieved through the use of electron donating (amines, methyl, etc.) and accepting (halogens, nitro, cyano, etc.) groups. For a linear molecule, it is apparent that placing these units at opposing ends of a conjugated molecule will impart a directionality to the induced dipole moment and a dependence of the induced dipole moment upon even order powers of the electric field (i.e. second and higher even order effects). Examples of the effects of structure on the magnitude of β as determined by the solvatochromism method is given in Table 1.

The ability to elucidate molecular structural control of β through synthesis is clearly outlined in Table 1 where the effects of changing group, length, and planarity are elucidated. The control material, para- nitro aniline (PNA), has a measured β of 5.7. Replacing the nitro group with a more effective electron attractor increases the β to 21.4. While an effective donator, the dimethylamino group in this compound is not quite coplanar with the rest of the molecule. Thus, the maximum conjugation cannot be achieved. Forcing the nitrogen into coplanarity through covalent bonding to the benzene ring (compound 3) results in a marked improvement in β to 41.8. The effect of increasing conjugation length on β is clearly seen from compounds 1, 4, and 5. An order of magnitude increase in β can be realized in going from one benzene ring (PNA) to three. Planarity also plays an important role. The azomethine linkage is not planar and will disrupt conjugation in the molecule shown. However, incorporation of an ortho hydroxyl group forces the molecule into planarity through hydrogen bonding resulting in a threefold increase in β.

Combining all three effects into a single molecule produces a material whose β is almost 20 times the control.

Table 1. Molecular Structural Control of β

Structure	β at 1.9 μm (x 10^{-30} esu)	Comment
H_2N—〇—NO_2	5.7	Standard
$N(CH_3)_2$—〇—C(CN)=C—CN, NC	21.4	Group
NH—〇—C(CN)=C—CN, NC	41.8	Group
NH_2—〇—〇—NO_2	20.1	Length
NH_2—〇—〇—〇—NO_2	50.7	Length
$N(CH_3)_2$—〇—CH=N—〇—NO_2	23.4	Planarity
$N(CH_3)_2$—〇—CH=N—〇—NO_2, OH	61.6	Planarity
$N(CH_3)_2$—〇—CH=CH—CH=CH—〇—NO_2	111.2	All

NLO SIDE CHAIN LIQUID CRYSTALLINE POLYMERS

Design criteria for low molecular weight organics with exceptionally high molecular hyperpolarizability have been established, as described in the previous section. A number of organic molecules have been prepared in accordance with these criteria and have been shown to have large hyperpolarizabilities demonstrating the general applicability of the criteria.

While some of these compounds have been grown as single crystals with a proper crystal symmetry for NLO application, the single crystal approach for NLO materials suffers from inherent difficulties. Most importantly, the resulting symmetry and orientation of molecular units packed within a unit cell cannot be controlled by chemical synthesis or the application of external forces. In addition, formation of the requisite noncentrosymmetric symmetry for $\chi^{(2)}$ materials is generally opposed by the highly dipolar character of high molecules. Further large, defect-free single crystals of organic molecules are extremely difficult to grow and therefore become very expensive. Finally, organic crystals are soft solids with high vapor pressures and do not lend themselves to fabrication into complicated shapes.

Guest/host materials in which the NLO molecules are dissolved in a polymeric matrix have been extensively studied at Celanese and possess many attractive properties. However, guest/host materials suffer from limited solubility of the active species in a polymer host which lowers the achievable level of the nonlinear optical properties. Other problems such as phase separation, nonuniformity of dispersion and evaporation of the active species are also prevalent.

Many of these difficulties are removed when the NLO active species are chemically attached to polymer chains. Polymers allow for high concentration of the active moiety; can be manipulated by various processing techniques into different sizes, shapes, and degree of orientation; and, moreover, their long relaxation times can be utilized to maintain the desired structure at ambient temperature for an indefinite period.

Due to the tensorial character of NLO coefficients, alignment and extent of noncentrosymmetry of the NLO moieties in NLO materials are critically important for maximizing the NLO response of the materials. For example, the value of $\chi^{(3)}$ for an unoriented polymer containing long, axially symmetric NLO units is increased by a factor of 5 by orienting all of the NLO moieties along the same axis. Further, unoriented polymers are by definition centrosymmetric and have no $\chi^{(2)}$ properties, whereas orienting high β moieties in the same direction produces high $\chi^{(2)}$ values. Therefore, the ability to control the orientation and symmetry of a polymeric material is critically important to its final response.

Many of the properties of active NLO organic moieties are shared by liquid crystalline mesogens. Therefore, considerable research has been performed to create NLO polymers in which the NLO moiety acts to induce a liquid crystalline phase. By achieving liquid crystallinity, oriented $\chi^{(3)}$ materials may be fabricated. Further, electrical poling of oriented liquid crystalline polymers will produce an enhanced degree of parallel alignment for $\chi^{(2)}$ applications.

Some of the materials currently under investigation are side chain liquid crystal polymers. A general structural scheme for forming a side chain liquid crystalline polymer is illustrated in Figure 2. The polymer chains are typically composed of three parts: backbone, side chain spacer groups and mesogens[4-10]. Intermolecular interaction between the mesogens and, to a lesser extent, between the spacer groups results in mesophase formation in melts of these polymers. The function of the spacer group is to decouple the side chain motion from that of the backbone, otherwise the latter will effect phase stability. It is most fortunate that the same requirements for a highly active NLO molecule (rigidity, polarizability, etc.) are the same as those for mesogens in a liquid crystalline polymer. Thus, we can use NLO molecules as the mesogens while maintaining the other molecular structure unchanged and an LC polymer with NLO activities results.

By appropriate manipulation of the backbone, spacer and NLO group, a variety of mesophases can be realized in these polymers. Nematic, cholesteric, smectic A and smectic C mesophases have been observed in side chain LCPs.

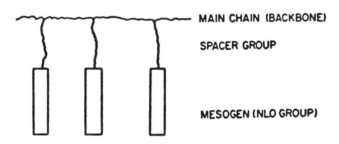

MAIN CHAIN (BACKBONE)

SPACER GROUP

MESOGEN (NLO GROUP)

Figure 2. Liquid Crystal Polymers
Containing NLO Active Groups

The smectic phase is a layered structure consisting of sheets of molecules with orientational order and translational order between the smectic layers. Many different types of smectics exist displaying varying degrees of order both within an individual smectic layer and between layers. Unoriented smectics show a domain texture in which the layers are aligned within each domain on a micron or submicron scale. Upon application of external forces, smectics are able to form large monodomains in which very high orientational order approaching that of a single crystal is achieved (i.e. orientational order parameters exceeding 0.85). With domain alignment, light scattering from defects is eliminated and the sample becomes optically transparent. The transition from a highly scattering polydomain state to a transparent monodomain state has been used as a method for optical storage.

Liquid crystalline polymers can display all of the same phases observed for small molecule liquid crystals. In addition, as polymers, these materials possess a glass transition temperature so that the molecular order for a given phase may be frozen for use over a wide range of temperatures.

For NLO applications, side chain polymers are favored because of their high orientational order in the melt phase. Furthermore, utilizing the slow relaxation characteristics of polymers, we may generate three different NLO samples from these polymers: (1) unoriented $\chi^{(3)}$ material, (2) oriented $\chi^{(3)}$ material by cooling an aligned smectic A structure, (3) oriented $\chi^{(2)}$ material by poling under a high electrical field of the sample.

The first series of side chain polymers to be investigated were those shown in Figure 3. The mesogen/NLO unit was based on 4-hydroxy-4'-nitro biphenyl and the spacer lengths were varied as shown. Although this particular moiety has a β only 80% that of MNA, it was chosen because of its relative ease of synthesis through known literature techniques. This would allow preparation of sufficient quantities of material for fabrication, poling, and device research. These "first generation" polymers would then provide the data base by which future higher activity materials would be investigated.

The synthesis of these materials is outlined in Figures 4 and 5. Hydroxy biphenyl is esterified with benzoyl chloride to form the benzoate ester. This compound is then nitrated with fuming nitric acid. Recrystallization afforded the 4-4' isomer. Hydrolysis of the ester, followed by neutralization, gave 4-hydroxy-4'-nitro biphenyl. Etherification, of this biphenyl, with the appropriate halo alkanol, followed by reaction with methacroyl chloride, produced the desired monomer.

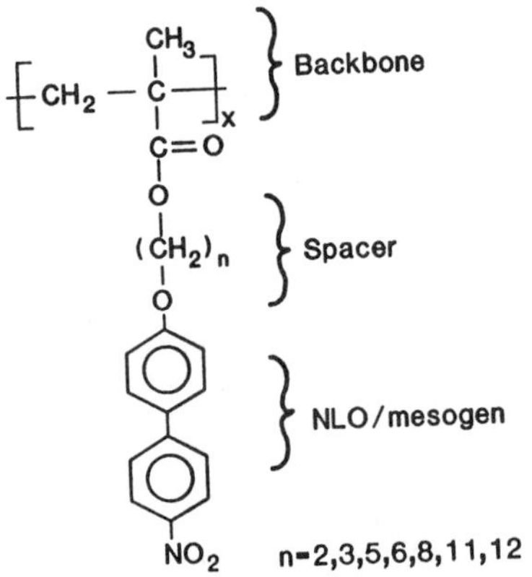

Figure 3. NLO Side Chain Polymers
First Generation

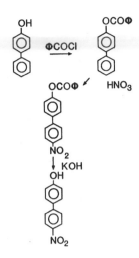

Figure 4. Synthesis of 4-Hydroxy-4'-Nitro Biphenyl

Polymerizations were conducted in toluene, with 1% AIBN, at 60-80 C for approximately 16 hours.

The results for the biphenyl polymers are reported in Table 2. For this series, improvements in the LC order were observed as the spacer length increased. Moving further away from the backbone reduces its influence and allows for more organizational freedom of the side chains, thus producing a more ordered phase. In fact, motion is so decoupled for the 11 and 12 carbon cases that crystallization of the side chain occurs when the polymers were cooled to ambient temperature. Unfortunately, as one moves away from the main chain, Tg decreases drastically. Although the desired smectic phase forms in the long spacer samples, the Tg is too low. Externally induced orientation will decay over a period of time. Figure 6 summarizes the properties of these polymers.

It appears that at moderate spacer lengths (5-6) where the Tg has a desirable value, the methacrylate backbone is stiff enough to interfere with the order of the LC phase. A more flexible backbone at the same spacer lengths should allow a more ordered phase to form. To test this hypothesis, a side chain polymer composed of a nitro biphenyl group with a 5 carbon spacer attached to a siloxane backbone was prepared (Figure 7). Although the Tg was low as expected, the LC phase was a classic smectic A and the clearing temperature was much higher than the corresponding methacrylate (165 vs. 72 C) indicating greater order and stability.

Table 2. NLO Side Chain Polymer Properties

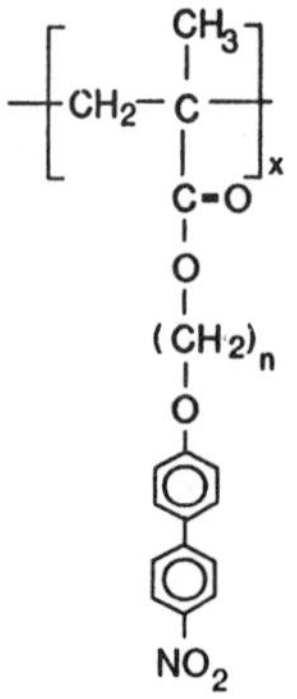

n	Phase	T_G (°C)	T_{cl} (°C)
2	I		
3	N	85	100
5	N	45	72
6	N	40	64
8	N	35	75
11	S_A	20	95
12	S_A	10	80

I - Isotropic
N - Nematic
S - Smectic
T_G - Glass Temperature
T_{cl} - Clearing Temperature (LC → Isotropic)

One of the primary premises of the NLO active polymer development effort was that the NLO activity of the polymers is the sum of NLO activity of the individual chromophores despite the fact that the chromophores are now chemically attached to the polymer chains rather than being physically dissolved into the polymer matrix. To validate this hypothesis, the solvatochromic response of the NLO moieties and the NLO active polymers were determined and their NLO properties were evaluated.

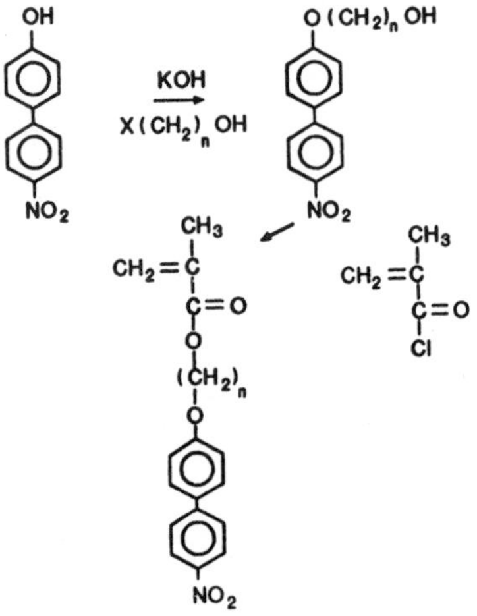

Figure 5. Monomer Synthesis

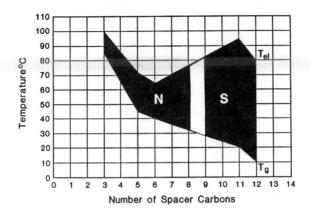

Figure 6.   Dependence of Glass and Clearing
Temperatures on Carbon Spacer Length

Comparison of the monomer and polymer data demonstrates that the NLO response was not altered to an appreciable extent in the polymerization. For example, 4-(12-dodecyloxy)-4'-nitrobiphenyl, its methacrylate and the polymer all showed an absorption band at about 29,500 cm^{-1} and a solvatochromic shift between methanol and n-hexane of about 1200 cm^{-1}. While there was observed a small decrease of the oscillator strength during the polymerization, this may be ascribable to the low solubility of the polymer in these solvents and to possible hydrolysis of the NLO compound (Table 3).

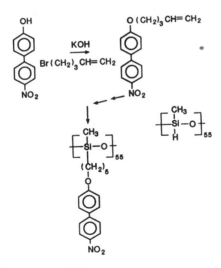

Figure 7.   Siloxane Based NLO Polymer

NLO POLYMER FABRICATION AND POLING

The poling experiments were largely carried out on free-standing film samples of thickness 50-100 μm. The samples were prepared by compression molding. Liquid crystalline polymer and other low MW polymer samples were prepared by pressing the polymer melt between two conductive glass surfaces. Here, the sample thickness was controlled by spacers (Figure 8).

Table 3. Covalent Attachment of an NLO
Unit to a Polymer Chain

	λmax (nm)	β(x10^{-30}esu)
NLO UNIT	340	11
Polymer	338	8

The free-standing samples were poled while held between 2.5 cm diameter brass electrodes. Typically, two layers of the samples were used to prevent premature dielectric breakdown due to pinholes in the samples. The samples and the electrodes were immersed in an oil bath maintained at a constant temperature for temperature control and to minimize surface dielectric breakdown. After the sample temperature was equilibrated to the bath temperature, an electrical field was applied across the samples by slowly raising the voltage to a desired level. The samples were kept at this poling condition, typically, for several minutes, taken out of the bath and rapidly cooled down to the ambient temperature before the electrical field was switched off.

Orientation of NLO liquid crystalline polymers is demonstrated in Figure 9. Here, an applied electric field is used to orient the liquid crystalline polymer creating an optically transparent monodomain material. As a result of the long range order of the liquid crystalline phase, defects are removed and the sample is transformed from a highly scattering opaque material to an aligned, optically clear $\chi^{(3)}$ material useful for nonlinear optics. $\chi^{(2)}$ materials may be created by applying higher DC fields as previously described[11].

NLO polymeric materials offer several distinct advantages for devices which can be divided into two general areas: intrinsic material properties and fabrication options. The intrinsic material properties have been well defined by single crystal NLO studies[12,13] and the generally known properties of organics. As has been highlighted, the NLO process for organics involves the polarization of loosely bound electrons in an organic molecule. This process is very fast (femtosecond) and lossless[14]. Both properties are quite unique. For example, nonlinear polarization of electrons in inorganic crystals is created in part by nuclear motion so that the tightly bound electrons move the with nuclei. This process is relatively slow. At high frequencies (e.g. optical frequencies), the electronic nonlinearities are much smaller. For GaAs or GaAs/AlGaAs multiple quantum well materials, a nonlinear change in the index of refraction is observed due to a nonlinear absorption process which involves the creation of electron-hole pairs[15]. However, there are significant absorption losses which lead to thermal heating.

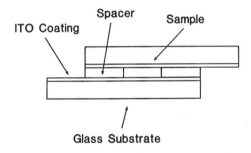

Figure 8. Microscopic Poling Cell for Perpendicular
Alignment of Polymers

NLO polymeric materials show a very slight dispersion in index of refraction (or dielectric constant) as a function of wavelength. For polymers, the dielectric constant at low frequencies is very nearly approximated by the square of the index of refraction at optical frequencies. This property is due to the disassociation of polarizable electrons from nuclei. For inorganics, the dielectric constant is much larger than this value. Thus, essentially all inorganic NLO materials have much higher dielectric constants than NLO polymers at electrooptical modulation frequencies. The implications for electro-optics are very significant.

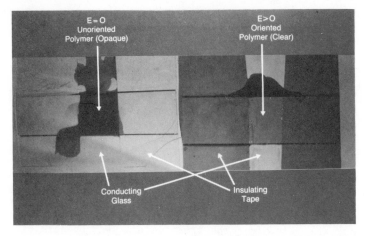

Figure 9. Electrical Poling of Liquid Crystalline Polymers

The ultimate advantages of NLO polymeric materials are yet to be fully realized. Polymers may be processed in very unique ways. Thus, fibers may be created by spinning, thin films for waveguiding may be prepared by spin coating, varied objects may be formed by molding, etc. As an example, cylinders of NLO polymeric material are shown in Figure 10. These materials possess novel electro-optical properties and show low absorption loss and an optical surface flatness (as determined by interferometry) which is acceptable for many applications. As device activities increase and become more equal to those activities for other NLO materials, the real merit of NLO polymer processing options should become readily apparent.

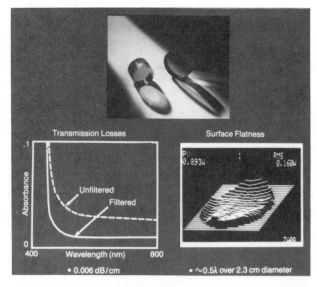

Figure 10. Optical Quality of Guest/Host Cylinders

EXPERIMENTAL

Synthesis

Polymethacrylate Backbone

4-Hydroxy-4'-Nitro Biphenyl

(a)   4-benzoyloxy biphenyl

To 500 mls of pyridine in a 1000 ml three necked flask is added
170 g. of 4-hydroxy biphenyl. The mixture is cooled to 10 C
whereupon 155 g. of benzoyl chloride is added dropwise keeping the
temperature below 20 C. After complete addition, the mixture is
gradually heated to reflux and held there for 30 minutes. The
reaction is then cooled to 100 C and poured into a large beaker to
cool. The hardened product is subsequently broken up with 250 ml
HCl and 250 ml water. Then, an additional 250 ml HCl and 500 ml
water was added and the whole mass mixed thoroughly in a blender.
The solid is then filtered, washed with water to a neutral pH, and
air dried overnight. Recrystallization is accomplished in n-
butanol: mp 149-150 C.

(b)   4-benzoyloxy-4'-nitro biphenyl

The above product (40 g) was mixed with 310 mls of glacial acetic
acid and heated to 85 C. Fuming nitric acid (100 mls) was slowly
added such that the temperature is kept between 85-90 C. After
complete addition, the reaction is cooled to room temperature.
The solid is filtered and washed with water and methanol. The
product is then recrystallized from glacial acetic acid: mp
211-214 C.

(c)   Hydrolysis of 4-benzoyloxy-4'-nitro biphenyl

The above product (60 g) is mixed with 300 mls ethanol and heated
to reflux. A solution of 40 g KOH in 100 mls water is then slowly
added dropwise at reflux. After complete addition, the mixture is
refluxed 30 minutes and cooled overnight. Next day, blue crystals
of the potassium salt are filtered, washed with tetrahydrofuran
until the wash is colorless, and dried. The salt is utilized in
all subsequent reactions. Pure 4-hydroxy-4'-nitro biphenyl can be
obtained by dissolving the salt in a minimum amount of boiling
water and adding 50/50 HCl/water until an acidic Ph is obtained.
The yellow solid is then filtered, washed with water, dried and
recrystallized from ethanol: mp 203-204 C.

Typical Prep of a Polymer with a 6 Carbon Spacer

4-(6-Hydroxyhexyloxy)-4'-Nitro Biphenyl

To 500 ml of toluene in a 1000 ml round bottom flask, fitted with
a condensor and magnetic stirrer, is added 7.6g. (0.03M) of the
potassium salt of 4-hydroxy-4'-nitro biphenyl, 4.9g. (0.035M) of
6-iodo-1-hexanol, and 0.5g. of 18-crown-6. The mixture was refluxed overnight. The next day, practically all of the potassium
salt had reacted as evidenced by the absence of blue crystals.

The solution was filtered hot and cooled to room temperature. After solvent removal, the solid residue was recrystallized from ethanol: MP 117-119 C.

4-(6-Methacryloxyhexyloxy)-4'-Nitro Biphenyl

The above product (22 g) is dissolved in 500 mls of dry dioxane and heated to 45 C. Triethylamine (14 g) is then added all at once. A solution of 10.5 g methacryloyl chloride in an equal volume of dioxane is slowly added dropwise keeping the temperature at 45 C. The reaction is then stirred at 45 C. At the end of the day, an additional 10.5g. of methacryloyl chloride is added and the mixture stirred overnight. The next day, the dioxane is removed under vacuum and the residue dissolved in water in a blender. The solid is filtered, washed with water, and air dried. The monomer is recrystallized from ethanol: mp 53-56 C.

Poly(6-[4-Nitrobiphenyloxy]-Hexyl)Methacrylate

The monomer (2 g) is dissolved in 20 ml degassed toluene and 1 mole% AIBN is added. The container is placed in a 60 C oil bath and heated 1 day. During this time, polymer forms and collects at the bottom of the container. After the required time, the supernatant is poured off and the lower layer mixed with methanol in a blender. The solid polymer is filtered, washed with methanol and vacuum dried.

Polysiloxane Backbone

4-(4-Penteneoxy)-4'-Nitro Biphenyl

To 400 ml ethanol is added 21.5 g of 4-hydroxy-4'-nitro biphenyl and the mixture heated to reflux. A solution of 7.1 g KOH in 30 mls water is added dropwise at the boil. After complete addition, 18 g of 5-bromo-1pentene is added all at once and the reaction heated under reflux overnight. The next day, ethanol is removed under vacuum from the cooled mixture. The residue is mixed with water in a blender, filtered, washed with water, and air dried. Recrystallization from 90/10 hexane/toluene afforded straw colored needles: mp 74-76 C.

Poly[(4-Nitrobiphenyloxypentyl)Methyl]siloxane

Reactions are conducted in dry toluene at 60 C. Both materials are added such that there is a 10 mole% excess of biphenyl relative to the Si-H bonds. To this mixture is added 1-2 drops of chloroplatinic acid catalyst (5% weight/volume in isopropanol). After overnight heating at 60 C, the polymer is precipitated into methanol, filtered, and dried. Further purifications are carried out by dissolving in chloroform and precipitation into methanol.

PHASE IDENTIFICATION

The phases exhibited by the side chain liquid crystal polymers were characterized in the following way. The temperature at which the iso-tropic to liquid crystal phase occurs is determined first. This is accomplished by optical microscopy using a Leitz Laborlux 12 pol polar-izing microscope fitted with a Mettler FP-80 82 controller-hot stage combination. The Mettler controls the temperature to within 0.1 C and can be programmed to heat or cool at rates from 0.1 to 20.0 C per min-ute. The temperature is raised at 10.0 C per minute until the birefrin-gent texture of the polymer, prepared on a microscopy slide, disappears. The temperature is then lowered at 5.0 C per minute until birefringence is regained. The Mettler is thus cycled up and down in temperature at slower rates of heating and cooling until the transition temperature is determined exactly. The oven is then set to cool into the liquid crys-tal phase from the isotropic state at 0.1 C per minute. Once some bire-fringence is seen optically, the temperature is held for long periods of time to allow the phase to anneal into its natural texture. Often, it is necessary to cool the oven in 0.2 C intervals every half hour to allow large domains to form in order to identify the phase. The tex-tures that the side chain liquid crystal polymers exhibit have been found to be analogous to those shown by small molecule liquid crystal materials.

By taking these precautions, the probability of misclassifying the liquid crystal phase is low and problems of misidentification, as is sometimes found in the literature, is avoided.

SUMMARY

The molecular structural characteristics which give rise to large optical nonlinearities have been delineated. This has been accomplished by several key steps and has led to the development of a structural al-gorithm for the design of molecules with high susceptibilities. The molecular characteristics are:

a. Increased molecular length

b. Molecular planarity

c. Delocalized electronic structure (including connector units along the molecular long axis)

d. Asymmetric electronic environment using electron donating and attracting groups

Molecules combining all important structural features have been synthesized and characterized.

High activity molecular units have been covalently attached to form polymers which can contain up to 60% of the NLO moiety, and the result-ing molecular optical properties have been characterized.

The second order molecular susceptibility was found to be undimin-ished by properly designed covalent bonding to form a polymer, and spec-tral characteristics were found to closely follow that of the NLO unit.

For the same backbone, Tg is influenced by the mesogen and the length of the spacer. Short spacers inhibit crystallizing ability of the side chains, lead to less ordered structures and raise Tg. This is due to the large influence of the backbone. At these short spacer levels, no decoupling of main chain and side chain motions is possible. A less but still profound influence is exerted by the main chain on mesogen organization at intermediate spacer lengths. More rigid chains (polymethacrylate and polyacrylate) lead to less ordered structures. Flexible chains (polysiloxanes) lead to more ordered states. Tg is lowered in methacrylate series over polymers with shorter spacers. Complete decoupling of the main chain and side chain motions occurs in polymers containing long spacers resulting in the most ordered structures. However, Tg is the lowest for this series. Due to this decoupling, side chain crystallization occurs more readily.

The NLO activity of the polymers was the sum of the NLO activity of the individual NLO moieties despite the fact that the NLO moieties were chemically attached to the polymer chains rather than being physically dissolved into the polymer matrix.

Control of orientation has been achieved by applying external fields.

ACKNOWLEDGEMENTS

This work was carried out under Air Force Contract No. F49620-85-C-0047. We gladly acknowledge the valuable technical guidance given to our effort by Professor A.F. Garito. We thank Professor I.M. Ward, F.R.S., for his technical insight, especially regarding poling experiments which he conducted at the University of Leeds. The support, guidance and encouragement of D. Ulrich of the Air Force Office of Scientific Research, J. Neff of the Defense Advanced Research Projects Agency, and W. Elser of the Center for Night Vision and Electro-Optics are gratefully acknowledged.

REFERENCES

1. A. F. Garito, K. D. Singer, and C. C. Teng, in D. J. Williams, Ed., "Nonlinear Optical Properties of Organic and Polymeric Materials", ACS, New York, pp. 1-26 (1983).
2. For a detailed description of DCSHG method, see K. D. Singer and A. F. Garito, J. Chem. Phys., 75, 3572 (1981).
3. A. Buckley, E. Choe, R. DeMartino, T. Leslie, G. Nelson, J. Stamatoff, D. Stuetz, H. Yoon, ACS Symposium on Solid State Polymerization, April 13-18, 1986, New York.
4. H. Finkelman, in "Mesomorphic Order in Polymers", ACS Symposium, 74, p. 22 (1978).
5. H. Finkelman, in "Polymeric Liquid Crystals", Academic Press (1982), p. 35.
6. V. P. Shibaev and N. A. Plate, Advances in Polymer Science 60/61, 354 (1984).
7. H. Finkelman, Makromol. Chem., 179, 2541 (1978).
8. H. Finkelman, Makromol. Chem., 179, 273 (1978).
9. H. Ringsdorf, Makromol. Chem., 184, 253 (1983).

10. V. P. Shibaev, Eur. Polym. J., 18, 651 (1982).
11. J. B. Stamatoff, A. Buckley, G. Calundann, E. W. Choe, R. N. DeMartino, G. Khanarian, T. Leslie, G. Nelson, D. Stuetz, C. C. Teng, H. N. Yoon, SPIE Symposium on Molecular and Polymeric Opto-Electronic Materials: Fundamentals and Applications, August 21-22, 1986, San Diego, Vol. 682, p. 85-92.
12. A. F. Garito, C. C. Teng, K. Y. Wong, and O. Zammani'Khamiri, Mol. Cryst. Liq. Cryst., 106, p. 219-258 (1984).
13. J. Zyss, J. of Mol. Electronics, 1, p. 25-45 (1985).
14. G. M. Carter, J. V. Hryniewicz, M. K. Thalsur, Y. J. Chen, S. E. Meyler, Appl. Phys. Lett., 49, 998 (1986).
15. H. M. Gibbs, S. L. McCall, T. N. C. Venkatesan, A. C. Gossard, A. Passner, and W. Weigmann, Appl. Phys. Lett., 35, p.451-453 (1979).

ORIENTATIONALLY ORDERED ELECTRO-OPTIC MATERIALS

K. D. Singer, M. G. Kuzyk, and J. E. Sohn

AT&T, Engineering Research Center
P.O. Box 900, Princeton, NJ, USA 08540

1. INTRODUCTION

Suitable materials for processing optical information are critical for implementation of all-optical and electro-optical communications and data processing systems. These technologies require materials with large optical nonlinearities, which certain organic materials have been shown to possess.[1] [2] [3] [4] Guided wave electro-optic systems require not only large optical nonlinearities, but also materials suitable for waveguiding and integration. Material considerations include manufacturability, that is, capability of fabrication into reproducible devices; and integrability with sources, electronics, detectors, and interconnects. Additionally, materials should possess favorable dielectric properties: a low dielectric constant and dielectric loss, and the requisite optical quality for producing low-loss optical waveguides.

Several approaches to the fabrication of bulk organic nonlinear optical materials have been pursued.[2][4] These include crystals, crystalline polymers, Langmuir-Blodgett films, liquid crystals and liquid crystal polymers. We have demonstrated that molecularly-doped poled polymer glasses possessing reasonably large second-order nonlinear optical susceptibilities can be produced.[5] The nonlinear optical properties of many organic materials can be traced to the constituent molecules, with the bulk nonlinear optical properties determined by the degree of orientational order of the molecules. This order can be described by statistical physical methods and related to microscopic order parameters, such as those common in the liquid crystal literature.[6] [7] This statistical model can be used to predict nonlinear optical properties in materials where the microscopic order parameters have been measured independently, or conversely, to deduce material order from nonlinear optical measurements.[7] This model is applicable to poled polymer glasses which are described below, as well as to other orientationally ordered systems. Results of second harmonic generation and electro-optic coefficient measurements are presented and discussed in terms of the statistical model. In addition to the effects of order, material properties pertinent to device fabrication of polymer glasses are also discussed, including high optical quality, low dielectric constant and dielectric loss, and manufacturability. Fabrication flexibility (low temperature processing and metallization), integration with semiconductor substrates, optical sources, detectors, and drive electronics, and the inherent material diversity of polymer glasses are advantageous for application to integrated optics.

2. INTRODUCTION TO NONLINEAR OPTICS

Nonlinear optical effects arise when the movement of charge in a material is nonlinear in the strengths of the applied electric fields. The dipole per unit volume, or polarization density is the lowest order measure of charge distribution and is sufficient to describe a large number of nonlinear optical processes. In the dipole approximation, and assuming a Taylor series expansion, the polarization density, P, is related to the electric field, E, through

$$P_i = P_i^{(0)} + \chi_{ij}^{(1)} E_j + \chi_{ijk}^{(2)} E_j E_k + \chi_{ijkl}^{(3)} E_j E_k E_l + ..., \tag{1}$$

where P_i is the ith component of $\overrightarrow{P}$, E_i the ith component of $\overrightarrow{E}$, $\chi^{(n)}$ the nth-order nonlinear optical susceptibility tensor, and summation notation is implied. In general, the response of a material may depend on electric field values at previous times and at other parts of the material. In Eq. (1) these effects are ignored.

The first term in Eq. (1) is the spontaneous polarization and the second term describes linear optics including propagation of light in dielectric media, absorption and surface reflection. The third term leads to second harmonic generation, parametric mixing and the linear electro-optic effect, while the fourth term results in third harmonic generation and self-focusing. The susceptibilities, $\chi^{(n)}$, depend on the frequencies of light present in the material. If the electric fields are decomposed into Fourier components labeled by frequency, ω, the nth order susceptibility will lead to mixing of $n+1$ fields. For example, the second-order contribution to the polarization, $P_i^{\omega_3}$ at frequency ω_3 is given by

$$P_i^{\omega_3} = \chi_{ijk}^{(2)}(-\omega_3;\omega_1,\omega_2) E_j^{\omega_1} E_k^{\omega_2}, \tag{2}$$

where energy conservation requires that $\omega_3=\omega_1+\omega_2$. The susceptibility quantifying the second harmonic intensity corresponds to the susceptibility $\chi_{ijk}^{(2)}(-2\omega;\omega,\omega)$ where two field quanta at frequency ω combine through the nonlinearity to form a field at the second harmonic frequency 2ω. Similarly, the linear electro-optic effect results from the mixing of one optical field, $\omega_1=\omega$, and one dc field, $\omega_2=0$, resulting in a phase change in the output optical beam, $\omega_3=\omega$.

In analogy to Eq. (1), the polarization per molecule, p_I, can be expanded in terms of the molecular nonlinear response in powers of electric field, E_I,

$$p_I = \mu_I^0 + \alpha_{IJ}E_J + \beta_{IJK}E_JE_K + \gamma_{IJKL}E_JE_KE_L + \cdots, \tag{3}$$

where μ_I^0 is the molecular ground state dipole moment, α_{IJ} the linear polarizability and β_{IJK} and γ_{IJKL} the two lowest-order molecular nonlinear optical susceptibilities, or hyperpolarizabilities. Given the molecular nonlinear optical susceptibilities, the bulk nonlinear susceptibility of a material can be predicted if the density and ordering of the system is known and if the effects of the neighboring molecules on the local fields are properly taken into account.

3. EFFECTS OF ORDER ON NONLINEAR OPTICS

The second-order nonlinear optical susceptibility of a centrosymmetric material can be shown to vanish by symmetry. Imparting orientational order that breaks the center of inversion symmetry will thus greatly affect the second-order response. The orientational order can be described by a distribution function in systems where the ordering energies are comparable to or less than the thermal disordering energies. In the crystalline limit, the distribution function takes the form of a delta function where all the molecular axes are fixed within the unit cell.

In general, the nonlinear optical susceptibility depends on the intramolecular order which determines the molecular susceptibility, as well as the positional and orientational order of the molecules. In systems where the nonlinear molecules are weakly interacting, the macroscopic susceptibility can be assumed to be additive over the microscopic susceptibilities. In this case, only the orientational order must be considered in detail since positional ordering effects can be described in an average sense by local field factors. Under these assumptions, the macroscopic second-order susceptibility, $\chi_{ijk}^{(2)}$, can be related to the microscopic lowest-order hyperpolarizability, β_{IJK}, by[6]

$$\chi_{ijk}^{(2)}(-\omega_3;\omega_1,\omega_2)=N<\beta_{IJK}^*(-\omega_3;\omega_1,\omega_2)>_{ijk}, \qquad (4)$$

where IJK refers to the components in the molecular frame, ijk the laboratory frame, N the number density of molecules, β_{IJK}^* the local field corrected molecular hyperpolarizability, and the angle brackets denote an orientational average weighted by a distribution function $G(\Omega)$:

$$\chi_{ijk}^{(2)}(-\omega_3;\omega_1,\omega_2)=N\beta_{IJK}^*(-\omega_3;\omega_1,\omega_2)\int d\,\Omega\, a_{iI}a_{jJ}a_{kK}G(\Omega), \qquad (5)$$

where the a_{iL}'s are rotation matrices in terms of three Euler angles. All the information about the effects of order resides in the distribution function where the form of this function is determined by the orienting process.

For many molecular organic crystals, the nonlinearity has been shown to be additive over the substituent molecules.[4] With a delta function orientational distribution function arising from the intermolecular interactions where binding energies are greater than kT, the relationship between the microscopic and macroscopic second-order susceptibilities reduces to[8]

$$\chi_{ijk}^{(2)}(-\omega_3;\omega_1,\omega_2) = N_u f_i^{\omega_3} f_j^{\omega_1} f_k^{\omega_2} \sum_{i,j,k} \sum_{s=1}^{n} a_{iI(s)} a_{jJ(s)} a_{kK(s)} \beta_{IJK}(s), \qquad (6)$$

where $f_l^{\omega_m}$ are the local field factors along the l direction at frequency ω_m, the a_{iL}'s are rotation matrices expressed in terms of the three Euler angles relating the molecular to the crystallographic axes, N_u the density of unit cells, and where the sum is over the n molecules in the unit cell indexed by s. All crystalline materials can be described by 32 point groups of which 21 are noncentrosymmetric. These exhibit second-order nonlinear optical effects. The bulk second-order nonlinear optical susceptibility is thus given in terms of geometrical considerations and has been applied to a series of organic crystalline materials.[9]

Other material systems, such as liquid crystals, liquid crystal polymers, isotropic polymers, and Langmuir-Blodgett films possess less order than crystals.

Nonetheless, the degree of orientational order can be described with a knowledge of the distribution function of Eq. (5). Since it is not usually known *a priori*, it is convenient to expand the distribution function in terms of orthogonal functions so that it can be approximated by a finite number of terms. Since the systems listed above usually possess azimuthal symmetry, and since many nonlinear optical molecules possess axially symmetric optical properties, the distribution function will mainly depend on only one Euler angle, θ, and can be expanded in terms of Legendre polynomials,[7]

$$G(\theta) = \sum_{l=0}^{\infty} \frac{(2l+1)}{2} A_l P_l(\cos\theta),\tag{7}$$

where the coefficients A_l are determined by

$$A_l = \int_0^{\pi} \sin\theta\, d\theta\, G(\theta) P_l(\cos\theta).\tag{8}$$

The coefficients A_l (the ensemble averages $<P_l>$), can be independently determined and are the microscopic order parameters that characterize the moments of the orientation with respect to P_l. Upon expanding $G(\theta)$ and employing the orthonormality of the Legendre polynomials, Eq. (5) yields for the even- and odd-order nonlinear optical susceptibilities

$$\chi_{ijk...}^{(2n)} = N \sum_{m=0}^{n} \xi_{ijk...}^{(2m+1)} <P_{2m+1}>$$

and

$$\chi_{ijkl...}^{(2n+1)} = N \sum_{m=0}^{n+1} \xi_{ijkl...}^{(2m)} <P_{2m}>,\tag{9}$$

respectively, where $\xi_{lmn}^{(k)}$ is a function of the tensor components of the kth order susceptibility, N the number density and where the angular average $<P_{2m}>$ depends only on the orientational order.[7]

As seen in Eq. (9), order plays a crucial role in the magnitude of the nonlinear optical properties. In guest-host systems, where these properties arise from the guest molecules, centrosymmetric ensembles exhibit no even-order nonlinear optical effects, although odd-order effects are present. Orientational order, imparted by the application of an electric field, yields materials that possess the symmetries of the ∞mm noncentrosymmetric point group where the unique axis is in the direction of the poling field. Such materials exhibit an infinite-fold rotation and an infinity of mirror planes about the unique axis.

Applying this symmetry operation to the third rank tensor representing the second-order optical susceptibility, $\chi_{ijk}^{(2)}$, results in the nonzero tensor components $\chi_{333}^{(2)}$, $\chi_{113}^{(2)}=\chi_{223}^{(2)}$, $\chi_{131}^{(2)}=\chi_{232}^{(2)}$, and $\chi_{311}^{(2)}=\chi_{322}^{(2)}$. The polarizations are then given by

$$P_1^{(2)}(\omega_3) = \chi_{131}^{(2)}(-\omega_3;\omega_1,\omega_2)E_3(\omega_1)E_1(\omega_2) + \chi_{113}^{(2)}(-\omega_3;\omega_1,\omega_2)E_1(\omega_1)E_3(\omega_2)$$

$$P_2^{(2)}(\omega_3) = \chi_{131}^{(2)}(-\omega_3;\omega_1,\omega_2)E_3(\omega_1)E_2(\omega_2) + \chi_{113}^{(2)}(-\omega_3;\omega_1,\omega_2)E_2(\omega_1)E_3(\omega_2)$$

$$P_3^{(2)}(\omega_3) = \chi_{311}^{(2)}(-\omega_3;\omega_1,\omega_2)E_1(\omega_1)E_1(\omega_2) + \chi_{311}^{(2)}(-\omega_3;\omega_1,\omega_2)E_2(\omega_1)E_2(\omega_2)$$

$$+\chi^{(2)}_{333}(-\omega_3;\omega_1,\omega_2)E_3(\omega_1)E_3(\omega_2). \tag{10}$$

In a guest-host system, the equilibrium distribution function of an ensemble of molecules of species u under the influence of a poling field, $\mathbf{E}_p$, is of the form[7]

$$G_u(\Omega,E_p) = \sum_v \frac{\exp[-\frac{1}{kT}(U_{uv}-\mathbf{m}_u^*\cdot\mathbf{E}_p)]}{\int d\,\Omega\exp[-\frac{1}{kT}(U_{uv}-\mathbf{m}_u^*\cdot\mathbf{E}_p)]}, \tag{11}$$

where U_{uv} is the interaction potential between species u and v, $1/kT$ the Boltzmann factor, Ω the angular coordinates, and $\mathbf{m}_u^*$ the local field-corrected molecular dipole moment of species u. The nonlinear susceptibilities then take the form

$$\chi^{(2n)}_{ijk\ldots} = \frac{NE_p}{kT}^{n+1} \sum_{m=0} \zeta^{(2m)}_{ijk\ldots} <P_{2m}>,$$

$$\chi^{(2n+1)}_{ijkl\ldots} = \frac{NE_p}{kT}^{n+1} \sum_{m=0} \zeta^{(2m+1)}_{ijkl\ldots} <P_{2m+1}>, \tag{12}$$

where now the order parameters describe the system before poling and where the molecular dipole moments appear with the molecular susceptibilities in the tensors $\zeta^{(k)}_{lmn\ldots}$.[6] The nonlinear coefficients for a poled material can then be determined with knowledge of the molecular susceptibility and dipole moment, number density, and zero-field order parameters.

Assuming a one dimensional molecule, where β_{zzz} and m_z are the only nonvanishing components, Eq. (12) for the second-order susceptibility possesses two independent components given by

$$\chi^{(2)}_{333} \sim N\beta_{zzz}^* \frac{m_z^* E_p}{kT} \times \left\{ \frac{1}{5} + \frac{4}{7}<P_2> + \frac{8}{35}<P_4> \right\} \tag{13}$$

and

$$\chi^{(2)}_{311} = \chi^{(2)}_{113} = \chi^{(2)}_{131} \sim N\beta_{zzz}^* \frac{m_z^* E_p}{kT} \times \left\{ \frac{1}{15} + \frac{1}{21}<P_2> - \frac{8}{70}<P_4> \right\}. \tag{14}$$

In the isotropic phase, both $<P_2>$ and $<P_4>$ vanish and $\chi^{(2)}_{333}$ is three times larger than the other components. In liquid crystalline phases, the order parameters can be substantial. As the order parameters approach unity, the $\beta^{(2)}_{333}$ component becomes a factor of five bigger than the isotropic case whereas the other components vanish, as the bulk symmetries reflect the molecular single dimensionality. The nonlinearities are thus determined by the order imposed by the poling field as well as order arising from intermolecular forces as quantified by the order parameters $<P_l>$.

4. NONLINEAR OPTICAL PROPERTIES

We have prepared orientationally ordered electro-optic glassy polymer thin films using the technique of electric-field poling.[5] [10] This is accomplished by dissolving the optically-nonlinear azo dye, Disperse Red 1, (Figure 1) in the glassy polymer poly(methyl methacrylate) (PMMA).

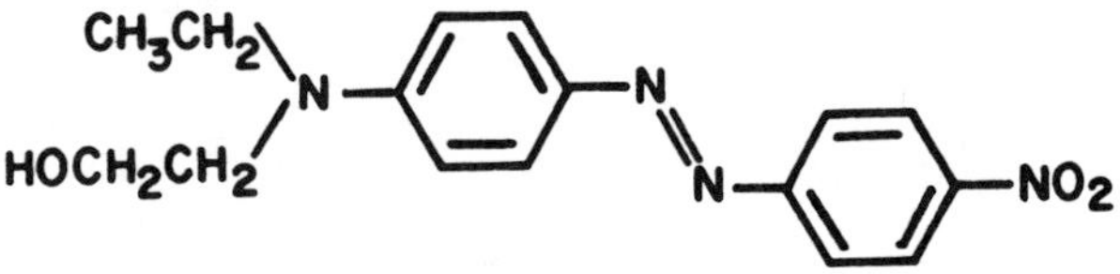

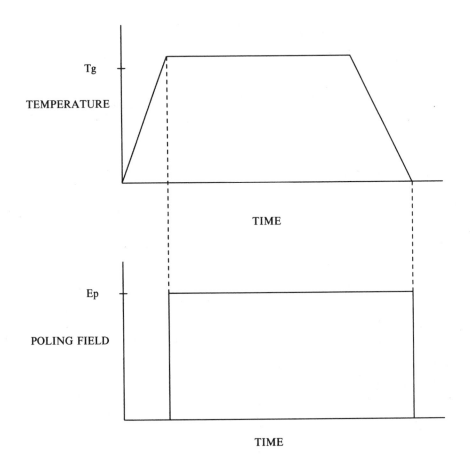

Figure 1. Disperse Red 1.

Figure 2: Schematic of the poling process.

Thin films of the azo dye in PMMA are prepared by spin coating onto indium tin oxide (ITO) coated glass; then a semi-transparent layer of gold is deposited on the film. The gold and ITO layers act as electrodes for poling. The film is heated above its glass-rubber transition temperature. In the rubbery state of the polymer, molecular motion is enhanced. Next, a strong electric field is applied to the sample which tends to align the molecules. The sample is cooled and the electric field removed. The induced polarization is then locked in, resulting in a noncentrosymmetric material. This process is shown schematically in Figure 2.

4.1 SECOND HARMONIC GENERATION

As mentioned previously, the point group of a poled glassy film is ∞mm. For second harmonic generation, Eq. (10) then reduces to[11]

$$
\begin{aligned}
P_x^{2\omega} &= 2d_{15}E_xE_z \\
P_y^{2\omega} &= 2d_{15}E_yE_z \\
P_z^{2\omega} &= d_{31}E_x^2 + d_{31}E_y^2 + d_{33}E_z^2,
\end{aligned}
\tag{15}
$$

where d_{uv} is the second harmonic coefficient, and standard contracted notation is used.[12] Owing to Kleinman symmetry, $d_{15}=d_{31}$.[13] And, as mentioned above, the thermodynamic model in Eqs. (13) and (14) yields $d_{31}=d_{33}/3$.

Second harmonic generation was measured in the films at a fundamental wavelength of $\lambda=1.58\mu m$.[5] The second harmonic coefficients are determined by measuring the second harmonic intensity as a function of incident angle, θ, in transmission through he film. In general, the second harmonic power in a uniaxial material deposited on a glass slide with the incident light falling on the film is given by[14] [15]

$$
P^{2\omega} = (512\pi^3/A)t_\omega^4 T_{2\omega}d^2t_o^2p^2P_\omega^2[1/(n_\omega^2-n_{2\omega}^2)^2]\sin^2\Psi(\theta),
\tag{16}
$$

where A is the area of the laser beam spot, d the appropriate second harmonic coefficient, P_ω the incident laser power, t_ω and $T_{2\omega}$ Fresnel-like transmission factors, t_o the Fresnel transmission factor of the second harmonic light through the glass substrate, p an angular factor that projects the nonlinear susceptibility tensor onto a coordinate frame defined by the propagating electric field, the $n's$ the refractive indices at the indicated frequencies, and $\Psi(\theta)$ the angular dependence of the second harmonic power resulting from interference between free and bound waves. When the coherence length, $l_c = \lambda/[4(n_\omega-n_{2\omega})]$, is much larger than the film thickness, $\sin^2\Psi(\theta)$ can be expanded for small $\Psi(\theta)$ leading to

$$
\sin^2\Psi(\theta) \sim \left[\frac{\pi}{2}\frac{L}{l_c}\frac{\bar{n}}{(N^2-\sin^2\theta)^{1/2}}\right]^2
\tag{17}
$$

where

$$
\bar{n}=\frac{n_\omega+n_{2\omega}}{2} \quad \text{and} \quad N^2=\bar{n}^2+\left[\frac{n_\omega-n_{2\omega}}{2}\right]^2
\tag{18}
$$

Since $(n_\omega-n_{2\omega}) << \bar{n}$, the second harmonic intensity is independent of the coherence length. For p-polarized second harmonic light, the Fresnel-like factor $T_{2\omega}$ is given by

$$T_{2\omega} = 2n_{2\omega}\cos\theta_{2\omega} \times \frac{(n_\omega\cos\theta + \cos\theta_\omega)(n_{2\omega}\cos\theta_\omega + n_\omega\cos\theta_{2\omega})}{(n_{2\omega}\cos\theta_{2\omega} + \cos\theta)(n_{2\omega}\cos\theta_{2\omega} + n_o\cos\theta_o)^2}, \tag{19}$$

where the angles θ_ω, $\theta_{2\omega}$, and θ_o are given by $n_\omega\sin\theta_\omega = \sin\theta$, $n_{2\omega}\sin\theta_{2\omega} = \sin\theta$, and $n_o\sin\theta_o = n_{2\omega}\sin\theta_{2\omega}$, and where n_o is the index of refraction of the glass substrate. The factor t_o is given by

$$t_o = \frac{2n_o\cos\theta_o}{\cos\theta_o + n_o\cos\theta}. \tag{20}$$

The components d_{33} and d_{31} can be measured independently by measuring p-polarized second harmonic with s- and p-polarized fundamental light. When the incident laser light and the transmitted second harmonic are both p-polarized (p-p), and when $d = d_{33} = 3d_{31}$, then

$$t_\omega = \frac{2\cos\theta}{n_\omega\cos\theta + \cos\theta_\omega}, \tag{21}$$

and

$$p = \left(\frac{1}{3}\cos^2\theta_\omega + \sin^2\theta_\omega\right)\sin\theta_{2\omega} + \frac{2}{3}\cos\theta_\omega\sin\theta_\omega\cos\theta_{2\omega}. \tag{22}$$

With the incident laser light s-polarized and the second harmonic light p-polarized (s-p), and with $d = d_{31}$,

$$t_\omega = \frac{2\cos\theta}{n_\omega\cos\theta_\omega + \cos\theta}, \tag{23}$$

and

$$p = 2\sin\theta_\omega\cos\theta_\omega\cos\theta_{2\omega}. \tag{24}$$

Using Eqs. (16)-(24), the data depicted in Figure 3 are consistent with $3d_{31} = d_{33}$. Further, the data indicate that $d_{33} = 6.0 \pm 1.3 \times 10^{-9}\,esu$, which compares favorably with that of KDP ($d_{36} = 1.1 \times 10^{-9}\,esu$).

For poled glasses, the integrals in Eq. (5) can be evaluated since $U_{\mu\nu}$ in Eq. (11) is negligible compared to kT, so that, for a one-dimensional molecule[5],[11]

$$d_{33}(-\omega_3;\omega_1,\omega_2) = N\beta_{zzz}(-\omega_3;\omega_1,\omega_2)\,f^{\omega_3}f^{\omega_1}f^{\omega_2}\,L_3(q), \tag{25}$$

where

$$L_3(q) = \frac{q}{5} - \frac{q^3}{105} + \cdots \tag{26}$$

and

$$q = \left[\frac{\epsilon(n^2 + 2)}{n^2 + 2\epsilon}\right]\frac{\mu E_p}{kT}. \tag{27}$$

The factors β_{zzz} and μ are the hyperpolarizability and dipole moment uncorrected for local fields. The f's, then, are local field factors at the appropriate frequencies, which, at optical frequencies are well-approximated by Lorentz-Lorenz expressions,

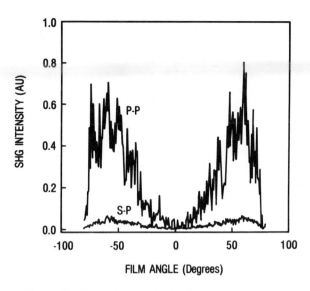

Figure 3. Data from polarization measurements.

$$f^\omega = \frac{n_\omega^2 + 2}{2}. \tag{28}$$

The poling local field has been included in Eq. (27). Taking $L_3(q)$ linear in q, Eq. (25) is consistent with Eq. (13).

Using the parameters for the film given in Table 1, where $\beta\mu$ has been measured independently using electric field induced second harmonic generation, and the thermodynamic model, d_{33} is calculated to be $d_{33} = 6.6 \pm 1.5 \times 10^{-9}$ *esu*. The agreement between the experimentally determined value of d_{33} and the calculated value, and that $d_{33} = 3d_{31}$ indicate that Eqs. (13)-(14) adequately describe the poled polymer.

TABLE 1. Properties of azo-dye/PMMA film used in second harmonic and electro-optic measurements. N is the number density of azo dye molecules in PMMA, E_p the magnitude of the poling field, n the refractive index, ϵ the static dielectric constant, and $\beta\mu$ the results of DC induced second harmonic generation measurements.[5]

N $(10^{20}/cm^3)$	E_p (MV/cm) SHG EO	n	ϵ	$\beta\mu$ $(10^{-30} cm^5 D/esu)$
2.42	0.62 0.37	1.52	3.6	525 ± 100

The thermodynamic model, Eqs. (25)-(27), predicts that the second order susceptibility depends on the processing parameters: N, the number density, E_p, the poling field, T, the poling temperature, ϵ, the bulk dielectric constant, and n, the refractive index. The remaining parameters are microscopic values of the molecular susceptibility, β, the molecular dipole moment, μ, and the local field factors, f^{ω_i}. A series of films were fabricated, varying the processing parameters N and E_p. The results of second harmonic measurements of the number density and poling field dependences are shown in Figures 4 and 5 respectively. The shaded areas were determined with no adjustable parameters from solution measurements of the molecular susceptibility and dipole moment, where the spread is due to experimental uncertainties. The thermodynamic model is consistent with the measured values of the nonlinear susceptibility in the two processing parameters. The linearity of the number density dependence suggests that the nonlinear molecules remain noninteracting in this range and that aggregation has probably not occurred.

4.2 ELECTRO-OPTIC MEASUREMENTS

Unlike second harmonic generation, the electro-optic effect can arise not only from electronic states, but also from molecular vibrations and rotations. It has been established experimentally, however, that the origin of the electro-optic effect in organic crystals is largely electronic.[11] In doped polymer glasses, this has not been established. The nonelectronic contributions can arise from both the polymer and the dopants, thus, low frequency measurements provide a means for determining the magnitude of the other processes leading to the electric field dependent change in the refractive index.

A two level model can be used to relate the second harmonic coefficient, $d_{kij}(-2\omega';\omega',\omega')$, to the electronic contribution of the electro-optic coefficient, $r^{el}_{ij,k}(-\omega;\omega,0)$ by accounting for dispersion in the second-order susceptibility and in the local field factors:

$$r^{el}_{ij,k}(-\omega;\omega,0) = \frac{4d_{kij}(-2\omega';\omega',\omega')}{n_i^2(\omega)n_j^2(\omega)} \times \frac{f^\omega_{ii}f^\omega_{jj}f^0_{kk}}{f^{2\omega'}_{kk}f^{\omega'}_{ii}f^{\omega'}_{jj}}$$
$$\times \frac{(3\omega_0^2-\omega^2)(\omega_0^2-\omega'^2)(\omega_0^2-4\omega'^2)}{3\omega_0^2(\omega_0^2-\omega^2)^2}, \tag{29}$$

where f^ω_{ll} is the local field factor at frequency ω in the l direction, $n_i(\omega)$ is the refractive index along the principle axis indexed by i, and where ω_0 is the characteristic frequency of the molecule. The difference between this value and the measured electro-optic coefficient is due to the slower vibrational and rotational processes.

Since the nonlinear properties of a poled material depends on the processing parameters such as poling field and number density, a comparison of different samples is possible by defining a parameter which is independent of these parameters.[6] As seen in the previous section, the nonlinear susceptibility is linearly proportional to both the number density and poling field. These effects can therefore be divided out by comparing values of r/NE_p. The parameters of the film used in the electro-optic experiment are given in Table 1. The electro-optic coefficient was measured by modulating the phase in one arm of a Mach-Zehnder interferometer.[16] The results of the measurements are summarized in Table 2.

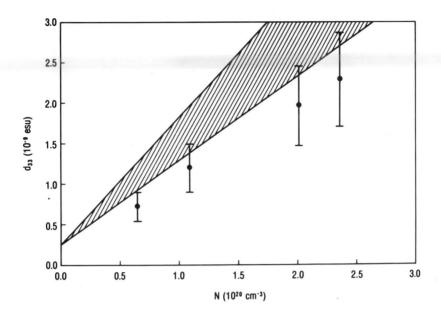

Figure 4. Number density dependence of the second harmonic coefficient.

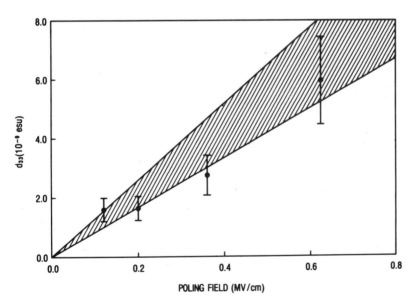

Figure 5. Poling field dependence of the second harmonic coefficient.

TABLE 2. Results of electro-optic measurements.

$r_{33}\,(expt)\,(10^{-12}\,m/V)$	$\dfrac{r_{33}\,(expt)}{E_p}(10^{-20}\,m^2/V^2)$	$\dfrac{r_{33}^{el}\,(expt)}{E_p}(10^{-20}\,m^2/V^2)$
2.5±0.4	6.7±0.9	10.5±2.1

The electronic contribution to the electro-optic coefficient as measured by second harmonic generation and corrected for dispersion by the two level model is larger than the electro-optic coefficient measured at a modulating frequency of $35kHz$ and wavelength of $633nm$. The nonelectronic contributions appear out of phase with the electronic process. The sign and percentage difference between r^{el}/E_p and r/E_p observed in the film is comparable to those found in organic crystals.[17] The difference may also be due to screening of the modulating field by trapped charge as well as inadequacies of the two-level model.

Poled doped polymer glasses thus possess substantial nonlinear optical bulk coefficients that can be understood in terms of the additivity of the microscopic nonlinear susceptibilities and molecular order. Using the thermodynamic model, the bulk susceptibility can be optimized by increasing order induced by both external and intermolecular forces, and by molecular engineering of dipole moments and hyperpolarizabilities.

To take full advantage of the nonlinear optical properties of poled doped polymer glasses in application to optical or integrated optical devices, other material considerations must be addressed. In contrast to the molecular origin of the nonlinear effects, properties such as optical quality, dielectric constant and loss, and manufacturability, are mostly determined by the bulk properties of the polymer glass.

5. OPTICAL PROPERTIES

Low-loss polymer channel waveguides have been produced by several techniques.[18] [19] These techniques include photolocking,[20] rf discharge polymerization,[21] mechanical embossing,[22] selective photopolymerizaton,[23] [24] [25] [26] and fiber drawing.[27] The materials used were mainly acrylates, although polystyrene and poly(organosilanes) have also been fabricated into low-loss waveguide structures. The low optical losses of these glassy polymer waveguides compares favorably with the propagation loss for devices fabricated in titanium-indiffused lithium niobate $(Ti:LiNbO_3)$, which is $\sim 0.1dB/cm$. $Ti:LiNbO_3$ is the current technology for fabrication of electro-optic devices.[28]

The factors contributing to optical loss in polymer glasses have been enumerated by Kaino. [29] Substantial improvements in loss in plastic optical fibers fabricated from polystyrene or PMMA are realized by controlling both the extrinsic processes (controllable through fabrication) and the intrinsic processes (material specific) that contribute to optical loss. It was determined that vibrational absorption owing to carbon-hydrogen absorption is the most significant source of optical loss in the near-infrared region.[27] Using a carefully controlled closed polymerization process with purified materials and a fiber drawing process, Kaino has fabricated the highest optical quality polymer waveguide material reported.[27] At $1.3\mu m$, plastic optical fibers of perdeutero-PMMA core and poly(fluoroalkyl methacrylate) cladding have propagation losses of ∼ 0.03dB/cm.

Loss measurements for organic crystals have also been reported. Tomaru *et al.* have produced optical channel waveguides of single crystal *m*-nitroaniline.[30] Scattering losses for a 5 *mm* long waveguide were reduced to 5dB/cm with a laser zone-melting technique. Nayar has prepared single-crystalline, void-free benzil fibers (diameter 2-10μm) with a propagation loss estimated to be 2dB/cm at 633nm.[31] The quoted value was about 1dB/cm greater than the attenuation in the bulk crystal. These values for crystalline materials are considerably higher than those measured in amorphous polymer glasses and in *Ti*:*LiNbO*$_3$; if crystals are to see application to guided-wave devices, these losses must be reduced substantially.

In addition to high optical quality, refractive index differences between the guiding region and the surrounding medium can be tailored depending on the fabrication process, materials, and structures fabricated. These differences can be used to control bending losses in guided wave configurations. The index differences obtained from several techniques range from 0.01 to > 0.5.[18] The refractive indices of typical glassy polymers range from 1.48 to 1.6. These values are relatively close to the index of optical fibers (∼ 1.45), thus mode mismatch between fibers and optical structures comprised of glassy polymers can be minimized. The data demonstrate that glassy polymers possess the required optical properties for application in integrated optics.

6. DIELECTRIC PROPERTIES

The speed of lumped-element electro-optic devices is determined by the RC time constants associated with the device, and thus are greatly determined by the dielectric properties of the electro-optic material. For instance the bandwidth per modulating power ($\Delta f/P$) of a guided-wave modulator is given by[32]

$$\frac{\Delta f}{P} \propto \frac{1}{1 + \epsilon/\epsilon_0},$$
(30)

where ϵ is the static dielectric constant. For a traveling wave modulator, the rise time of the modulator is given by $t_r \propto (\epsilon/\epsilon_0)^{1/2} - n$.[33] Thus a lower dielectric constant results in a faster device.

The dielectric properties of organic electro-optic materials are one of their more attractive properties. In contrast to ferroelectric crystals, the dielectric constants of organic molecular crystals are mostly determined by the molecular electronic structure. The dielectric constant of organic materials is not substantially larger than n^2. The dielectric constant of PMMA ranges from 2.5-3.6, depending on frequency, and the dielectric constant of polystyrene is ∼ 2.5 and remains flat to a frequency of 1GHz. The incorporation of organic electro-optic molecules does not substantially increase the dielectric constant.

Dielectric loss will also degrade device performance and affect the switching power through its dissipative effects. The dielectric losses in PMMA and polystyrene are also low and nondispersive.

7. MANUFACTURABILITY

Manufacturability implies that a reproducible device/system fabrication process can be developed. Since electro-optic polymer glasses have only recently been demonstrated, no statements can be made as to their ultimate manufacturability. However, experience with the fabrication of polymer glasses in the electronics and optics industries does provide some relevant data. Polymer glasses are widely used in the fabrication of electronic devices and device interconnects. Polymers are used as photoresists and as dielectric interlayers for electrical interconnects. As a result, a body of knowledge exists concerning planarization methods of polymers on substrates, the definition of microscopic features, and the fabrication of microstructures in planar polymer structures.

Polymer glasses are also widely used in the optics industry, as bulk and micro optical components.[34] They have been investigated as optical waveguide materials for short-haul optical communications,[35] and are already in wide use in consumer applications such as corrective lenses, and embossed holograms. In addition, polymer glasses have been investigated for integrated optical components. Kurokawa *et al.* have used selective photopolymerization of methyl acrylate in polycarbonate, and subsequent removal of unreacted monomer to fabricate optical dividers and low-loss couplers.[24] [25] [26] [29] Six-port star coupler circuits were reported with mixing guide lengths of 10-30*mm*, and a branching guide length of 12*mm*. With attached input and output fiber arrays, the insertion loss was 2.6*dB*. It was found that the insertion loss was independent of the optical fiber core and cladding dimensions used.

Tien *et al.* described a method using polymer films to form light guide interconnections for an integrated optical circuit.[36] A light-guiding film bridges the input terminal of one device and the output terminal of another device. The lightwave is found to propagate in the same mode in the devices and the film, by tapering the terminals of the devices and the ends of the interconnecting film.

A variety of optical structures have been fabricated from glassy polymers, some reaching the marketplace. Based on progress to date, the prospects for manufacturable devices and systems for application in integrated optics seem good.

The ability to integrate an electro-optic material with other optical devices, *e.g.* light sources and detectors, and with electronic drive circuits is important. Integrability implies that the electro-optic materials and the processing of these materials are compatible with the other components, and that electrical and optical interconnects can be fabricated. Polymer glasses show much promise in this respect. The use of polymers in electronic circuit fabrication and interconnects, planarization onto electronic substrates, electrical interconnection, and structure definition have been well investigated.

8. SUMMARY

The applicability of glassy polymer systems to integrated optics was discussed, including their manufacturability, integrability with sources, detectors, and drive electronics, and their optical and dielectric properties. Further, it was shown that orientational order plays a crucial role in the magnitude of the second-order nonlinearity. A model nonlinear optical glassy polymer system was fabricated and evaluated, and its nonlinear optical properties compare favorably with those of other materials. The combination of ordered large molecular hyperpolarizabilities and the advantageous bulk characteristics of glassy polymers affords the future optimization of bulk nonlinear optical properties, along with integrability, and high optical and dielectric quality, making this material class a promising one for use in integrated optics.

9. REFERENCES

1. A. F. Garito and K. D. Singer, *Laser Focus*, **80**, 59 (1982).

2. D. J. Williams, *Angew. Chem. Int. Ed. Engl.*, **23**, 690 (1984).

3. D. J. Williams, ed., *Nonlinear Optical Properties of Organic and Polymeric Materials*, ACS Symposium Series No. 233, American Chemical Society, Washington D. C., (1983).

4. D. S. Chemla and J. Zyss, eds., *Nonlinear Optical Properties of Organic Molecules and Crystals* (Academic Press, Orlando, 1987).

5. K. D. Singer, J. E. Sohn, and S. J. Lalama, *Appl. Phys. Lett.*, **49**, 248 (1986).

6. K. D. Singer, M. G. Kuzyk and J. E. Sohn, *J. Opt. Soc. B*, **4**, No. 6, (1987), in press.

7. J. D. Legrange, M. G. Kuzyk, and K. D. Singer, submitted to *Mol. Cryst. Liq. Cryst.* and references cited therein.

8. J. Zyss and J.L. Oudar, *Phys. Rev. A* **26** , 2028 (1982).

9. C. W. Dirk and R. J. Twieg, unpublished.

10. K. D. Singer, S. J. Lalama, and J. E. Sohn, *SPIE Proc.*, **578**, 130 (1985).

11. K. D. Singer, S. J. Lalama, J. E. Sohn, and R. D. Small, in *Nonlinear Optical Properties of Organic Molecules and Crystals*, D. S. Chemla and J. Zyss, eds., (Academic Press, Orlando, 1987).

12. J. F. Nye, *Physical Properties of Crystals* (Oxford University Press, London, 1957).

13. D. A. Kleinman, *Phys. Rev.* **126**, 1977 (1962).

14. P.D. Maker, R.W. Terhune, M. Nisenoff, and C.M. Savage, Phys. Rev. Lett. **8**, 21 (1962).

15. J. Jerphagnon and S.K. Kurtz, J. Appl. Phys. **41**, 1667 (1970).

16. M. G. Kuzyk, J. E. Sohn, S. J. Lalama, and K. D. Singer, to be published.

17. M. Sigelle and R. Hierle, *J. Appl. Phys.*, **52**, 4199 (1981).

18. S. J. Lalama, J. E. Sohn, and K. D. Singer, *SPIE Proc.*, **578**, 168 (1985), and references cited therein.

19. W. J. Tomlinson and E. A. Chandross, "Organic Photochemical Refractive-Index Imaging Recording Systems," in *Advances in Photochemistry*, John Wiley & Sons, 1980, and references cited therein.

20. E. A. Chandross, C. A. Pryde, W. J. Tomlinson, and H. P. Weber, *Appl. Phys. Lett.*, **24**, 72 (1974).

21. P. K. Tien, G. Smolinsky, and R. J. Martin, *Appl. Opt.*, **11**, 637 (1972).

22. G. D. Aumiller, E. A. Chandross, W. J. Tomlinson, and H. P. Weber, *J. Appl. Phys.*, **45**, 4557 (1974).

23. T. Kurokawa, N. Takato, S. Oikawa, and T. Okada, *IOOC*, A8-3 (1977).

24. Y. Katayama, N. Takato, T. Kurokawa, and S. Oikawa, *Org. Coat. Plast. Chem.* **40** 374 (1979).

25. T. Kurokawa, N. Takato, and Y. Katayama, *Org. Coat. Plast. Chem.*, **40**, 368 (1979).

26. T. Kurokawa, N. Takato, and Y. Katayama, *Appl. Opt.*, **19**, 3124 (1980).

27. T. Kaino, *J. Polym. Sci., Polym. Chem.*, **25**, 37 (1987).

28. Optical Society of America Proceedings of the Topical Meeting on Photonic Switching, Incline Village, March 1987.

29. Kaino, T; *Jpn. J. Appl. Phys.*, **1985**, *24*, 1661.

30. S. Tomaru, M. Kawachi, and M. Kobayashi, *Opt. Commun.*, **50**, 154 (1984).

31. B. K. Nayar, in *Nonlinear Optical Properties of Organic and Polymeric Materials*, D. J. Williams, ed., ACS Symposium Series No. 233, American Chemical Society, Washington, D. C., (1983).

32. I. P. Kaminow, J. R. Carruthers, E. H. Turner, and L. W. Stulz, *Appl. Phys. Lett.*, **22**, 540 (1973).

33. G. White and G. M. Chin, *Opt. Commun.*, **5**, 374 (1972).

34. S. Musikant, "Optical Materials: An Introduction to Selection and Application," Marcel Dekker, Inc., New York, 1985.

35. S. J. Lalama, J. E. Sohn, and K. D. Singer, *SPIE Proc.*, **578**, 168 (1985), and references cited therein.

36. P. K. Tien, R. J. Martin, and G. Smolinsky, *Appl. Opt.*, **12**, 1909 (1973).

CONFORMATIONAL TRANSITIONS IN POLYDIACETYLENE SOLUTIONS

D. G. Peiffer, T. C. Chung, D. N. Schulz, P. K. Agarwal,
R. T. Garner, M. W. Kim, and R. R. Chance

Corporate Research Laboratories
Exxon Research and Engineering Company
Annandale, NJ 08801

ABSTRACT

Visible absorption, infrared absorption, Raman, and kinetic measurements are all strongly suggestive of a single-chain origin for the conformational transitions observed for polydiacetylenes in the solution phase, poly4BCMU and poly3BCMU being the prototypical examples. Intramolecular association between substituent groups on the polydiacetylene backbone has been assigned as the driving force for the transition; in the case of the BCMU polymers, hydrogen bonding provides the association mechanism. In this paper we review recent work on polydiacetylene solutions with emphasis on rheological and time-temperature dependent quasielastic light scattering measurements for poly4BCMU in the dilute-semidilute regime. Light scattering measurements during the coil-to-rod transformation (yellow-to-red color change) demonstrate that dramatic color changes take place without any measurable change in hydrodynamic radius, confirming the single-chain origin of the conformational change. On aging, aggregation is observed as an order-of-magnitude increase in hydrodynamic radius.

INTRODUCTION

High molecular weight polydiacetylenes are generally insoluble even in exotic organic solvents. The first exceptions to this rule were the butoxycarbonylmethyleneurethane (BCMU) substituted polydiacetylenes.[1] The substituents for these polymers, commonly referred to as poly3BCMU and poly4BCMU, are shown as follows:

$$3BCMU \quad -(CH_2)_3-O-\overset{\overset{\displaystyle O}{\|}}{C}-\underset{\underset{\displaystyle H}{|}}{N}-CH_2-\overset{\overset{\displaystyle O}{\|}}{C}-O-C_4H_9$$

$$4BCMU \quad -(CH_2)_4-O-\overset{\overset{\displaystyle O}{\|}}{C}-\underset{\underset{\displaystyle H}{|}}{N}-CH_2-\overset{\overset{\displaystyle O}{\|}}{C}-O-C_4H_9$$

These polydiacetylenes (PDA) are soluble in common organic solvents, such as $CHCl_3$, due to the high entropy content of the complicated urethane substituent groups. The synthesis of these soluble polymers has led to a number of interesting developments,[1-4] perhaps the most remarkable being the discovery of a conformational transition in the polymer solutions, referred to as a "visual conformational transition."[1]

The conformational transition in BCMU polymers is induced by a change in solvent or temperature. Though conformational transitions are commonly observed in polymer solutions, the PDA system is unique because of the sensitivity of the electronic properties to backbone conformation, which results in dramatic color changes during the conformational transition. This sensitivity to conformation originates from the conjugated structure of the PDA backbone [=RC-C≡C-CR=]. π-Electron conjugation requires a planar structure. Conformational disorder limits conjugation by disrupting planarity to produce a distribution of chromophores of various "conjugation lengths" on the individual chains. The electronic properties vary with conjugation length, usually with a length^{-1} dependence.[4] Short conjugated PDA segments absorb in the blue spectral region (yellow solutions) while long segments absorb in the red (blue solutions). Therefore, the conformation transition in soluble PDAs is thought to involve a significant change in conjugation length. In poly4BCMU, the transition is accompanied by a yellow-to-red color change, a 2000-cm^{-1} shift in optical absorption. In poly3BCMU the color changes is yellow to blue, an absorption shift of more than 5000 cm^{-1}. Two photon absorption, which dominates the nonlinear response in the solutions, shows similar spectral shifts. In the blue solution phase of poly3BCMU the optical properties are very similar to those of its crystalline form,[5] a result which suggests an effectively infinite conjugation length in the blue solutions.

Additional soluble polydiacetylenes have been discovered which show temperature- or solvent-induced color changes with absorption shifts similar to poly4BCMU.[6-8] These solutions are believed to consist of aggregates in the red phase.[6-9] Evidence is mixed as to whether polyBCMU solutions in their red or blue phases consist of aggregates or single chains. Lim and Heeger,[3] using light-scattering measurements, argue for the single chain interpretation based on the lack of variation of the hydrodynamic radius with polymer concentration. Wenz et al.[6,9] suggest the solutions are composed of aggregates as large as 700 chains. These light-scattering measurements on the solutions cannot address the important question of the molecular origin of the conformational transition, and in particular whether it is intra- or intermolecular. Previous spectroscopic[1] and kinetic[10] work has suggested that the driving force for the transition is purely intramolecular in origin (involving the hydrogen-bonding network of the urethane substituent groups) and suggests that, if aggregation takes place, it must be subsequent to the conformation transition. Recent experiments involving field-induced birefringence in poly4BCMU solutions strongly support the intramolecular interpretation of the color transition.[11]

In this paper we discuss the dilute-semidilute properties of poly4BCMU solutions, via visible absorption, rheological measurements, and temperature-time dependent quasielastic light scattering.[12] The results offer definitive evidence that the conformational transition is intramolecular in origin. In particular, the light scattering results demonstrate that aggregation occurs subsequent to the conformational transition but has no observable effect on solution optical properties.

EXPERIMENTAL

Polydiacetylene (4BCMU) was prepared by a similar synthetic route described by Patel et al.[1] 5-Hexyn-1-ol, obtained from FARCHAN Laboratories, was oxidatively coupled by two ways, either Hay's method using copper (I) and tertiary amine as catalyst or stoichiometry oxidation procedure using copper (II) in pyridine-methanol-ether (1,1,4) solution. The corresponding diacetylene diol was then reacted with n-butyisocyanatoacetate in dibutyltin-bis-(2-ethyl hexanote) and tri-ethylamine catalyst system to obtain the desired diacetylene diurethane monomer. Further purification was carried out by recrystallization of monomer in acetone-hexane mixed solvent. The solid state polymerization of monomer proceeded by 1,4 reaction of the diacetylene units which was ini-tiated by high intensity UV radiation to give the conjugated polymer chain. Hot acetone was used to remove unreacted monomers and oligomers. The polydiacetylene structure was confirmed by IR. GPC measurements showed the weight-average molecular weight above 3 million (degree of polymerization of 6000) by comparing the calibration curve from 13 samples of polystyrene. The molecular weight distribution, $M_w/M_n \doteq 8$, was quite broad. However, it might be due to the instrumental limitation, such as high molecular weight degradation in GPC column. In addition, it should be noted that the average molecular weight of these materials are markedly higher than previously reported.[2]

The polymer solutions for all studies were prepared by dissolving the polydiacetylene in tetrahydrofuran (THF)-toluene mixed solvents (or the pure solvents) at temperatures between 60 and 65°C. The resultant homo-geneous yellow solution was slowly cooled to room temperature over a period of five hours. As will be described shortly, the polymer upon cooling behaved differently depending on the solvent composition, i.e. red, orange, and yellow colors are observed.

Reduced viscosities were measured with a standard Ubbelhode capillary viscometer placed into a temperature controlled water bath, typically from 25 to 65°C. The solutions were temperature equilibrated for approximately 15 minutes prior to viscosity measurement. The measurement was repeated several times (typically three measurements) until reproducible flow rate times were obtained.

A Contraves Rheometer (LS-30) was used to measure the shear viscosity of the solutions as a function of shear rate, polymer concentration and any time dependent changes at specific shear rates.

The visible wavelength absorption regions were obtained on a Perkin-Elmer Lambda 7 double beam UV/VIS spectrophotometer at room temperature, spanning typically the wavelength region from 300 to 700 nm. A bandpass of 2 nm, scan speed was 60 nm/min and path lengths of 1 mm were used for all spectra reported here. All spectra are plotted with the solvent or solvent mixture as a reference.

Quasielastic light scattering was used to measure the hydrodynamic radius as a function of temperature (typically from 10°C to 65°C) and time. A BI-2030 Brookhaven Instrument Correlator equipped with Model 124B Spectra Physics Helium-Neon laser operating at approximately 15 mW power and at a wavelength of 632.8 nm. The hydrodynamic radius were calculated from the well-known Stokes-Einstein relation for particle diffusion where the so-called Guinier conditions apply.

RESULTS AND DISCUSSION

Solvent Quality Effects

In Figure 1 is presented the absorbance as a function of wavelength for 400 ppm poly(4BCMU) solutions covering the complete range of solutions compositions comprising the solvents tetrahydrofuran (THF) and toluene. In both the pure solvents or their respective mixtures, the solution

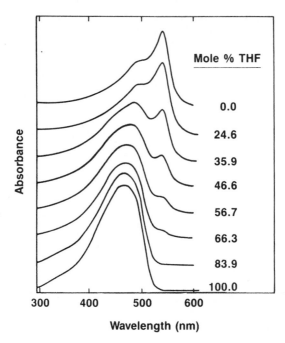

Fig. 1. Absorption spectra for poly(4BCMU) in THF, toluene and their
corresponding mixtures. Spectrum were taken at 25°C with the
polymer concentration at 400 ppm.

characteristics were quite stable for long periods of time. Even under
relatively nonvigorous stirring, shaking or elongation flow did the solvent
mixtures (with less than approximately 30 mole % THF) show phase separation
with time. In these instances, aggregation followed by precipitation of
the poly(4BCMU) occurred in a few minutes. Therefore, in this work, we
concentrated our efforts on the 46.6 mole % THF solution mixtures. A
distinctive red color appears only at THF levels less than 24.6 mole %;
otherwise, a pronounced yellow-orange or reddish-orange color is observed.
A brilliant red color is noted in nearly pure toluene solutions. A close
examination of the spectra in Figure 1 conforms these visual observations.
As the level of THF is decreased, the peak in the yellow region, i.e.
having λ_{max} at about 470 nm is gradually reduced and reappears at longer
wavelengths. Furthermore, a peak in the red region (λ_{max} at ~540 nm
appears when the THF level is between 70 and 80 mole %. These results
closely approximate previously observed spectra 1 (CHCl₃/hexane solvent
system) except in our experiments there is less variation in solvent
quality. (Hexane is a nonsolvent for poly4BCMU. As a result, it appears
that as THF level is decreased, a rather gradual transformation of the
local chain conformation is occurring, compared to the rather abrupt
transformation in the CHCl₃/hexane system.[1]

We have further noted that the stability of the poly(4BCMU) solutions
are inversely related to the absorbance in the orange-red region. This
could again follow to increase in the conjugation length within the chain
structure itself. It is well-known that rigid, noncharged macromolecules
tend to strongly aggregate making them extremely difficult to melt or
solution process. This is due, in part, to the difficulty in overcoming
the high glass transition temperature and crystallinity of these materials.
As a result of these considerations, improved processability and solubility
occurs when some flexible units are introduced into the rigid chain. In
this respect, poly(4BCMU) in THF/toluene solutions behave as an example of

these latter polymeric materials. The longer the conjugation lengths (rigid elements in the chain backbone), the more facile is the aggregation process.

However, at this point, it is of interest to briefly examine the yellow solutions since these are quite stable with regard to outside perturbations. In addition, the yellow solution can be cooled and subsequently transformed into red solution or solutions intermediate between the yellow and red phases. In this way, the properties of the complete range of solution characteristics can be examined especially if the rate of cooling or heating is carefully controlled.

100 mole % THF Solution (Yellow Solution)

Figure 2 shows the reduced viscosity-concentration profiles of poly(4BCMU) from room temperature to 60°C. In all instances, the solutions were yellow in color and did not display any time-dependent effects over a broad concentration. As a result, the data show that above about 400 ppm, the reduced viscosity rises rapidly due undoubtedly to the overlapping of polymer chains. At lower concentrations, the intrinsic viscosity can be easily obtained via extrapolation to zero concentration with the results presented in Figure 3. A comparison of these results with that of Wenz and Wegner[6] confirms the poly(4BCMU) solutions consists of coiled macromolecules in a good solvent. The intrinsic viscosity changes with temperature as expected with no indications of a phase transition or any aggregation effects.

Throughout the remainder of this paper, we focus on the concentration regime from 100 to 400 ppm. This latter concentration appears to be the region where the dilute state begins to reach the chain overlap concentration. In addition since solution stability is a problem at the THF concentrations less than about 30 mole %, we focus our attention on

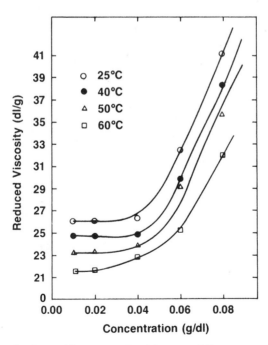

Fig. 2. Reduced viscosity-concentration profiles as a function of temperature of poly(4BCMU) in THF.

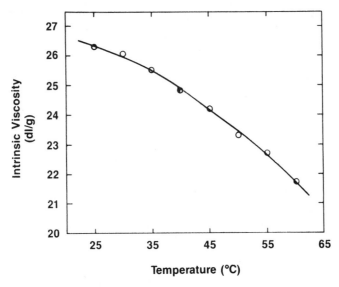

Fig. 3.   Intrinsic viscosity of poly(4BCMU) in THF as a function of
          temperature.

solutions containing 46.6 mole % THF, which from Figure 1 is intermediate
between the yellow and red solutions. (We will refer to this solution as a
red-orange solution.) Our results show that even though there is still a
yellow component to this latter solution, the behavior found appears to be
readily translated to the less stable red solution systems.

<u>Yellow to Red Transition</u>

     As described previously, a broader range of solubilization conditions
are found if the nature of the good and poor solvent for poly(4BCMU) is
changed.   The effect of change in solvent conditions is reflected in the
change in hydrodynamic radius as a function of mole % THF (Table 1).   At
low THF levels R_H is large. However, an order of magnitude decrease is
observed at approximately 56 mole % THF.   A similar observation occurs with
a good solvent, chloroform, upon the addition of hexane, a nonsolvent.
However, precipitation of the polymer does not occur in THF/toluene solvent
mixtures even though the polymer concentration closely approximates the
chain overlap concentration.   Furthermore, it is noted that the sharp
change in R_H occurs at about the same concentration at which the first hint

Table 1

Hydrodynamic Radius of poly4BCMU as a Function of THF Concentration
for THF-Toluene Solvent Mixtures (Polymer Concentration: 0.04 g/dl)

THF Concentration (Mole %)	Hydrodynamic Radius (Microns)
0.0	8.2
24.6	7.3
46.6	7.8
51.3	6.4
56.1	5.7
56.7	1.1
75.1	0.6
100.0	0.6

of the red peak appears in the visible spectrum of this material
(Figure 1). The color of the solution is, however, red-orange in nature.
Therefore, again the change in the absorption characteristics corresponds
to the change in R_H and the previously described solution stability
observations.

In order to more closely examine this transformation, the R_H was
examined as a function of temperature. This data is shown in Figure 4,
where the solution contains 46.6 mole % THF at the 400 ppm polymer level.
Above 40°C, R_H remains constant. It should be noted that the solution
transforms from a red-orange to a completely yellow solution in this
temperature regime. Upon slow cooling without any outside perturbations,
R_H is found to remain remarkably invariant from 70°C through the "transformation interval" to 10°C. The color, however, has changed from the yellow
color at high temperatures to the various "hues of orange" back to a
red-orange color. If, indeed, the color change is directly attributed to a
modest change in the conjugation length, then it is apparent that this can
occur without any significant degree of aggregation of chains occurring.
Furthermore, it is also unnecessary for the coil to transform completely
over its entire contour length into a rigid rod. Only relatively small
"stiffening" of a number of adjacent monomer units is apparently required
for this to occur. It is assumed that these minor changes in the local
persistence length should not have a dramatic influence on the overall
chain conformation as is observed. We observe also that the absorption
spectra of the material at the beginning of the heating cycle closely
approximately the slowly cooled solution at 10°C.

Careful aging of the slowly cooled solutions at 0°C results in the
transformation of R_H from 0.7 to approximately 6.0 microns (Figure 5) which
closely approximates the original size of the chains in solution. However,
the absorption spectrum does not change in any perceptible way during this
increase in R_H. Apparently, the chains are slowly aggregating in this
polymer concentration regime without influencing the spectral absorption
characteristics. We should note that the absorption peaks in the slowly
cooled samples are considerably sharper than those in Figure 1, an
observation which has also been made for poly4BCMU in $CHCl_3$/hexane
mixtures.[13] Improved chain packing during a slowed growth of the

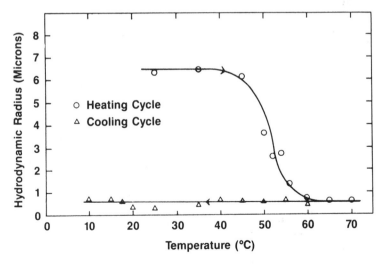

Fig. 4. Hydrodynamic radius of poly(4BCMU) with increasing temperature
followed by a slow cooling cycle. Polymer concentration is
400 ppm in a 46.6 mole % THF solution.

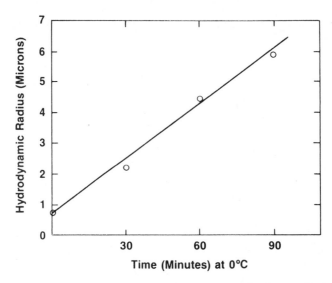

Fig. 5. Hydrodynamic radius of poly(4BCMU) of slowly cooled solution
thermally aged at 0°C. Polymer concentration is 400 ppm dissolved
in a 46.6 mole % THF solution.

aggregates may have a bearing on this observation. We are currently
fractionating this relatively broad molecular weight sample and repeating
the above spectral, QELS, conventional light scattering, and viscometry
measurements in order to more fully understand the polymer transformation
in solution.

Shear Rate Measurements

Throughout this study, we observed that careful handling of
poly(4BCMU) in a number of THF/toluene solution compositions was crucial in
obtaining non-aggregating systems. Therefore, it is of interest to examine
the shear rate dependence of the aggregation process. Preliminary results
for the 100.0 and 46.6 mole % THF solutions are shown in Figure 6. Due
primarily to the low polymer concentration, the yellow solution (100 mole %
THF) is Newtonian over the entire shear rate range examined. In addition,
the viscosity remained invariant with time at each shear rate. On the
contrary, the red-orange solution shows pronounced shear thinning characteristics over most of the shear rate except at $0.044S^{-1}$ where the
viscosity slowly rose with time. This behavior is indicative of the
formation of aggregates which can form at specific shear rates. Similar
behavior is observed even at 100 ppm polymer concentration, although the
magnitude of the aggregation effects are markedly reduced.

However, several groups have experimentally determined[14] and theoretically predicted[15] that coiled polymers, for example, can exhibit phase
separation in a flow field. Apparently the energy stored in the sheared
state is added to the Gibbs energy of the solution at rest. As a result,
demixing can occur under a shear stress especially if the solvent is of
poor quality. Moreover, if the energetics of the phase separation are
favorable, aggregation of the phase separation chains can occur. In
qualitative terms, this could be the reason for the formation of aggregates
of poly(4BCMU) as the quality of the solvent is reduced.

Polarizing Microscope Examination

Due to the fact that the molecular weight of the poly(4BCMU) is
relatively high and the size of the aggregates formed are large even in

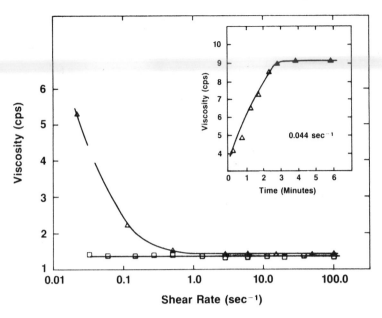

Fig. 6.   Viscosity-shear rate profiles of poly(4BCMU) in pure THF (□) and
46.6 mole % THF (Δ). Polymer concentration is 400 ppm.

dilute solution, the particles can be examined with optical techniques,
such as optical microscopy. With the aid of crossed polars, the anisotropy
and average size can be at least qualitatively determined. Typical
polarized photomicrographs of the red or red-orange solution are shown in
Figure 7. Gently placing a cover slip onto a small volume of solution does
not itself cause aggregation. Two points become immediately obvious: (1)
There are a large number of particles present in solution possessing a size
that is good agreement with the QELS measurements, and (2) the individual
particles are highly birefringent indicative of high chain alignment.
Therefore, as the aggregation phenomenon takes place, the individual
diacetylene chains self-align into an array of highly oriented chains. As
of the present time, little is known concerning the subtle details of this
alignment process (exact degree of chain concentration, degree of crystallinity, kinetics, etc.); however, as noted by Heeger et al.,[8,9] in the
context of film formation a single polymer chain can undergo a collective
coil-to-rod transition. They noted that it is the intermolecular mean
field interaction that markedly enhances the rigidity of the individual
rod-like polymers and tends to keep them self-consistently aligned.

     Furthermore, in this work, we note that if the conjugation length of
the chain segments are sufficiently long, i.e. red component to the absorption spectrum, then these can be the sites where aggregation occurs initially. Subsequently, the chains themselves could align utilizing the other
chains as a template. The growing oriented aggregate is now a nucleus for
further nucleation controlled growth. In this respect, a rod-to-coil
transition could occur if the individual diacetylene chains are supported
by other oriented chains.

     Finally, we note that in the yellow solution, birefringence is com-
pletely absent in dilute/semidilute solutions. However, highly birefringent films are formed when even the pure THF solution is dried.
Figure 8 shows some typical results in this regard where the films formed
have a preferred chain axis along the solvent evaporation front. Again we
observe that chain alignment can easily occur from solutions containing

Fig. 7. Photomicrographs with crossed polars of a thermally aged (0°C) poly(4BCMU) solution at 400 ppm with (right side) and without (left side) a cover slip.

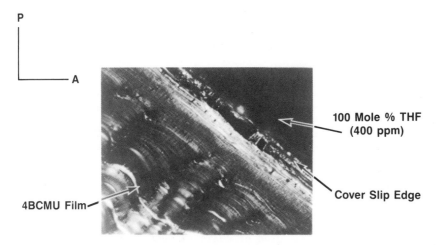

Fig. 8. Photomicrograph with crossed polars of the 100 mole % THF solution and an air dried film formed at the air-THF boundary.

coiled chains of polydiacetylenes due again to this preferential nucleation and "growth" of the individual chains acting as a template for each other.

CONCLUSION

Absorption spectra, rheological measurements, polarizing microscopy and temperature-time dependent hydrodynamic radius (R_H) were preformed in order to examine the semi-dilute solution properties of a model polydiacetylene, poly(4BCMU). Our preliminary measurements have confirmed that is is possible to have dramatic color changes with no change in R_H. As a result, it appears that the color changes are occurring via a modest changes in the "local" conjugation length. Apparently, any increase in the intramolecular hydrogen bonding can cause an increased delocalization of the π-electron cloud. This increased conjugation length also appears to be a site for aggregation especially as the solvent quality becomes poorer (i.e. absorption spectrum has an increasingly larger red peak absorption maximum). This aggregation phenomena is reflected in a variety of time-temperature and rheological measurements.

Polarized light microscopy examination of the thermally aged solutions confirms that the aggregation size closely corresponds to the hydrodynamic radius measurements. Moreover, the microstructure of the aggregates are highly anisotropic indicating that during the aggregation step a nucleation and growth process occurs. At the present time, the exact nature of this process from either dilute or semidilute solution is not well understood. In any event, this interesting orientation process permits the coiled chain to become highly extended, possibly a completely rigid rod. Further synthesis and physical measurements are being planned in order to confirm if these observations are a general phenomena of the polydiacetylene family over a broad concentration range. With moderate changes in molecular structure, the extent of aggregation could become even more manageable, thus making these soluble materials useful with regard to controlling the rheological properties of aqueous and nonaqueous solutions.

ACKNOWLEDGEMENTS

The authors gratefully acknowledge the visible spectroscopy measurements of Ms. K. Graf and Ms. J. Quodomine, and W. W. Graessley and D. S. Pearson for useful discussions.

REFERENCES

1. G. N. Patel, R. R. Chance and J. D. Witt, _J. Chem. Phys_. 70, 4387 (1979).
2. G. N. Patel and E. Walsh, _J. Polym. Sci. (Lett.)_ 17, 203 (1979).
3. K. C. Lim and A. J. Heeger, _J. Chem. Phys_. 82, 522 (1985).
4. R. R. Chance, M. L. Shand, C. Hogg and R. Silbey, _Phys. Rev_. B22, 2340 (1980).
5. R. R. Chance and G. Patel, J. Witt, _J. Chem. Phys_. 71, 206 (1979)
6. G. Wenz and G. Wegner, _Makromol. Chem. Rapid. Commun_. 3, 206 (1980); 3, 231 (1982).
7. C. Plachetta, N. O. Rau, A. Hauck and R. C. Schultz, _Makromol. Chem. Rapid. Commun_. 3, 249 (1982)
8. C. Plachetta and R. C. Schultz, _Macromol. Chem_. 3, 815 (1982).
9. G. Wenz, M. Müller, M. Schmidt and G. Wegner, _Macromolecules_ 17, 837 (1984).
10. R. R. Chance, M. W. Washabaugh and D. J. Hupe, _Polydiacetylenes_, R. R. Chance and D. Bloor, Ed. (Martinus Nijhoff, The Netherlands, 1985), p. 239; _Chemtronics_ 1, 36 (1986)
11. K. C. Lim, A. Kapitulnik, R. Zacher, and A. J. Heeger, _J. Chem. Phys_. 82, 516 (1985).
12. D. G. Peiffer, T. C. Chung, D. N. Schulz, P. K. Agarwal, R. T. Garner and M. W. Kim, _J. Chem. Phys_. 85, 4712 (1986).
13. R. R. Chance, J. M. Sowa, H. Eckhardt and M. Schott, _J. Phys. Chem_. 90, 3031 (1986).
14. G. Ver Strate and W. Philippoff, _J. Polym. Sci_., _Polym. Lett_. Ed., 12, 267 (1974).
15. B. A. Wolf, _Macromolecules_ 17, 615 (1984).

MOLECULAR ENGINEERING APPROACHES TO MOLECULAR AND MACROMOLECULAR NONLINEAR

OPTICAL MATERIALS: RECENT THEORETICAL AND EXPERIMENTAL RESULTS

D. Li[a], J. Yang[b], C. Ye[a], M. A. Ratner[a], G. Wong[b], and
T. J. Marks[a]

[a]Department of Chemistry and the Materials Research Center
[b]Department of Physics and the Materials Research Center
Northwestern University, Evanston, IL 60201

INTRODUCTION

As materials, polymers offer unexcelled diversity and tailorability in terms of light weight, strength, elasticity, plasticity, chemical resistance, toughness, thermal stability, friction resistance, and processability with regard to forming films, foils, fibers, coatings, etc. Traditionally, the application of polymers in optics technology has been limited largely to inexpensive lenses, prisms, fiber optics, anti-reflection coatings, filters, etc. It is now clear, however, that this picture is rapidly changing, and recent advances in the area of polymeric/organic substances with highly nonlinear optical (NLO) properties[1-4] signal that new generations of materials for optics and electronics technology await synthesis, characterization, understanding, and ultimate application. The latter sphere could well include unique new materials for optical telecommunications, optical information processing, integrated optics, electrophotography, and other applications.

The attraction of organic-based NLO materials lies in the very large nonlinear responses possible over a broad frequency range, rapid response times, high laser damage thresholds, and the intrinsic tailorability of organic structures. The scientific challenge lies in the design and synthesis of chromophores with high optical nonlinearities, in the rational construction of tunable multichromophore arrays, in the fabrication of new types of devices, and in understanding molecular architecture/electronic structure/materials performance relationships at a fundamental level. Research in this area is thus necessarily interdisciplinary in character and involves an interplay of synthesis, materials processing, theory, and optical characterization. The purpose of this article is to review recent progress in the area of tailored molecular and macromolecular NLO materials at Northwestern University. We begin with a discussion of basic physical requisites and synthetic strategies. We then focus upon recent quantum chemical developments which greatly facilitate the design and evaluation of new NLO chromophores and are based on π-electron calculations using the Pariser, Parr, Pople model Hamiltonian.[5,6] Finally, we discuss one successful synthetic approach to the construction of polymeric materials with efficient frequency doubling characteristics: chromophore-functionalized glassy polymers.[7,8]

BACKGROUND AND SYNTHETIC STRATEGIES

The fundamental relationship describing the change in molecular dipole moment (polarization upon interaction with an external electric field, as in the oscillating electric vector of electromagnetic radiation) can be expressed in a power series (eq.(1)).[1-4] Here P_I is the polarization

$$P_I = \sum_J \alpha_{IJ} E_J + \sum_{JK} \beta_{IJK} E_J E_K + \sum_{JKL} \gamma_{IJKL} E_J E_K E_L + \cdots \qquad (1)$$

induced along the Ith molecular axis, E_J is the Jth component of the applied electric field, α is the linear polarizability, β the quadratic hyperpolarizability, and γ the cubic hyperpolarizability. The even order coefficient, β, which is responsible for second harmonic generation (SHG), vanishes in a centrosymmetric environment. There are no environmental parity restrictions on the odd order tensors. The analogous macroscopic polarization arising from an array of molecules is given by eq.(2) where

$$P_i = \sum_i \chi_{ij}^{\omega} E_j + \sum_{jk} \chi_{ijk}^{2\omega} E_j E_k + \sum_{jk\ell} \chi_{ijk\ell}^{3\omega} E_j E_k E_\ell + \cdots \qquad (2)$$

the χ's are macroscopic susceptibilities. An important consequence of these relationships is that light waves of frequency ω passing through an array of molecules can interact with them in such a way to produce new light waves at double (2ω, SHG), triple (3ω, THG), etc., the fundamental frequency (a nonlinearity in the polarization). Significantly, it has been found that optical nonlinearities can be extremely high for certain classes of conjugated organic molecules--far higher than for traditional inorganic solids (e.g., $LiNbO_3$, KDP). Moreover, the π electron character of these phenomena in the organic materials implies response times (ca. 10^{-14} sec) which are far shorter than in conventional inorganic materials.

Considerable progress (much of it empirical) has been made in identifying those molecular characteristics which give rise to the highest non-linear responses; by far the greatest interest to date has been in SHG. Donor and acceptor functionalities connected by an appropriately conjugated π electron system can give rise to very large β values, and the contribution of charge transfer states (eq.(3)) to the nonlinearity has been empha-

$$D \!-\!\!\bigcirc\!\!-\! A \xrightarrow{\ h\nu\ } \overset{+}{D} \!-\!\!\bigcirc\!\!-\! \bar{A} \qquad\qquad (3)$$

A = acceptor substitutent   D = donor substituent

sized.[9] Nevertheless, criteria for the design of new chromophores are presently at a very primitive level. Efficient approaches to predicting the characteristics of new chromophores, and properties descriptions that embody conventional chemical electronic structure concepts, are greatly needed.

Beyond the above problem of designing molecules with appropriate optical characteristics lies the enormous challenge of rationally constructing supermolecular chromophore assemblies with high optical nonlinearities. In the case of SHG materials, the challenge is to design a non-centric (and ultimately phase-matchable) array in which the molecular dipoles are aligned in the same direction. Ideally, the noncentric

character should also be readily tunable. Criteria for organic THG
materials are at present less clear, however extended conjugation lengths
appear to be an asset.[10-12]

From the standpoint of enforcing molecular organization and orientation in a controlled fashion as well as offering a ready means for producing mechanically stable films, fibers, and other desirable geometries,
combining nonlinear optical chromophores with the properties of certain
polymeric materials is highly desirable. Undoubtedly the simplest approach
is a guest-host polymer/chromophore motif (e.g., I) in which a NLO chromo-

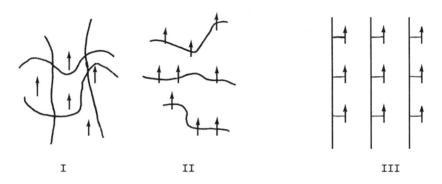

I   II   III

phore is dissolved in a suitable polymer matrix.[13-15] The success of such
an approach depends critically on the ability to preferentially orient the
chromophoric guest (e.g., by using a poling electric field near or above
the glass transition temperature) and to maintain this configuration for
extended periods of time. Major anticipated difficulties include ther-
mally-induced relaxation back to a centrosymmetric structure,[15] chromo-
phore-polymer immiscibility and phase separation, as well as chromophore
leaching/evaporation. Other structural arrangements (e.g., II,III) address
the second and third of these problems by fixing the chromophore to the
polymer backbone. Problems in achieving a noncentric structure[16] are still
a major consideration. Of course, all of these strategies rely upon proper
selection of the component chromophore.

π-ELECTRON THEORY FOR PREDICTING MOLECULAR NONLINEAR OPTICAL PROPERTIES

The accuracy with which semi-empirical methods can describe the linear
optical properties of relatively large molecules suggests application,
within a perturbation theoretical framework, to nonlinear optical properties. Indeed, all-electron CNDO methods have, in the past several years,
been successfully employed to calculate molecular NLO properties (β).[17-21]
An alternative approach, which capitalizes upon the fundamentally π-elec-
tron character of the NLO phenomena in the molecules of interest here,
employs the purely π-electron Pariser, Parr, Pople (PPP) model.[5,6,22-29]
The attraction of this approach is the far greater computational efficiency
of a calculation which only deals with the π electrons. For example, a
monoexcited configuration interaction PPP calculation on benzene would
involve 9 singlet states whereas a comparable all-valence electron CNDO
calculation would involve 900 singlet states. Thus, the advantages of the
PPP approach are that a greater number of π-electron configurations can be
included in the sums over states (affording greater accuracy), that more
elaborate chromophores and chromophore assemblies can be examined, and that
a greater number of chromophore geometries and polarization frequencies can
be explored. It will be seen that the present approach[5,6,30] yields
results in satisfactory agreement with experiment and with comparable or
greater accuracy than the all-electron CNDO method.

The nature of the basic PPP Hamiltonian[5,6,22-29] and the perturbation theoretical expressions for the β tensor[17,31,32] are discussed elsewhere. In the present calculations, we employ the monoexcited configuration interaction approximation[25,26] and standard PPP parameters for optical spectra. Gratifyingly, the results are not particularly sensitive to reasonable variations in these quantities. Chromophore metrical parameters are taken from standard sources.

In the following discussion, we focus first upon the accuracy of the PPP approach by comparing our results with experiment. A quantity of particular interest is β_{vec}, the β result obtained in solution electric field induced SHG experiments (eqs. (4), (5)). In Figure 1 is shown a plot

$$\beta_{vec}(-2\omega;\omega,\omega) \equiv \left[\beta_1^2 + \beta_2^2 + \beta_3^2\right]^{1/2} \tag{4}$$

$$\beta_i = 1/3 \sum_{k=1}^{3} \{\beta_{ikk} + \beta_{kik} + \beta_{kki}\} \qquad i = 1,2,3 \tag{5}$$

of experimental[4,5,17,18,19] versus calculated β_{vec} values for twelve conjugated organic chromophores. Structures and relevant data are set out in Table I. It can be seen that the agreement between PPP-derived and experimental β_{vec} values is very good. Additional verification is provided by Figure 2 in which a comparison is made of the calculated and experimental[33] frequency dependence of β_{vec} for p-nitroaniline. Again, the agreement is impressive.

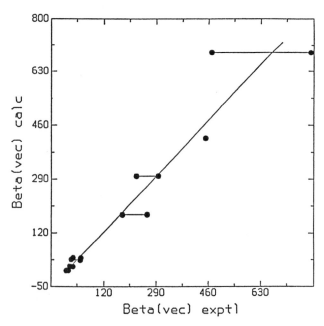

Fig. 1.   Plot of β_{vec} values in units of 10^{-30} cm^5 esu^{-1} calculated from the PPP model versus experimental β_{vec} values from the literature ($\hbar\omega = 1.17$ eV). Where two experimental β_{vec} values are given in the literature, both are shown in the plot.

Table I. Calculated and Experimental Values of β_{vec} for Various Molecular Chromophores[a,b]

Molecule	$\hbar\omega$ (eV)	$\beta_{vec}^{PPP}(\omega)$	β_{vec}(exp.)	$\beta_{vec}^{CNDO}(\omega)$
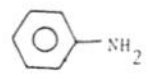 ⬡—NH$_2$	1.17	2.5	0.79–2.46	3.662
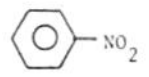 ⬡—NO$_2$	1.17	5.4	1.97–4.6	0.62
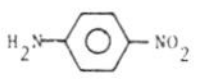 H$_2$N—⬡—NO$_2$	1.17	35.9	16.2–47.7	12.18
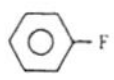 ⬡—F	1.17	0.4	1.06	0.84
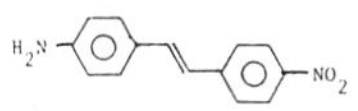 H$_2$N—⬡—CH=CH—⬡—NO$_2$	1.17	300.0	225–295	69.15
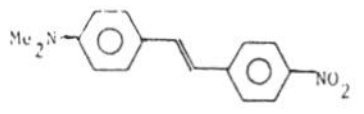 Me$_2$N—⬡—CH=CH—⬡—NO$_2$	1.17	420.0	450	
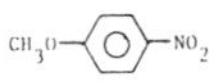 CH$_3$O—⬡—NO$_2$	0.656	13.5	14.3 – 17.5	7.35
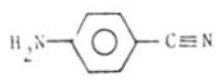 H$_2$N—⬡—C≡N	0.656	15.0	13.34	7.21
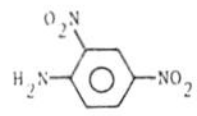 O$_2$N—, H$_2$N—⬡—NO$_2$	1.17	39.1	21	9.25
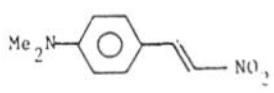 Me$_2$N—⬡—CH=CH—NO$_2$	1.17	179.3	180–260	126
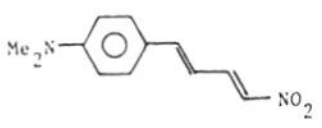 Me$_2$N—⬡—CH=CH—CH=CH—NO$_2$	1.17	687.9	470–790	355
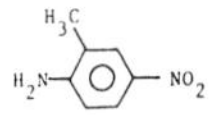 H$_3$C—, H$_2$N—⬡—NO$_2$	1.17	35.7	16– 42	

[a]β values are in units of 10^{-30} cm^5 esu^{-1}.
[b]Experimental values and CNDO results are from refs. 4, 5, 17, 18, 19.

One attractive application of the PPP model is in evaluating the properties of potential NLO chromophores prior to undertaking a laborious synthesis. Several examples of interesting hypothetical molecules and calculated β_{vec} values (in units of 10^{-30} cm^5 esu^{-1}; $\hbar\omega = 0.0$ eV) are shown below. The effects of incrementally varying donor and acceptor substituents is readily assessed. Also intriguing are "experiments" with multiple donors and acceptors as shown below. It is furthermore possible to probe the effects of polyene conjugation length on β_{vec} at differing

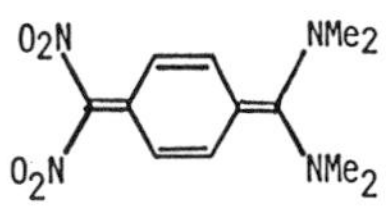

$\beta = 172$

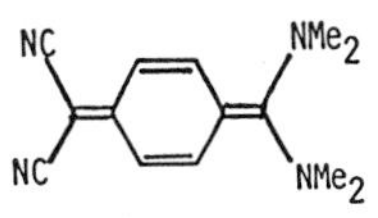

$\beta = 81$

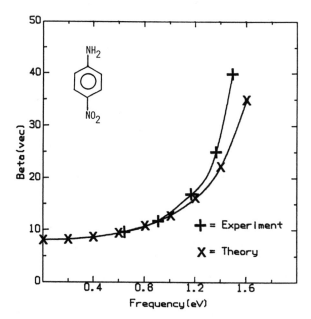

Fig. 2.   Plot of β_{vec} values in units of 10^{-30} cm^5 esu^{-1} at various frequencies calculated from the PPP model compared to experimental values from the literature (ref. 33).

$\beta = 267$

$\beta = 7$

$\beta = 57$

$\beta = 72$

$\beta = 71$

$\beta = 90$

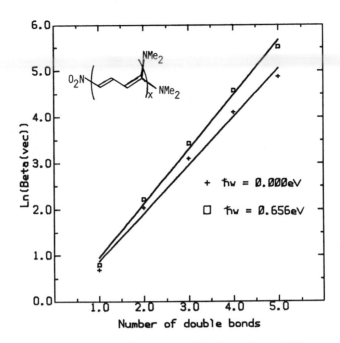

Fig. 3. Effect of conjugation length on PPP-
derived β_{vec} values for a donor,
acceptor-substituted polyene. Results
are shown at two frequencies. β_{vec} is in
units of 10^{-30} cm^5 esu^{-1}.

frequencies as shown in Figure 3. Although not discussed here, the present
PPP approach can be used to predict, with substantial accuracy, both the
linear optical spectral characteristics of a particular chromophore
(absorption maxima, oscillator strengths) and the dipole moment.[5,6] The
former information is crucial for anticipating laser/chromophore transparency characteristics while the latter provides insight into how the
chromophore will respond to a static, orienting electric field[13,14,15] used
in certain processing methodologies.

It has also been of interest to determine whether useful correlations
exist between NLO behavior and more conventional descriptors of organic
molecule π-electronic structure. The Hammett free-energy relationship
(eq.(6)) has long been used to relate thermodynamic and kinetic parameters

$$\log\frac{K}{K_0} = \sigma\rho \qquad (6)$$

to substituent effects in aromatic systems.[34] Here the K's are equilibrium
constants, σ's are substituent constants describing electron-donating or
accepting tendencies, and ρ describes the sensitivity of the process to
changes in substituents. The reference system is usually the ionization
constants of substituted benzoic acids where ρ is set equal to unity. We
have calculated β_{vec} for a series of para-substituted benzenes (IV) and the

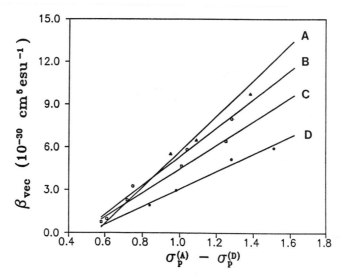

Fig. 4. PPP-derived β_{vec} values for <u>para</u>-substituted benzenes. Each line represents a series with fixed acceptor (A = NO_2, B = CHO, C = CO_2Me, D = CN) and donor varying in the order (right to left on the plot): NMe_2, NH_2, OMe, Me.

IV

results are related to σ values for <u>para</u>-substituents (σ_p) in Figure 4. It can be seen that for A held constant, there is a linear relationship between the difference in substituent constants and the NLO parameter β_{vec}.

These initial results indicate that perturbation-theoretic methods implemented within the framework of a semi-empirical π-electron PPP Hamiltonian should be extremely useful in understanding the NLO properties of known organic chromophores as well as in predicting the characteristics of new ones. Further results from this effort will be reported shortly.

CHROMOPHORE-FUNCTIONALIZED GLASSY POLYMERS

As noted above, the synthesis of a noncentric chromophore assembly for optimal SHG characteristics represents a considerable synthetic challenge. We have adopted a functionalized polymer approach (II above) to ensure maximum guest-host stability and compatibility. The ideal host polymer should be readily functionalizable to optimize synthetic flexibility, should have a relatively high glass transition temperature (T_g) to stabilize the assembly in a noncentric orientation subsequent to electric field poling, should have good optical characteristics (low crystallinity, λ_{max} at short wavelengths), should have good environmental stability, and should be easily processable into films, fibers, monoliths, etc. We have chosen polystyrene for our initial studies.

We find that polystyrene can be readily chloromethylated via the procedure shown in Scheme I.[7,8,35] This low temperature methodology circumvents the need for large quantities of chloromethylmethyl ether[36,37] and allows fairly extensive functionalization without detectable cross-linking. Reactions can be easily monitored by 400 MHz ^{1}H NMR. For a number of the functionalization reactions, it was found advantageous to convert the chloromethylated polymer to the corresponding, more reactive iodomethylated derivative[38] as shown in Scheme I. Two types of functional-ization pathways are shown in Scheme II. Thallium-mediated etherifica-tion[39] (Scheme IIA) can be used to introduce alcohols such as the well-characterized high-β chromophore, Disperse Red(**V**). Alternatively, a quaternization procedure (Scheme IIB) introduces another high-β chromophore (**VI**).[40,41] All of the new polymers were characterized by 400 MHz ^{1}H NMR spectroscopy, uv-visible spectroscopy, elemental analysis, and DSC. Most work to date has been carried out with polystyrene having $M_n \approx 22,000$ and ca. 7% chloromethylation. Measured T_g values are ~105-109°C for (PS)CH$_2$-**V** and ~103-107°C for (PS)CH$_2$-**VI**.

Scheme I

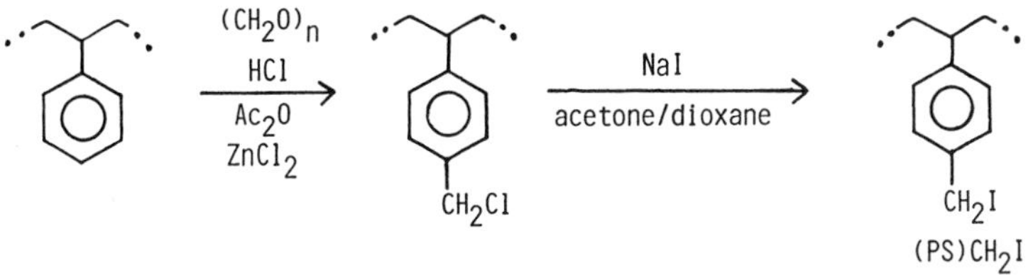

Scheme II

A. Alcohol Chromophores

$$ROH + TlOEt \xrightarrow{\text{benzene}} TlOR + EtOH$$

$$(PS)CH_2I + TlOR \xrightarrow{\text{DMF}} (PS)CH_2OR + TlI_{(s)}$$

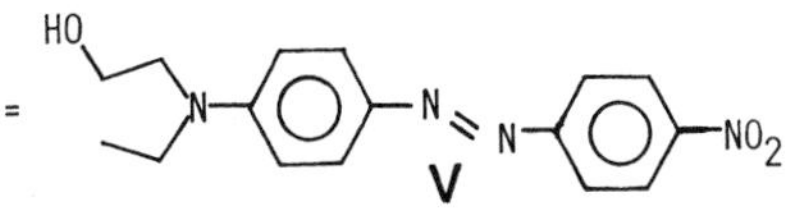

B. Pyridinium Chromophores

$$(PS)CH_2I + py \xrightarrow{\text{MeOH}} (PS)CH_2py^+I^-$$

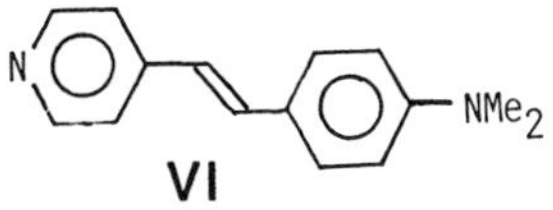

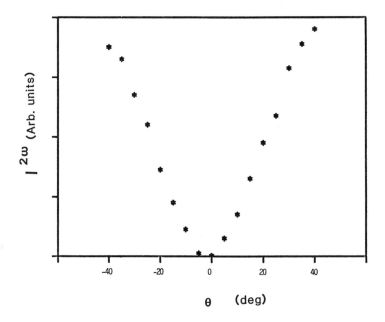

Fig. 5. Second harmonic intensity of p-polarized
fundamental (1.064 μm) and second
harmonic light versus angle of incidence
for a poled film of the chromophore-
functionalized polymer (PS)CH$_2$-**VI**).

Using dust-free conditions, multiply-filtered CH$_2$Cl$_2$ solutions of
(PS)CH$_2$-**V** and (PS)CH$_2$-**VI** have been spin-coated onto indium tin oxide (ITO)-
coated conducting glass to form thin films ($\leqslant$ 1 μm). Poling experiments on
these films were carried out at 115°C using a second piece of ITO glass as
the other electrode. A large electric field of 0.05-0.5 MV cm^{-1} was used
in the poling process.

The second harmonic coefficients of the films were measured in the p-
polarized geometry using the instrumentation described previously.[42]
Figure 5 shows the dependence of second harmonic intensity as a function of
incident angle for a typical sample. The angular dependence was analyzed
by assuming d$_{31}$ = d$_{21}$ = d$_{24}$ = d$_{15}$ = 1/3 d$_{33}$[13] to obtain d$_{33}$. By cali-
brating against d$_{11}$ of quartz, we obtained the preliminary results: d$_{33}$ =
2.7 x 10^{-9} esu for (PS)CH$_2$-**V** films poled at 0.3 MV cm^{-1} and d$_{33}$ = 1.4 x
10^{-9} esu for (PS)CH$_2$-**VI** films poled at 0.5 MV cm^{-1}. The small but non-
negligible absorption at 0.5320 μm of the thin (PS)CH$_2$-**V** films was
corrected for in the analysis using a procedure similar to that described
in reference 15. Detailed studies of d$_{33}$ as a function of poling conditions and fundamental wavelength are in progress.

Acknowledgements

We are grateful to the MRL program of the NSF and to the AFOSR for
support of this research (grants #DMR85-20280, AFOSR-86-0105). We thank
other members of the Northwestern Nonlinear Optics group, including J.
Torkelson and S. H. Carr, for helpful discussions.

REFERENCES

1. D. S. Chemla and J. Zyss, eds. "Nonlinear Optical Properties of Organic Molecules and Crystals," Vols. 1,2, Academic Press, New York (1987).
2. G. Khanarian, ed., "Molecular and Polymeric Optoelectronic Materials: Fundamentals and Applications," *SPIE*, 682 (1986).
3. D. J. Williams, *Angew. Chem. Intl. Ed. Engl.* 23:690 (1984).
4. D. J. Williams, ed., "Nonlinear Optical Properties of Organic and Polymeric Materials," American Chemical Society, Washington, 1983; Symp. Ser. No. 233.
5. D. Li, T. J. Marks, and M. A. Ratner, *Chem. Phys. Lett.* 131:370 (1986).
6. D. Li, T.J. Marks, and M. A. Ratner, submitted for publication.
7. C. Ye, T. J. Marks, J. Yang, and G. Wong, submitted for publication.
8. C. Ye, T. J. Marks, J. Yang, and G. Wong, manuscript in preparation.
9. A. F. Garito, K. D. Singer, and C. C. Teng in ref. 4, pp. 1-26.
10. F. Kajzar, J. Messier, J. Zyss, and I. Ledoux, *Optics Commun.*, 45:133 (1983).
11. A. F. Garito, K. Y. Wong, Y. M. Cai, H. T. Man, and O. Zamani-Khamiri in ref. 2, pp. 2-12.
12. P. N. Prasad in ref. 2, pp. 120-123.
13. K. D. Singer, J. E. Sohn, and S. J. Lalama, *Appl. Phys. Lett.*, 49:248 (1986).
14. K. D. Small, K. D. Singer, J. E. Sohn, M. G. Kuzyk, and S. J. Lalama in ref. 2., pp. 160-169.
15. G. R. Meredith, J. G. Van Dusen, and D. J. Williams, *Macromolecules*, 15:1385 (1982).
16. A. R. McGhie, G. F. Lipscomb, A. F. Garito, K. N. Desai, and P. S. Kalyanaraman, *Makromol. Chem.* 182:965 (1981).
17. S. L. Lalama and A. F. Garito, *Phys. Rev.* A20:1179 (1979).
18. A. F. Garito, C. C. Teng, K. Y. Wong, and O. Enmmankhamiri, *Mol. Cryst. Liq. Cryst.* 106:219 (1984).
19. V. J. Docherty, D. Pugh, and J. O. Morley, *J. Chem. Soc. Faraday Trans. II* 81:1179 (1985).
20. S. Allen, J. O. Morley, D. Pugh, and V. J. Docherty in ref. 2, pp. 20-26.
21. J. A. Morrell and A. C. Albrecht, *Chem. Phys. Lett.* 64:46 (1979).
22. J. A. Pople, *Trans. Faraday Soc.* 49:1375 (1953).
23. R. Parser and R. G. Parr, *J. Chem. Phys.* 21:466 (1953).
24. R. Pariser, *ibid.* 21:568 (1953).
25. J. Linderberg and Y. Öhrn, "Propagators in Quantum Chemistry," Academic Press, London, (1974).
26. J. Linderberg and Y. Öhrn, *J. Chem. Phys.* 49:716 (1967).
27. J. Koutecky, *J. Chem. Phys.* 47:1501 (1967).
28. P. Jørgensen and J. Linderberg, *Inter. J. Quantum Chem.* 4:587 (1970).
29. J. N. Murrell and A. J. Harget, "Semi-Empirical Self-Consistent-Field Molecular Orbital Theory of Molecules," Wiley, London (1972).
30. C. W. Dirk, R. J. Twieg, and G. Wagniere, *J. Am. Chem. Soc.*, 108:5387 (1986).
31. J. F. Ward, *Revs. Mod. Phys.* 37:1 (1965).
32. N. Bloembergen, H. Lotem, and R. T. Lynch, *Indian J. Pure Appl. Phys.*, 16:151 (1978).
33. C. C. Teng and A. F. Garito, *Phys. Rev. Lett.*, 50:350 (1983).
34. F. A. Carey and R. J. Sundberg, "Advanced Organic Chemistry," Second Ed., Part A, Plenum Press, New York (1984), pp. 179-190.
35. An adaptation of: *Jpn. Kokai Tokkyo Koho JP 59 89,307* (23 May 1984). Chem. Abs. 101:152574g (1985).
36. G. D. Jones, *Ind. Eng. Chem.* 44:2686 (1952).
37. R. G. Feinberg and R. B. Merrifield, *Tetrahedron*, 30:3209 (1974).
38. A modification of: V. V. Korshak, S. V. Rogozhin, and V. A. Duvankov, *Izvest. Akad. Nauk. SSSR Ser. Khim.* 1912 (1965).

39. General procedure: H.-O. Kalinowski, D. Seebach, and G. Crass, *Angew. Chem. Int. Ed. Engl.*, 14:762 (1975).
40. T. H. Lu, T. J. Lee, C. Wong, and K. T. Kuo, *J. Chin. Chem. Soc.*, 26:53 (1979).
41. A. N. Kost, A. K. Sheinkman, and A. N. Rozenberg, *J. Gen. Chem. USSR*, 34:4106 (1964).
42. S. J. Gu, S. K. Saha, and G. K. Wong, *Mol. Cryst. Liq. Cryst.*, 69:287 (1981).

POLED COPOLY(VINYLIDENE FLUORIDE-TRIFLUOROETHYLENE) AS A HOST FOR
GUEST NONLINEAR OPTICAL MOLECULES

Philip Pantelis, Julian R. Hill and Graham J. Davies

British Telecom Research Laboratories
Martlesham Heath, IPSWICH, IP5 7RE, U.K.

Optical frequency doubling has been observed in a partly
crystalline copolymer of vinylidene fluoride and trifluoroethylene
in which an optically nonlinear guest was dissolved. The nonlinear
behaviour was induced in the composite by a room temperature corona
poling process which both aligned the guest and caused cooperative
alignment of the chains forming the crystalline regions of the host
polymer. These ordered crystallites created stable electric fields
within the film which were spectrophotometrically measured to be as
high as 1.5×10^8 V/m, which was sufficient to maintain the partial
alignment of the guest.   The angular dependance of second harmonic
radiation produced in these poled films was analysed by the method
of Maker. A composite containing 10% of an aminoazo compound
exhibited a second harmonic coefficient of 6.1 $\pm$0.3 $\times 10^{-9}$ esu.
Thermal annealing and the incorporation of other guests resulted in
composites with larger nonlinear coefficients.

INTRODUCTION

Since the publication four years ago of the ACS Symposium Series
book on nonlinear optical properties of organic and polymeric
materials [1], there has been an increased awareness and interest in the
use of organic materials for a number of signal processing applications.
Polymeric materials provide an alternative to single crystals as their
advantages lie in their variety and ease of fabrication, both at a
molecular level and in device structures [2]. A noncentrosymmetric
ordering of the optically nonlinear moieties is a prerequisite for
accessing second order nonlinear effects in these materials.   Williams
has reviewed approaches based on guest-host polymer structures and has
in particular addressed electrical field induced alignment (poling) of
liquid crystal polymers[3][4]. Recently, thermal poling and simultaneous
freezing-in of the induced alignment of guest molecules in a glassy
polymer host has been demonstrated [5].

In this paper, we describe a new composite which comprises of a
ferroelectric host in which an optically nonlinear low molecular weight
guest is dissolved. This guest is subsequently aligned using a room
temperature corona poling process [6].

The polymer studied was Foraflon-7030 (Atochem UK Ltd). It is a copolymer of vinylidene fluoride and trifluoroethylene in the ratio 70:30 mol percent. It has advantages over the homopolymer PVDF as a host, as it can be easily and conveniently solvent cast to give thin transparent films which possess good optical, as well as electrical quality. The copolymer is partially crystalline but importantly, unlike the homopolymer, it does not need stretching to promote its crystalline regions to be in the non-centrosymmetric crystal form required for promoting a piezoelectric response [7][8].

The manifestation of ferroelectric behaviour on poling is the key to our new host system. The intense corona poling field causes the carbon-fluorine dipoles of the molecular chains in the crystalline regions to be permanently brought into alignment in the direction of the applied field. This imparts piezoelectric and pyroelectric properties to the whole film, which after an initial decay in the first day, remain stable over a period of years [9]. The poled polymer is also weakly optically nonlinear [10] because of the slight hyperpolariziblity of the sigma-electrons forming the carbon-fluorine bonds. Guest molecules selected for their highly nonlinear properties residing in the amorphous region of the host would also align with the intense electric field caused by corona poling. When this applied field is removed there still remains within the composite a permanent internal electric field caused by the ordered crystalline regions. It is this field, we propose, will maintain the alignment of the guest molecules and cause the composite to remain as a stable optically nonlinear material.

There is, however, no direct knowledge of the exact magnitude of this internal field in the amorphous region of the poled polymer. A rough estimate of it can be inferred from the coercive fields obtained from electrical displacement versus applied electric field hysteresis curves. For thin PVDF copolymer films, coercive fields as high as 125MV/m have been recorded [11].

In the initial part of this paper we describe spectrophotometric experiments which directly determine the average of the internal fields within Onsager cavities in the poled amorphous region by using probe molecules. Later, we describe the analysis of Maker fringe experiments to measure the second harmonic coefficient (d_{33}) of poled PVDF copolymer composites containing aminoazobenzene guest molecules.

FILM FORMING

Foraflon-7030 together with the guest compound of interest was dissolved in acetone to give a solution containing up to 15% total solids by weight. These solutions were cast as films onto glass at 30°C under dry nitrogen using an ICI design automatic film spreader (Sheen Instruments). The dried films of about 8μm in thickness were either electrically poled in this 'as cast' state or thermally annealed and then poled.

GUEST MATERIALS

A number of low molecular weight guest materials have been investigated and the properties of two well characterised compounds are listed in Table 1.

Table 1. Properties of Nonlinear Optical Guest Molecules

Compound	Melting point(°C)	Refractive index n_D	Ground-state dipole moment (Debye)	β (10^{-30} esu) at 1.064μm
(I)	101-2	1.647	7.5 (a)	~16
(II)	153-4	1.674	9.3 (a)	360 (b)

(a) Measured by A.H. Price, University College of Wales, Aberystwyth.
(b) Measured by G.R. Meredith, du Pont Central R & D Dept. Wilmington.

We have concentrated our nonlinear optical investigation on 4-(4'-cyanophenylazo)-NN-bis-(methylcarbonylmethyl)-aniline(II) because of its high molecular hyperpolarisablity relative to methyl-N-(4'nitrophenyl)-2-aminopropionate (I).

(I)   (II)

These compounds were designed to have a high solubility in the host matrix by the incorporation of ester groups and to have their ground state and transition dipole moments in approximately the same direction. This combination of properties made them suitable as probe molecules for estimation of internal fields in the host polymer and also as guests for nonlinear optical studies.

CORONA POLING

In the poling process, the polymer film is supported on a PTFE mount between two tungsten wires of diameter 0.1mm, which are each spaced approximately 1cm from the film surfaces. These wires are held at +10kV and -10kV respectively creating corona discharges. This sets up a large electric field across the film causing the dipolar alignment of the various molecular species in the polymer. The poling process is effectively completed within 15 seconds.

LINEAR OPTICAL AND WAVEGUIDING PROPERTIES

The refractive index and dispersive properties of the composites were obtained using an Abbé refractometer. Film thicknesses were accurately determined by analysis of the optical interference fringes of the films in the visible and near infrared [12].

The bulk optical loss for the pure copolymer throughout the visible and near infrared spectrum was determined spectrophotometrically and is shown in Figure 1. The main source of loss in the visible is that due to scatter rather than absorption. In the near infrared there are two optical 'windows' at the wavelengths of greatest interest for telecommunications (1.3 and 1.55μm) where again the predominant loss is due to scatter. For the majority of guest species, additional absorption losses in these near infrared windows is negligible.

Waveguiding structures composed of a 5μm overlay of a guest/host composite on a pure polymer base layer have also been prepared. The total optical loss of these waveguides at 633nm has been found to be typically 5 to 6dB/cm, which is in good agreement with the measured bulk losses of about 4dB/cm. Total waveguide loss in the near infrared windows is expected to be less than these values since the bulk optical loss in these regions is only about 1.5dB/cm.

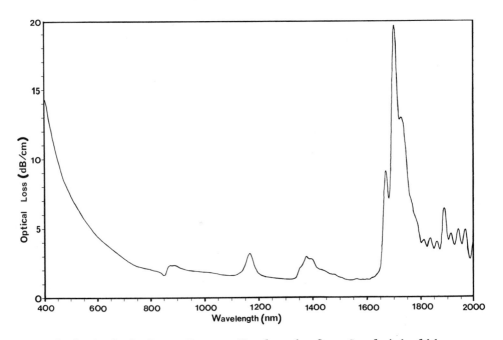

Fig.1. Optical Loss Versus Wavelength for Copoly(vinylidene fluoride - trifluoroethylene)

DETERMINATION OF INTERNAL FIELDS

From experiments in electric field induced second harmonic generation the relationship between the electric field used to align the optically nonlinear species and the observed nonlinear coefficient is known [13]. Prior to quantitative nonlinear optical measurements, we spectrophotometrically determined the internal field in our composites to predict the magnitude of the nonlinear coefficients.

The theoretical relationship between the field (Fi) within an Onsager cavity within the amorphous regions of the host matrix in which the guest resides and the contribution of this guest to the nonlinear second harmonic coefficient (d_{33}) of the composite is given by Equation(1).

$$d_{33} = N. f_\omega^2 . f_{2\omega} . \beta . \mu . Fi/(10.k.T)   \qquad EQUATION(1)$$

Where, (N) is the number density of nonlinear guest molecules, (β) is the microscopic nonlinear coefficient of each guest molecule vectored onto its ground state dipole moment and f_ω, $f_{2\omega}$ are Lorenz-Lorenz local field factors of the form $(\epsilon+2)/3$ [14]. The value of ϵ has been taken as the square of the refractive index of the composite at either the fundamental or second harmonic frequency. From this expression it follows that the second harmonic coefficient should be linear with both internal field and guest concentration.

The average internal electric field was determined by an analysis of the changes in the visible absorbance spectra of the composites. Films about $8\mu m$ thick were prepared from Foraflon-7030 containing either 0.6% (w/w) of (I) or 0.4% (w/w) of (II). Their absorbance was measured by a Perkin-Elmer Lambda-9 UV/VIS/NIR spectrophotometer and recorded on an integral PE3600 data station. Ten of the samples were kept as an experimental control while ten were corona poled at room temperature. Remeasurement of the absorbance spectra of the films after poling showed a decrease in their absorbance due to a partial dipolar alignment of the guest.

The ratio of the absorbance prior to poling (Ao) and that observed after poling (Ap) was analysed using Equation(2) which was derived by Havinga and Van Pelt [15] for polymers above their glass transition temperature.

$$Ap/Ao = 1-(3 Cos^2(\o)-1)(1-3Coth(x)/x + 3/x^2)/2   \qquad EQUATION(2)$$

Where
$\qquad$ $x = \mu.Fi/(k.T)$.
$\qquad$ μ = Ground state dipole moment.
$\qquad$ Fi = The electric field in an Onsager cavity containing the guest.
$\qquad$ k = Boltzmann's constant.
$\qquad$ T = Absolute temperature.
$\qquad$ $\o$ = The angle between the ground state and transition dipole moments for the guest.

233

This expression provides an average value for the field acting on the guest molecules and is strictly only valid when the ratio Ap/Ao changes linearly with field strength, approximately over the range x<5.

Equation (2) does not differentiate between those internal fields which are parallel to the bulk remanent polarisation of the sample and those which run anti-parallel. The field value obtained is therefore an average of field intensity which neglects the sign of the various local fields. If anti-parallel dipolar alignments are present then these predictions will over-estimate the bulk nonlinear optical coefficient of the composite.

Using the PPP molecular orbital method [16] we have calculated the angle (ø) between the ground state dipole moment and the transition dipole moment to be approximately zero for (I) and 4.8 degrees for (II). The error due to the use of a calculated value for this angle is small since Cos(ø) tends to unity for small angles.

Figure 2. shows the changes in the spectrum of a sample containing the aminoazo guest. Immediately after corona poling the absorbance ratio Ap/Ao was small but this ratio increased with time to converge to a constant value. The spectra for the poled aging samples were taken at 20 minute intervals. It is interesting to note that in addition to a decrease in the absorption intensity of the sample, there was also observed a slight spectral shift and broadening. This can be ascribed to a field induced change in the position of the energy levels of the guest [15].

After an initial period of 24 hours, the absorbance of the films showed no further change over a period of several weeks. This suggests that fields in the polymer due to unstable states such as trapped charges decay rapidly, leaving only the internal fields due to permanent crystallite polarisation. Our value for this stabilised field is given in Table 2.

The degree of crystallinity in our dried, as cast, pure polymer is low and has been estimated to be 18% from its density of $1.83g/cm^3$ [17]. An increase in crystallinity occurs with thermal annealing prior to poling [18]. Ten films containing (II) were annealed at 80°C for 30 minutes prior to measuring their initial absorption spectrum and their subsequent corona poling. As shown in Table 2. this annealing process leads to an increase in their measured internal fields.

TABLE 2. Onsager Cavity Field Obtained from Absorbance Changes on Poling Guest/Host Composites.

SAMPLE	GUEST	Ap/Ao	Fi (10^8 V/m)
As cast	(I)	0.970	1.10 ± 0.1
As cast	(II)	0.954	1.12 ± 0.1
Annealed	(II)	0.918	1.54 ± 0.2

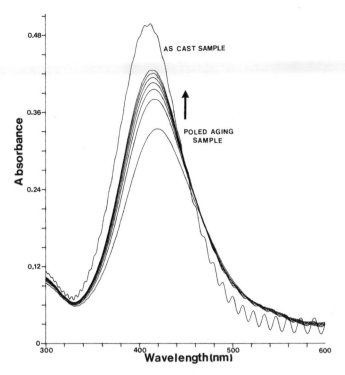

Fig.2. Changes in the Absorbance Spectrum of a Guest/Host Composite on Poling.

The large internal fields measured by this method suggested that the second harmonic coefficient (d_{33}) for non-annealed poled films containing (II) could be as high as 1.9×10^{-10} esu per percent of guest. Annealing the composite prior to poling increases the internal field by a factor of 1.38 and from Equation (1) this should result in a corresponding increase in the optical nonlinearity due to the guest. As discussed later, these results compare well with our experimentally determined nonlinear coefficients.

NONLINEAR OPTICAL MEASUREMENTS

Nonlinear optical measurements were performed on the composites at least a day after poling allowing time for unstable internal electric fields in the samples to fully decay. The second harmonic coefficients of the poled composites were determined by Maker fringe analysis. These measurements were made relative to a Y-cut quartz plate ($d_{11} = 1.1 \times 10^{-9}$ esu). In this technique, a plane parallel sample of the nonlinear optical material is rotated in the path of a fundamental beam of wavelength 1.064μm from a Q-switched Nd/YAG laser. For samples with thicknesses of millimeters, the power of the 532nm radiation produced by optical frequency doubling in the film is found to oscillate in a periodic fashion as the incidence angle is varied. Analysis of the periodicity and intensity of these fringes provides information on the coherence length of the radiation within the sample and also on the value and tensorial nature of the second harmonic coefficient. For films a few microns thick, only one maximum of the second harmonic is observed when the incidence angle is varied from zero to 90°.

Theoretical analysis of Maker fringes has shown the second harmonic power in uniaxial materials [19] and aligned polymers [5] to be given by Equation(3).

$$P_{2\omega} = \frac{512\pi^3.d^2.t_\omega^4.T_{2\omega}.p(\emptyset)^2.P_\omega^2.Sin^2(\psi)}{a.c.[n_\omega^2 - n_{2\omega}^2]^2}$$ EQUATION(3)

Where

$$\psi = 2.\pi.L.(n_\omega.Cos(\emptyset_\omega) - n_{2\omega}.Cos(\emptyset_{2\omega}))/\lambda$$

and where

$P_{2\omega}$ = the transmitted second harmonic power.
P_ω = the incident fundamental power.
c = the speed of light in a vacuum.
a = the area of the beam.
d = the nonlinear coefficient (d_{33} for our films).
L = the sample thickness.
n_ω = the refractive index at the fundamental.
$n_{2\omega}$ = the refractive index at the second harmonic.
$\emptyset$ = the angle of incidence.

The transmission factors t_ω and $T_{2\omega}$ are functions of n_ω, $n_{2\omega}$ and ($\emptyset$) and are defined in the literature [19]. Projection of the incident laser light onto the film nonlinear polarisation components gives p($\emptyset$) as

$$p(\emptyset) = (Cos^2(\emptyset_\omega)/3 + Sin^2(\emptyset_\omega)).Sin(\emptyset_{2\omega}) +$$ EQUATION(4)

$$(2/3).Cos(\emptyset_\omega).Sin(\emptyset_\omega).Cos(\emptyset_{2\omega})$$

Figure 3. gives the experimental points showing the variation of the second harmonic intensity with incidence angle for an 8μm thick composite containing 6% by weight of (II). The line drawn gives the theoretical power predicted by Equation(3).

236

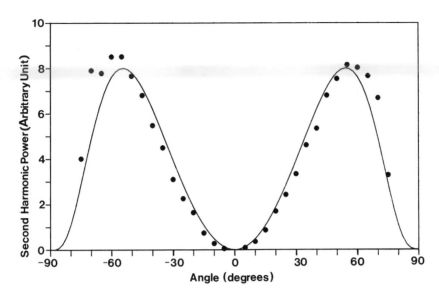

Fig.3. Second Harmonic Power Versus Incidence Angle for a Poled Polymer Film Containing 6% by Weight of Guest (<u>II</u>).

Frequent oscillation of the second harmonic power generated with incidence angle is not observed in thin films. Therefore the gradual change in second harmonic power with angle in our samples enabled simple nonlinear measurements on them to be made at a fixed angle. This was experimentally convenient as only one power measurement was needed from each sample to obtain the second harmonic coefficient. The ensuing reduction in accuracy was acceptable in comparative experiments.

Samples containing one to four percent of (<u>II</u>) were placed at a fixed angle of 45° to the incident beam and the second harmonic power produced from them was measured. The resulting relationship between power and concentration had a quadratic form as was expected from Equations (1) and (3). This relationship is shown in Figure 4. which gives the average powers obtained from six samples at each concentration.

The second harmonic output passed through zero as the concentration was raised and then increased rapidly and continuously with further additions of the guest. This can be interpreted in terms of the nonlinear coefficient of the composite having a constant negative contribution from the host polymer together with a positive contribution from the guest material which was linear with concentration. As the concentration of the guest was raised there was a threshold point where these two effects cancelled to give a zero nonlinear optical coefficient. The low concentration at which this effect occured demonstrated the greater hyperpolarisability of the guest molecules compared to the host.

This interpretation is supported by the observation of a negative molecular hyperpolarisability for difluoromethane [20] where back donation of an electron from the fluorine atom to an antibonding sigma-orbital occurs on electronic excitation [21]. The sign of the molecular hyperpolarisability of (<u>II</u>) should however be positive [22].

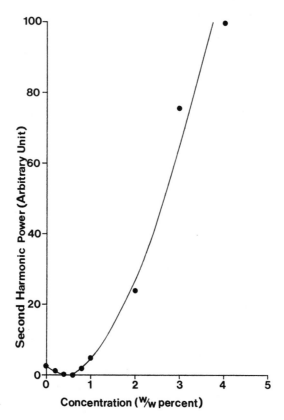

Fig.4. Second Harmonic Intensity at 45° Versus Guest (II) Concentration.

The complete angular dependent analysis of our composites using Equation(3) over the range $-55° \leqslant \emptyset \leqslant 55°$ gave a value for d_{33} but did not reveal its absolute sign. Taking the value of d_{33} below the threshold concentration to be negative and all values above this point to be positive, a linear experimental relationship between d_{33} and concentration of (II) was obtained. This is shown in Figure 5.

The value of d_{33} for the pure host polymer is obtained from the intercept of the line with the ordinate. This value of $d_{33} = -4.1 \pm 2 \times 10^{-10}$ esu is the first quantitative value for a PVDF system. Earlier work on homopolymer PVDF only stated that the second harmonic coefficient was comparable to quartz [10]. The gradient gives the contribution of the guest material to the coefficient of the composite as $6.5 \pm 0.3 \times 10^{-10}$ esu per percent of (II) by weight. This is lower than the value calculated by Equation(1) using the measured internal field in the matrix. However, the prediction of the nonlinear coefficient to within an order of magnitude is satisfactory because of the complexity of the film morphology and consequent uncertainty in parameters required for the theoretical analysis.

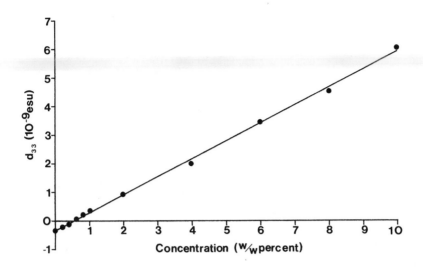

Fig.5. Second Harmonic Coefficient Versus Guest (<u>II</u>) Concentration.

The effect of thermal annealing on the nonlinear behaviour of composites containing 2% of (<u>II</u>) was investigated at a fixed incidence angle of 45°. It was found that samples annealed prior to poling exhibited second harmonic coefficients which were on average 1.4 times greater than their equivalent non-annealed composites. This is in good agreement with the value of 1.38 as predicted from the increased internal field measured in annealed samples.

The nonlinear behaviour of the composite is dominated by the guest and so enhancement of the nonlinear coefficient may be readily effected by substitution of an alternative guest compound. For example, substitution of 5% of (<u>II</u>) by 5% of 4-amino-NN-bis- -(methylcarbonylmethyl)-4'-nitro-trans-stilbene (<u>III</u>) increased the second harmonic coefficient observed from the composites by an average ratio of 1.3 in fixed 45° angle experiments.

(<u>III</u>)

This enhancement is due to the greater product of dipole moment and molecular hyperpolarisability for the stilbene derivative relative to the aminoazo compound. Other molecules with a further increase in this $\beta.\mu$ product are under current investigation. These compounds are also designed to have enhanced solubility in our copolymer system.

CONCLUSION

We have demonstrated for the first time, second harmonic generation from a poled composite composed of a nonlinear guest dissolved in a copolymer of vinylidene fluoride and trifluoroethylene. Nonlinear properties of the composite are dependent on internal electric fields in the polymer which retain the alignment of the guest compound. Guest materials have been used to probe the Onsager cavity field in the amorphous region of the poled polymer and have confirmed the presence of a stable electric field as high as 1.5×10^8 V/m. The second harmonic coefficient for the pure host polymer has been found to be about -4.1×10^{-10} esu whereas the coefficient for composites containing 10% of an aminoazo compound has been measured as $6.1 \pm 0.3 \times 10^{-9}$ esu. The nonlinear behaviour of the composite is dominated by the guest and so enhancement of the nonlinear coefficient may be readily effected by substitution of an alternative guest compound. The optical loss of a waveguide constructed from the composite has been found to be 5 ± 0.5 dB/cm at 633nm. Losses at the longer wavelengths used in telecommunications are expected to be significantly lower.

ACKNOWLEDGMENT

Acknowledgment is made to the Director of British Telecom Research Laboratories for permission to publish this work, which was carried out with the support of the UK Department of Trade and Industry under the JOERS programme. The authors would also like to thank their colleagues P.L.Dunn and J.D.Rush for their laser measurements, S.N.Oliver for chemical synthesis of the guests and R.Heckingbottom for useful discussions.

REFERENCES

1. "Nonlinear optical properties of organic materials and devices", D.J. Williams, ed., (A.C.S.Symposium Series 233) , Washington,D.C., (1983)

2. G.J.Bjorklund, R.W.Boyd, G.M.Carter, A.F.Garito, R.S. Lytel, G.R. Meredith, P.N.Prasad, J.Stamatoff and M.Thakur, _Appl. Optics_, 26:227, (1987).

3. D.J. Williams, _Angew. Chem. Int. Ed. Engl._, 23:690, (1984).

4. D.J. Williams, Nonlinear Optical Properties of Guest-Host Polymer Structures _in_ "Nonlinear Optical Properties of Organic Molecules and Crystals", vol 1, D.S.Chemla, J.Zyss eds., Academic Press, Orlando, In Press.

5. J.E. Sohn, K.D.Singer, S.J.Lalama and M.G.Kuzk, _Polymeric Materials Science and Engineering_, 55:532, (1986).

6. P. Pantelis and G.J. Davies, Patent GB Appl.84/31682.

7. A.J. Lovinger, _Jap.J.Appl.Phys_, 24, suppl 24-2:18,(1985).

8. H. Ohigashi, _Jap.J.Appl.Phys_, 24, suppl 24-2:23,(1985).

9. P. Pantelis, _Phys.Technol._, 15:239, (1984).

10. J.H. McFee, J.G. Bergman and G.R. Crane, _Ferroelectrics_, 3:305, (1972).

11. K. Kimura and H. Ohigashi, _Jap.J.Appl.Phys._, 25:383, (1986).

12. J.C. Manifacier, J. Gasiot and J.P. Fillard, _J.Phys.Sci.Inst._, 9:1002, (1976).

13. B.F. Levine and C.G. Bethea, _Appl.Phys.Lett._, 24:445, (1974).

14. J.L. Oudar, _J.Chem.Phys._, 67:446, (1977).

15. E.E. Havinga and P. van Pelt, _Ber.Bunsenges.Phys.Chem._, 83:816, (1979).

16. J. Griffiths, "Colour and Constitution of Organic Molecules", Academic Press, London, NY, San Francisco, (1976).

17. M. Jimbo, T. Fukada, H. Takeda, F. Suzuki, K. Horino, K. Koyama, S. Ikeda and Y. Wada , _J. Polym. Sci., Polym. Phys._, 24:909, (1986).

18. K. Koga and H. Ohigashi, _J.Appl.Phys._, 59:2142, (1986).

19. J. Jerphagnon and S.K. Kurtz, _J.Appl.Phys._ , 41:1667, (1970).

20. G.R. Meredith, Prospect of New Nonlinear Organic Materials, _in_ "Nonlinear Optics: Materials and Devices", C. Flytzanis and J.L. Oudar eds., Springer-Verlag,Berlin, Heidelberg, NY, Tokyo, (1986).

21. J.A. Barltrop and J.D. Coyle, "Excited States in Organic Chemistry", John Wiley & Sons, London, N.Y., Sydney, Toronto, (1975).

22. J.L. Oudar and H. Le Person, _Optics Commun._, 15:258, (1975).

ELECTROACTIVE POLYMERS: CONSEQUENCES OF ELECTRON DELOCALIZATION

Larry R. Dalton

Department of Chemistry
University of Southern California
Los Angeles, California 90089-0482

INTRODUCTION

Linear polyenes and heteroaromatic polymers (which we shall hereafter refer to collectively as delocalized electron polymers) have received considerable attention during the past decade because of their interesting electrical conductivity properties.[1] As summarized in Fig. 1, pristine polymers routinely exhibit electrical conductivities typical of insulators or semiconductors. However, upon doping by chemical, electrochemical, or ion implantation methods, conductivities are observed to increase to metallic levels. Such electrical conductivity behavior has attracted theoretical attention because of the contribution of solitonic and polaronic species to electrical conductivity.[2-8] In order to verify theoretical predictions, considerable experimental effort has been focused upon definition of "intrinsic" domain-wall defect (highest occupied molecular orbital) delocalization and upon measuring (soliton/polaron) diffusion, including both intra- and intermolecular contributions to the diffusion coefficient. Substantial insight has been obtained concerning the detailed mechanisms of electrical conductivity, but questions remain, motivating the search for materials which will provide additional insight as well as practical utility.

More recently, significant nonlinear optical (NLO) activity[9-12] has been observed for delocalized electron polymers. Simple theoretical calculations[13,14] suggest that such activity may be strongly influenced by π-electron delocalization with the third order susceptibility, $\chi^{(3)}$, proportional to the sixth power of the delocalization length. Another important feature of these materials is that nonlinear optical activity is electronic in origin, so that the NLO switching times, τ, are typically less than a picosecond. Indeed, the figure of merit (defined as $\chi^{(3)}/\alpha\tau$, where α is the linear optical absorption coefficient) for delocalized electron polymers is comparable to that of the best inorganic materials. Although magnitudes of $\chi^{(3)}$ for electroactive polymers are currently too small for applications such as receiver protection and for the variety of components necessary for optical information processing, many avenues exist for improving the magnitude of $\chi^{(3)}$.

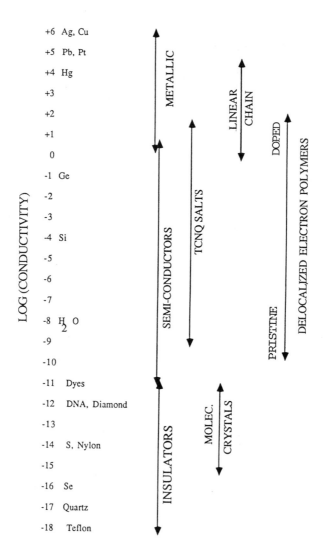

Figure 1. Electrical conductivity behavior for various materials

Another important consequence of electron delocalization is the effect upon physical properties such as solubility, mechanical strength, and thermal stability. Until recently, electroactive polymers have typically been prepared as insoluble films or powders. As such they cannot be effectively purified, characterized, or processed. At the International Conference on Ultrastructure in Organic and Inorganic Polymers, Amherst, October 21-25, 1985, we demonstrated that improved solubility could be achieved by derivatization which destabilized polymer-polymer interaction while enhancing polymer-solvent interaction. This route to soluble polymers, which can be effectively purified, characterized, and processed, has been pursued by ourselves[15-17] and by others.[18-21] The use of selective derivatization has complemented the use of soluble precursor polymers[22,23] and of charge-transfer solvents[24] to achieve successful processing of polymers. The strong advantage of derivatization is that lattice contamination by elimination products from the soluble precursor polymer is avoided and the electroactive polymer can be characterized and processed by standard techniques. The solvent dependence of optical properties can, for example, be conveniently investigated.

Improved understanding of the consequences of electron delocalization upon electrical, optical, magnetic, and physical properties (see Figure 2) has set the stage for the rapid development and application of electroactive polymers. In the following, we present an overview of our efforts toward this goal. Rather than undertake a comprehensive discussion of results for each polymer system, we shall focus upon selected findings which illustrate important concepts, particularly those useful in guiding future efforts. Moreover, we note that we have restricted our research to polymers capable of supporting solitonic or polaronic species.

We divide the following presentation into three sections: Polymer synthesis, characterization of electron intrinsic and dynamic electron delocalization, and discussion of electroactivity. In the section on synthesis, we review various methods for synthesizing polyacetylene. In the subsequent section on characterization, we discuss the consequences of various synthetic approaches. We then turn our attention to the preparation of soluble heteroaromatic polymers by derivatization. The preparation of fully cyclized ladder polymers via open-chain intermediates is discussed, as is the preparation of oligomer and polymer chains of defined length by sequential synthesis techniques.

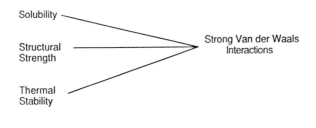

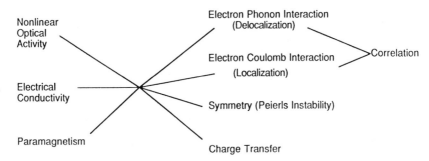

Fig. 2. Properties and principles relevant to delocalized electron polymers

The section on characterization of electron delocalization focuses upon the use of electron nuclear double resonance (ENDOR) and electron spin echo (ESE) spectroscopic techniques to define both intrinsic and dynamic electron delocalization. Intrinsic electron delocalization can be thought of as the extent of delocalization observed in the low temperature limit. Data presented in this section emphasize the importance of interchain mechanisms and of variable range hopping as a contributor to observed electrical conductivity. Intrinsic electron delocalization is considered as a function of polymer or oligomer length and as a function of polymer substituents.

The final section attempts to make predictions of nonlinear optical susceptibilities and of electrical properties based upon the data of the preceding section. We conclude with a discussion of promising future research directions.

POLYMER SYNTHESIS

Trans-polyacetylene prepared by the method of Shirakawa[25] (see Figure 3) is insoluble and intractable. Moreover, the fibril polymer lattice is contaminated by the incorporation of residual catalysis, crosslinks (including peroxide crosslinks), and cis-polyacetylene segments which are

246

incapable of supporting solitons. To address the problems associated with polymer insolubility, Feast[26] developed a route to trans-polyacetylene that included a soluble polymer precursor (see Figure 4).

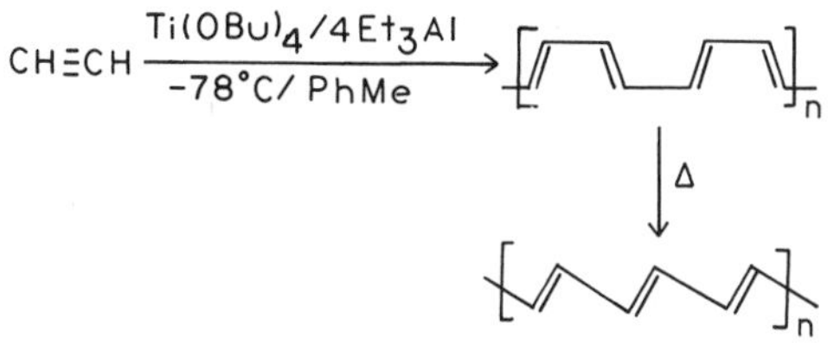

Fig. 3. Shirakawa synthesis of polyacetylene

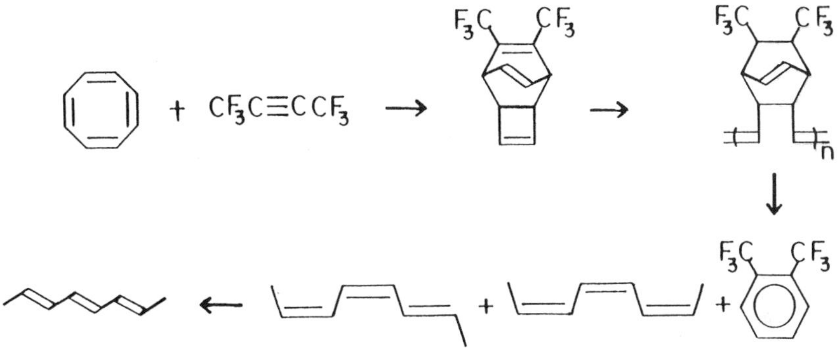

Fig. 4. Feast (or Durham) synthesis of polyacetylene

We have synthesized polyacetylene by both the Shirakawa[25] and Feast[26] methods. As will be discussed in the next section, we have characterized both intrinsic and dynamic electron (highest occupied molecular orbital) delocalization for these materials. Because problems of lattice contamination remain even for polyacetylene prepared by the soluble precursor route and because of the intrinsic atmospheric instability of polyacetylenes, we have more recently focused our attention upon heteroaromatic ladder polymers. Ladder polymers are characterized by excellent thermal and chemical stability and these materials frequently exhibit exceptional mechanical strength. Indeed, it was these properties that motivated the original synthetic interest in these polymers by Stille[27-29] and by Arnold and Van Deusen[30-33]. Our synthetic efforts can be viewed as a minor but important modification of the general synthetic procedure developed by these workers. In particular, our efforts focus upon the preparation of derivatized monomers and upon the investigation of the effects of solvent and temperature upon the rates of the various condensation reactions.

The general scheme for the production of high symmetry ladder polymers by the condensation of derivatized quinones with aromatic di- and tetra-substituted amines is depicted in Figure 5. Clearly, this is a very versatile approach, affording even the opportunity of preparing copolymers incorporating metallomacrocycles.

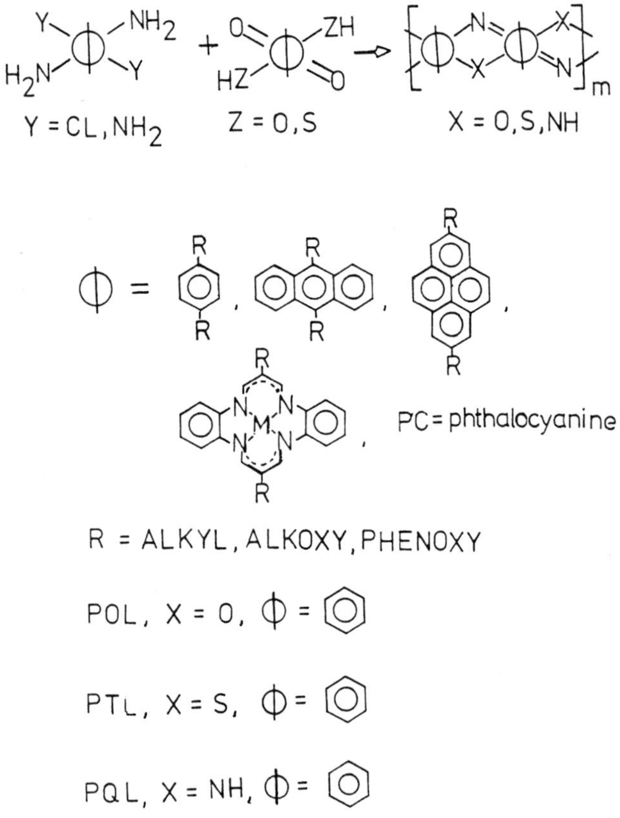

Fig. 5. General synthetic scheme of preparation of electroactive heteroaromatic polymers

Our major activity has been the preparation of derivatized monomers to be used in the preparation of polymers or oligomers. Alkyl derivatization has been accomplished by adaptation of the literature reactions, e.g.[34].

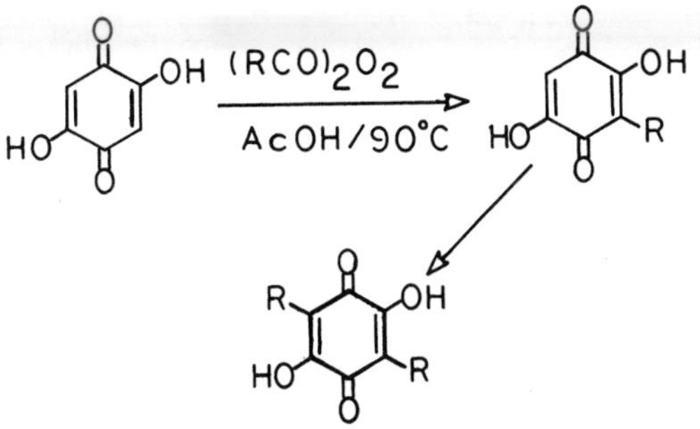

Fig. 6. Synthesis of mono and disubstituted alkylated monomers. The reaction shown proceeds by a free radical mechanism.

Alkylated monomers have also been prepared according to the scheme, Oxy derivatives have been prepared according to [35].

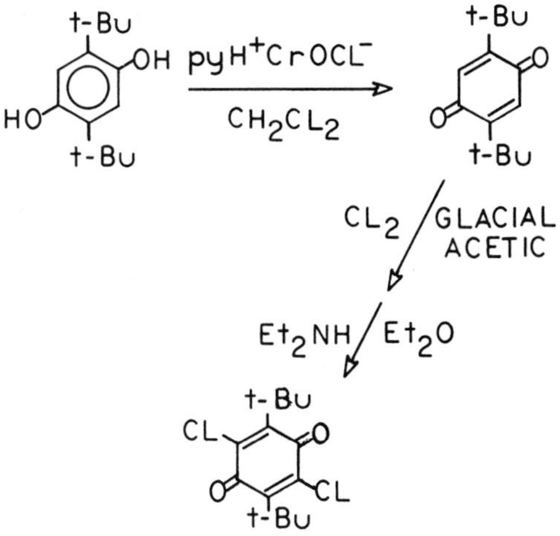

Fig. 7. Synthesis of the di-t-Butyl derivatized dichloroquinone monomer.

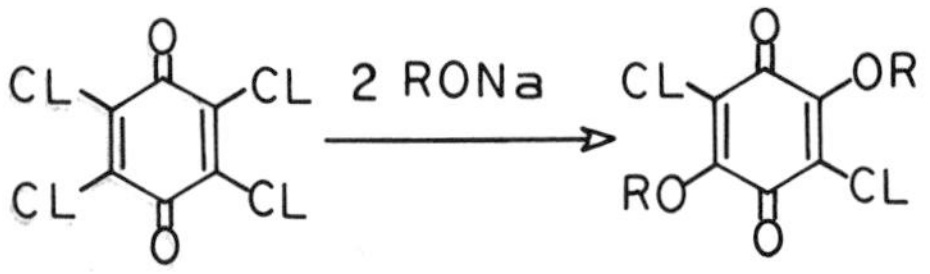

Fig. 8. Synthesis of the di-alkoxy derivatized dichloroquinone monomer.

In like manner, large ring systems can be prepared. For example, the alkylated anthacene analog can be prepared according to[36]

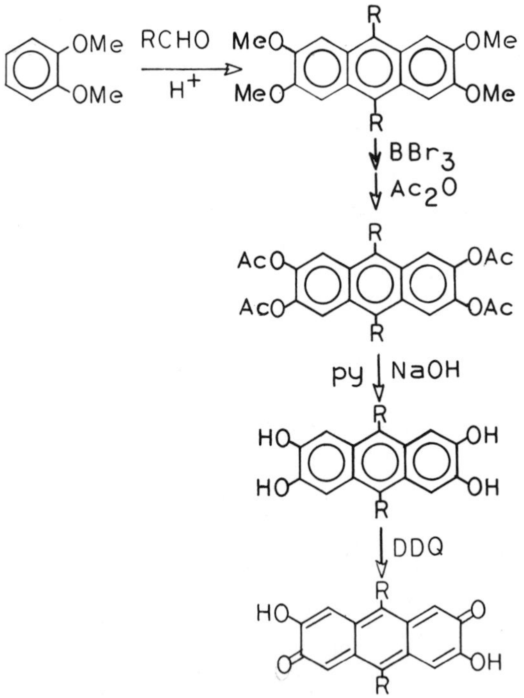

Fig. 9. Synthesis of the alkylated anthacene analog of the dihydroxyquinone monomer.

while 4,5,9,10-pyrenetetrone analogs can be prepared in several steps by ozonolysis of the appropriate pyrene starting material, e.g.,[37]

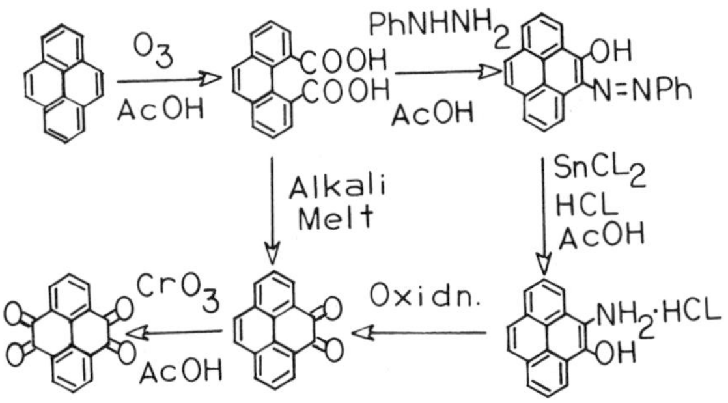

Fig. 10. Synthesis of 4,5,9,10-pyrenetetronone.

An important aspect of the synthesis of ladder polymers is that by adjustment of reaction conditions, open-chain precursors of the fully cyclized ladder polymers can be prepared. For example, reaction in polyphosphoric acid (PPA) at elevated temperature (e.g., 260°C) yields the fully cyclized PQL ladder polymer according to

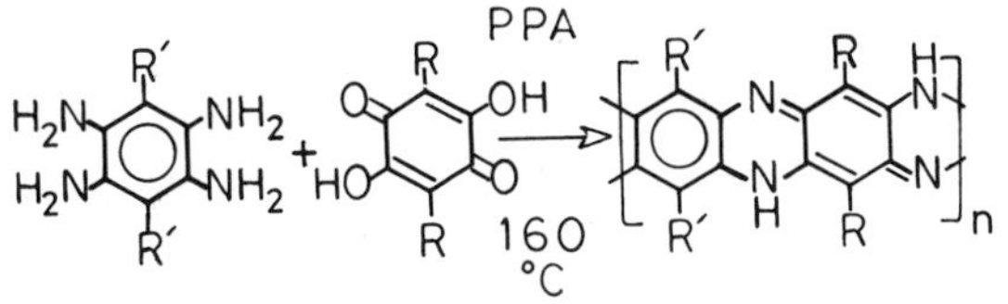

Fig. 11. Synthesis of the R = $C_{11}H_{23}$, R' = H fully cyclized analog of PQL. PPA is polyphosphoric acid.

On the other hand, the following reaction in a THF/H_2O solvent medium at ambient temperature yields an open-chain precursor

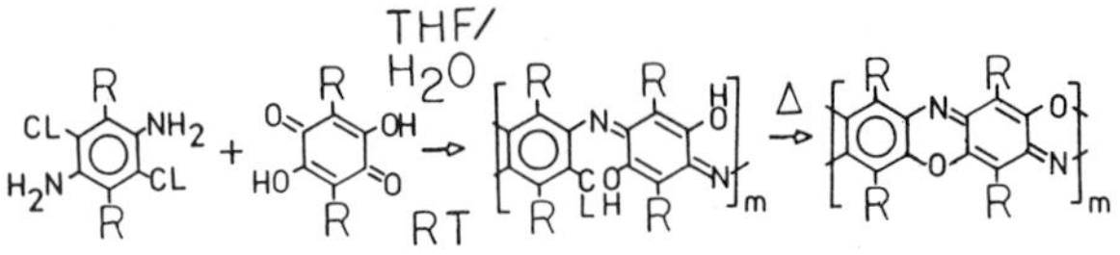

Fig. 12. Synthesis of derivatized analog of POL via an open-chain
          precursor polymer.

which can be converted to the fully cyclized POL (poly[2H,11H-bis[1,4]
oxazino[3,2-b:3',2'-m]triphenodioxazine-3,12-diyl-2,11-diylidene-11,12-bis
(methylidine)] analogs by heating. When dealing with derivatized polymers
care must be taken to avoid temperatures so high as to effect dealkylation.

        Another important advantage of the condensation synthesis scheme is
that it is easily adapted to the preparation of oligomeric model compounds
and polymers of defined chain length. In the most extreme case, one might
anticipate a Merrifield[38] approach to the sequential synthesis of polymers
of increasing chain length. A more simple approach, which we have
explored, is the use of end-capped monomers, together with appropriate
isolation of mixed products, to synthesize oligomers of defined length.
A typical reaction scheme is as follows:

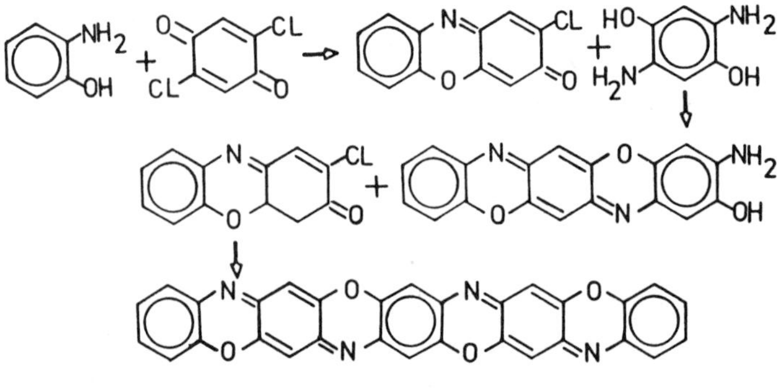

Fig. 13. Synthesis of oligomeric model compounds is illustrated. For
          the sake of simplicity, 1,4 substituents have been omitted.

        We have also prepared BBL, BBB, and several derivatized polythiophene
polymers following literature methods.[30-33, 18-21]

CHARACTERIZATION OF ELECTRON DELOCALIZATION

        In the broadest sense, electron delocalization is important for both
electrical conductivity and nonlinear optical activity, although these

252

properties depend upon substantially different aspects of electron delocalization. NLO activity is almost certainly dominated by short timescale
or intrinsic delocalization while electrical conductivity is dominated
by long range charge transport and thus will be dominated by rate limiting
(long timescale) diffusion processes.

If soliton or polaron species contribute to electrical conductivity
and/or nonlinear optical activity, then it will be important to define
the extent of intrinsic delocalization of the highest occupied molecular
orbital (HOMO) and to define the diffusion coefficient for the mid-gap
(solitonic or polaronic) defect species. Since solitonic and polaronic
species are predicted to be paramagnetic, electron paramagnetic resonance
(EPR) techniques provide a natural route to the investigation of HOMO
delocalization.

For high molecular weight oligomers and polymers where the electron
delocalization extends over a large number of nuclei, EPR spectra are unre-
solvable owing to the large number of resulting hyperfine associated
transitions. This problem of unfavorable spectral density has been discussed elsewhere.[39,40] In such a case, one must turn to techniques such
as electron nuclear double resonance (ENDOR) and electron spin echo (ESE)
spectroscopies to measure hyperfine interactions (and, in turn, the square
of the HOMO wavefunction at various nuclear positions). For low molecular
weight oligomers, sufficient resolution exists in the EPR spectra to
permit measurement of hyperfine interactions directly by spectra simulation. The effect of oligomer/polymer molecular weight upon EPR spectra
is illustrated in Figure 14.

We begin our discussion of electron delocalization with a brief overview of our work on polyacetylene. As is illustrated in Figure 15, the
EPR spectra of various polyacetylene samples always consist of a single
unresolved line, although at high amplification a second signal from
titanium (catalyst) can be detected (note: for the small concentrations
of residual catalyst, special ESE techniques are required for observation
in the presence of the stronger soliton signal). Recall that ENDOR transition frequencies are linearly related to hyperfine interaction frequencies
by

$$\nu_{\pm}(i) = \nu_N \pm \frac{a_i}{2}$$

where ν_N is the nuclear Larmor frequency and a_i is the hyperfine interaction. Since hyperfine interactions are in turn related to the square of
the HOMO wavefunction at various nuclear positions, an ENDOR spectrum
provides a convenient display of electron delocalization.

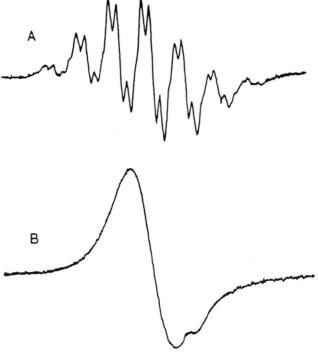

Fig. 14.   Upper.   The ambient temperature EPR spectrum of the 5-ring oligomer analog of derivatized POL.
Lower.   The EPR spectrum of the derivatized POL polymer.   In
both cases, the solvent is DMSO.

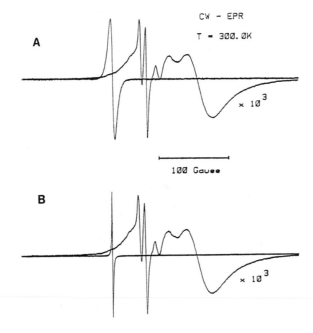

Fig. 15.   The EPR spectra of cis-(upper) and trans-rich (lower) samples of
Shirakawa polyacetylene.   The amplified portions to the right are
Ti EPR signals from residual catalyst.

Figures 16 and 17 indicate that ENDOR spectra for polyacetylene are very
sensitive to polymer lattice.

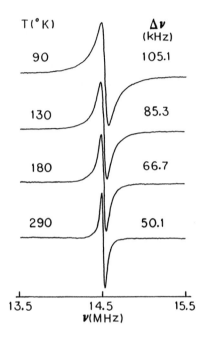

Fig. 16.   The ENDOR spectra of a trans-polyacetylene sample with minimum
           lattice contamination is shown as a function of temperature.

The ENDOR spectra shown in Figure 16 are suggestive of considerable
electron (in this case, soliton) delocalization. The strong temperature
dependence suggests that this delocalization is dynamic in nature and
determined by an activated process. If we continue to lower the temperature to 2K, an ENDOR spectrum analogous to that shown in Figure 17 is
obtained. At helium temperatures, the ENDOR spectra change only slightly
with temperature (see Figure 18). Thus, the spectra in this range can
be used to define the "intrinsic" or low temperature limit for the electron delocalization. This extent of delocalization can be viewed as the
short timescale delocalization after longer range diffusion has been
quenched by lengthening the diffusion time.

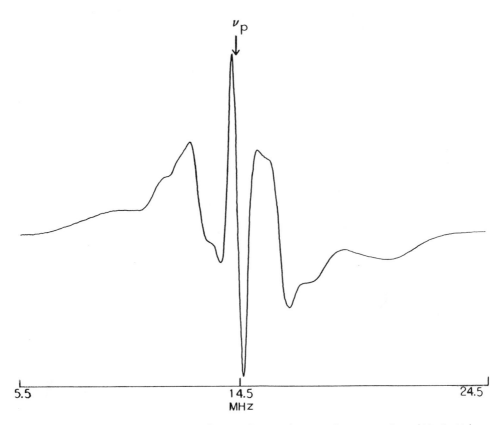

ν_P

5.5   14.5   24.5

MHz

Fig. 17. The ENDOR spectrum of a Durham polyacetylene sample with lattice
contamination by the fluoroxylene elimination product. The
temperature is 120°K.

As is illustrated in Figure 17, the "intrinsic" delocalization spectra can be observed at elevated temperatures if pinning potentials are
present. In the case of the spectra shown, the soliton pinning potential
is provided by residual fluoroxylene trapped in the trans-polyacetylene
lattice following the thermal elimination step (see Figure 4). Let us
now relate the experimentally observed ENDOR spectra reflecting intrinsic
electron delocalization to theory by employing quantum mechanics and
spectral simulation techniques. Employing unrestricted Hartree-Foch
methods with electron coulomb interactions treated by perturbation methods,
we have calculated the HOMO wavefunction and corresponding spin densities.[41] These results are summarized in Table I. These spin densities
can in turn be employed to calculate ENDOR transition frequencies and
probabilities (see Figure 19). By employing density matrix theory, ENDOR
spectra can be simulated which reflect molecular and applied time-dependent
interactions (see Figure 20). Comparison of the calculated and experimental results indicate that the shape of the wavefunction is not well
reproduced. This is not surprising in that the calculations have been

TABLE I. Spin Densities and Proton Hyperfine Tensor Elements for an
Effective Hubbard U = 3eV

Static Case

n	$\rho(n)$	A_{xx}	A_{yy} (MHz)	A_{zz}
0	+0.225	-8.19	-16.38	-24.57
1	-0.078	+2.84	+5.68	+8.52
2	-0.208	-7.57	-15.14	-22.71
3	-0.066	+2.40	+4.80	+7.21
4	+0.165	-6.01	-12.01	-18.02
5	-0.049	+1.78	+3.57	+5.35
6	+0.116	-4.22	-8.44	-12.67
7	-0.033	+1.20	+2.40	+3.60
8	+0.075	-2.73	-5.46	-8.19
9	-0.021	+0.76	+1.53	+2.29
10	+0.046	-1.67	-3.35	-5.02
11	-0.013	+0.47	+0.95	+1.42
12	+0.027	-0.98	-1.97	-2.95
13	-0.007	+0.25	+0.51	+0.76
14	+0.016	-0.58	-1.16	-1.75

Dynamically 1-D Averaged Case

n	$\rho(n)$	A_{xx}	A_{yy}	A_{zz}
even	+0.102	-3.71	-7.43	-11.14
odd	-0.038	+1.38	+2.77	+4.15

carried out for a completely static lattice, yet we know that lattice vibrations will cause fluctuations of the wavefunction within the confining potential. If we consider such a time averaging effect, we compute the spin densities shown in the lower part of Table I. These values yield theoretical spectra· in quantitative agreement with the experimentally observed spectra, providing a quantitative definition of intrinsic electron delocalization (see Figure 21).

Let us now turn our attention to the detailed consideration of soliton "diffusion". Analysis of ENDOR spectra, employing a simple jump model, for samples where lattice contamination has been minimized, yields activation barriers to diffusion on the order of 200-400°K and an ambient temper-

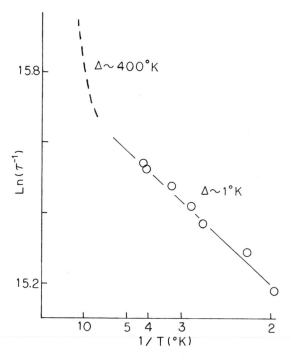

Fig. 18.   The correlation time for the modulation of hyperfine interactions (extracted from the temperature dependent shifts of ENDOR turning points) is plotted as a function of reciprocal temperature. The Δ indicate activation barriers extracted from the analysis of high temperature and helium temperature ENDOR data.

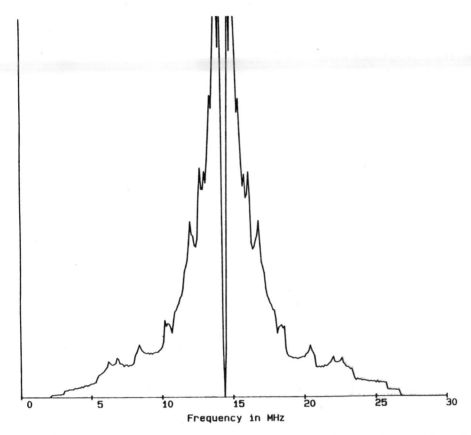

Fig. 19.   ENDOR transition probabilities calculated employing a "static" soliton wavefunction.

ature dynamic correlation time on the order of 0.1 nanosecond. Similar results are obtained from the analysis of ESE data shown in Figure 22. The minimum behavior observed for the plot of T_M versus temperature is characteristic of an activated dynamical process. At low temperatures, the experimentally measured echo phase memory decay time, T_M, is linearly related to the correlation time, τ, as every dynamical event produces spin dephasing. At high temperature a reciprocal dependence is expected because it is only the component of the spectral density function for the molecular dynamics which lies at the spin resonance frequencies that is capable of effecting dephasing.

The comparison of EPR linewidth and ESE T_M given in Figure 22 is also consistent with this picture. At low temperature where molecular dynamical frequencies are less than the frequencies of magnetic interactions, the EPR lines are inhomogeneous broadened and the widths will be greater than the intrinsic packet widths which are defined by T_M. Detailed analysis of ESE data yields an activation barrier of 200-400K consistent with ENDOR data.

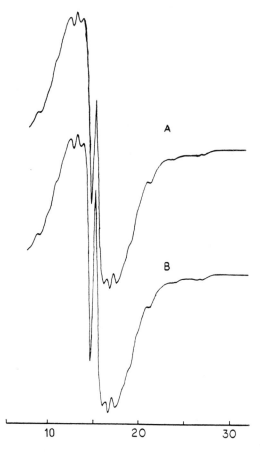

Fig. 20.   Computer simulation (calculated using density matrix theory) of the ENDOR spectrum of the static soliton.   Spectrum A was computed neglecting nuclear spin diffusion while this energy transfer mechanism is active in B.   The sweep is given in megahertz.

Studies of the effects of irreversible oxygen exposure provides insight into the nature of pinning potentials.   As is seen in Figure 23 (when considered with respect to Figure 22), the effects of oxygen exposure is to increase the activation barrier to soliton delocalization. Additional studies employing isotopically enriched molecular oxygen suggest

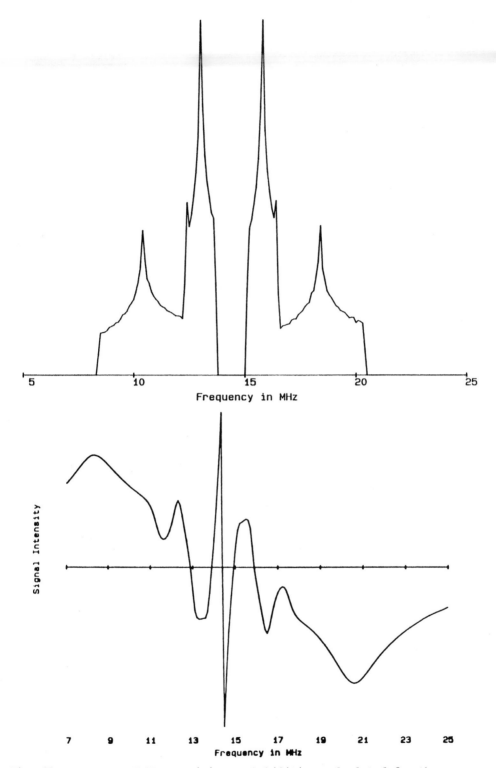

Frequency in MHz

Signal Intensity

Frequency in MHz

Fig. 21. Upper. ENDOR transition probabilities calculated for the averaged (phonon modulated) soliton.
Lower. Computer simulation, employing density matrix theory, of
the ENDOR spectrum of the 1-D averaged soliton.

261

that the irreversible oxygen effect can be attributed to the formation of peroxide crossbridges. A reversible oxygen exposure effect is also observed, which can be attributed to electron Heisenberg spin exchange between triplet oxygen and the doublet soliton.

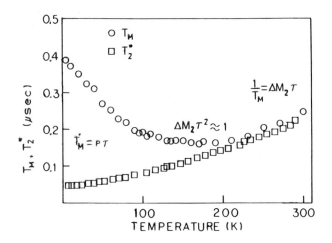

Fig. 22. Circles. Electron phase memory relaxation times (T_M) for a sample of perdeuterated trans-polyacetylene.
Squares. Effective spin-spin relaxation times from EPR linewidths of a perdeuterated trans-polyacetylene sample.

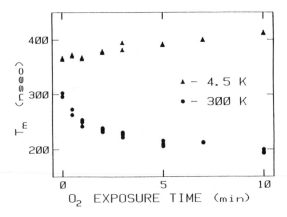

Fig. 23. The dependence of ESE phase memory time (T_M) upon transient exposure (followed by evacuation) to molecular oxygen.

Electron spin-lattice relaxation measurements provide insight into high frequency election dynamics. Such dynamics affect spin-lattice relaxation only through nonsecular processes; thus, only processes with frequencies on the order of the election Larmor frequency will contribute to electron spin-lattice relaxation. Moreover, analysis of the frequency dependence of electron spin-lattice relaxation times provides insight into the dimensionality of dynamical processes. The experimental data shown in Figure 24 clearly require the existence of fast, 1-dimensional electron dynamics.

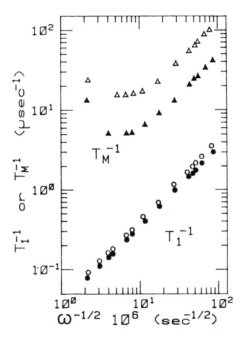

Fig. 24. The frequency dependence of electron phase memory time (triangles) and electron spin-lattice relaxation (circles) data. The solid triangles and circles are for deuterated trans-polyacetylene while the open triangles and circles are for protonated trans-polyacetylene.

Investigation of the dependence of electron spin-lattice relaxation times upon polymer isotopic composition and upon soliton concentration establish phonon modulation of electron-electron dipolar interactions as

the dominant electron spin-lattice relaxation mechanism. Analysis of the
temperature dependence (see Figure 25) of electron spin-lattice relaxation
rates establish that phonon modulation of the extended soliton wavefunction
is responsible for the short timescale (0.1 picosecond) soliton dynamics.
Note that phonon modulation does not have to produce soliton translational
diffusion to effect electron spin-lattice relaxation. A phonon passing
through an extended wavefunction of a pinned soliton can produce the
fluctuation of electron-electron interactions necessary to effect spin
relaxation. Note that the correlation time for spin-lattice relaxation
should also define the nonlinear optical switching time.

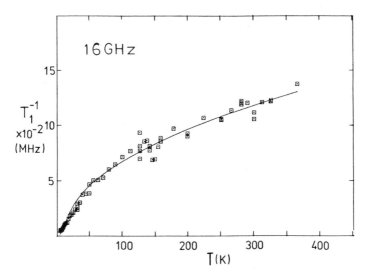

Fig. 25. Experimental (squares) and theoretical (solid line) 16 GHz
          electron spin-lattice relaxation data for trans-polyacetylene.
          Theoretical rates were computed assuming phonon modulation of
          an extended soliton wavefunction.

An important observation for understanding electrical conductivity is
the role of interchain soliton-soliton interaction. We have already observed that electron spin-lattice relaxation is dominated by such interaction. ENDOR studies of polyacetylene/polyethylene composites characterized
by different soliton concentrations show that the observed ENDOR spectra
vary with soliton concentration reflecting the influence of electron Heisenberg spin exchange.

The observed effect of soliton concentration upon ENDOR spectra suggests that an important contribution to electrical conductivity should be variable-range, interchain electron transfer (hopping). Unfortunately, for a given sample, the effects of electron Heisenberg spin exchange and soliton translational diffusion upon ENDOR spectra are indistinguishable. Discrimination can be accomplished by observation of multiple quantum NMR transitions observed by ESE methods. We have developed a modification of multiple quantum NMR to discriminate between electron translational diffusion and electron spin (Heisenberg exchange) diffusion. A sequence of microwave pulses producing electron spin polarization is used to effect nuclear spin polarization (via electron nuclear hyperfine interaction). The nuclear polarization of nuclei in contact with the paramagnetic electron is sampled after a time delay. Electron Heisenberg spin exchange (EHSE) does not affect nuclear polarization so if EHSE is the only dynamical process active then the same number and intensity of multiple quantum nuclear transitions will be observed by the sampling pulse sequence. On the other hand, electron translation diffusion brings a new pool of nuclei (unpolarized) into contact with the paramagnetic electron during the delay between the polarization and sampling pulse trains. In Figure 26, we demonstrate experimental and theoretically simulated multiple quantum transitions. At low temperatures, such multiple quantum spectra permit determination of intrinsic electron delocalization yielding results in good agreement with conclusions derived from the analysis of ENDOR data. Changes in multiple quantum NMR spectra with increasing temperature are determined solely by soliton translational diffusion. The temperature dependence of multiple quantum NMR spectra is much less dramatic than the temperature dependence of ENDOR spectra suggesting that interchain Heisenberg spin exchange is highly important. This again supports the contention that electrical conductivity is likely dominated by interchain variable range electron transfer.

ENDOR and multiple quantum spectra permit estimation of the intrinsic HOMO (soliton) delocalization length. It is this length that is relevant in the theoretical calculation of $\chi^{(3)}$ employing the method of Flytzanis.[12] Electron spin-lattice relaxation measurements permit determination of the electron-phonon correlation time which should, in turn, define the NLO switching time.

EPR, ENDOR, and ESE measurements on ladder polymers establish that the HOMO (polaron) exhibits intrinsic delocalization comparable to that found for polyacetylene (see Figure 27 and Table II). No long range translational diffusion is observed but evidence of electron Heisenberg spin exchange suggests that variable range hopping may dominate the electrical conductivity.

A typical ENDOR spectrum of polythiophene is shown in Figure 28. This spectrum establishes that the polaron defect is extensively delocalized (in terms of intrinsic delocalization) but is not undergoing long range translational diffustion. The temperature dependence of ENDOR spectra suggest the existence of Heisenberg spin exchange.

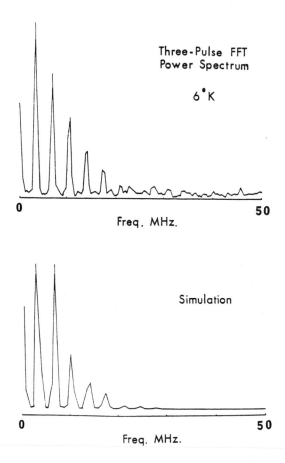

Fig. 26. The experimental (upper) and theoretical (lower) ESE-induced ^{13}C multiple quantum NMR spectra of trans-polyacetylene.

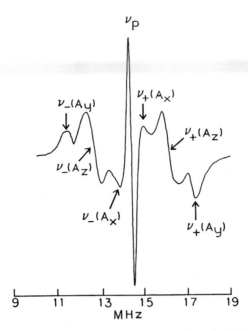

Fig. 27. Typical intermediate temperature proton ENDOR spectrum of a
ladder polymer.

Table II: Unpaired Spin Densities at Carbon (ρ_c) for π - Electron
Polymers and Corresponding Delocalization Lengths.

Polymer	$a(^1H)$, Gauss	ρ_C	N
Durham	−2.86	+0.110	15
t-PA	+0.99	−0.038	14 (29)
POL (X=O)	−1.970	+0.076	26-28
PTL (X=S)	−1.945	+0.075	26-28
PQL (X=NH)	−1.713	+0.066	26-28
BBL	−1.236	+0.048	
BBB	−1.156	+0.045	

$a(^1H)$ = isotropic proton hyperfine interaction from intermediate temperature ENDOR measurements
ρ_c = spin density at carbon determined from $\rho_c = a(^1H)/26$. Note $\rho_N \approx \rho_c$.
N = Number of atoms over which the electron delocalization extends.

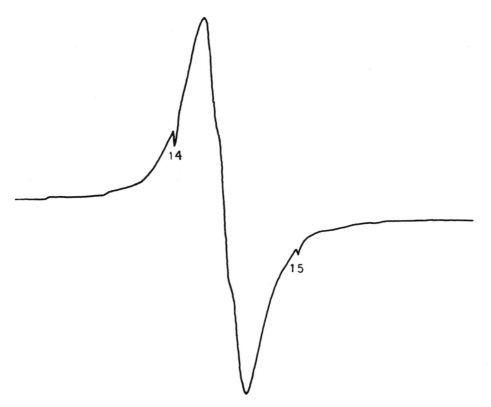

Fig. 28. ENDOR spectrum of polythiophene at 125°K.

DISCUSSION

Correlation of electrical conductivity/magnetic susceptibility studies suggest that the paramagnetic species in pristine polyacetylene samples is a soliton while in pristine heteroaromatic polymers the paramagnetic species is a polaron. The highest occupied molecular orbital is characterized by finite delocalization; substantial barriers exist to long-range, intrachain defect diffusion. The extent of intrinsic HOMO delocalization is surprisingly similar for the polymers investigated in this work. As we shall discuss shortly, this suggests that the magnitude of $\chi^{(3)}$ is not expected to vary dramatically from polymer to polymer. Electron Heisenberg exchange effects are observed for all of the polymers studied, suggesting the importance of interchain electron interactions (and hence of variable range hopping for electrical conductivity).

Chance and coworkers[7] have calculated the electronic energy structure of the fundamental ladder polymer with X = CH_2. They predict a bandgap of 0.7 eV and a bandwidth of 1.5 eV. Using these data, we calculate (employing the theory of Flytzanis[13]) a value for $\chi^{(3)}$ of 5×10^{-10} esu. This compares well with the magnitude of $\chi^{(3)}$ calculated for trans-polyacetylene.[12] The predicted comparable magnitudes of third order susceptibility for these polymers is consistent with the comparable intrinsic electron delocalization lengths observed by ENDOR for trans-polyacetylene and ladder polymer samples. The data in Table II further suggest that the X substituent does not strongly influence electron delocalization and thus is not expected to dramatically influence $\chi^{(3)}$. An important observation is that electron self-localization occurs in polymers. This observation, together with data on oligomers, indicates that after a given chain length, $\chi^{(3)}$ will no longer increase with increasing polymer length. This observation relaxes restrictions on the synthesis of polymers for electroactivity.

The estimate of third order susceptibility discussed above considers only nonresonant contributions. The theory of Flytzanis[13] predicts that $\chi^{(3)}$ will increase dramatically with increasing electron delocalization (or decreasing optical bandgap). Clearly, ladder polymers will be expected to be superior to open chain polyanilines because improved π-orbital overlap will lead to enhanced electron delocalization and reduced optical bandgap. Oxidation/reduction will affect optical properties, and a logical direction for future research is to examine the nonlinear optical activity of doped polymers. Using simple theory together with the theoretical data of Chance and coworkers[7], we calculate a value of $\chi^{(3)}$ of 5×10^{-8} esu.

Resonant contributions to $\chi^{(3)}$ for delocalized polymers have yet to be defined. Clearly, it is important to investigate $\chi^{(3)}$ values measured at the band edge. In this regard, the synthesis of oligomers affords the opportunity to tune the bandgap as the gap decreases approximately linearly with increasing oligomer length for short oligomers. Such experiments not only afford matching to available laser frequencies but also permit determination of the dependence on electron delocalization of the resonant contributions to $\chi^{(3)}$.

The present work suggests that Langmuir-Blodgett film techniques may usefully be applied to the fabrication of thin film samples of oligomers and short-chain ladder polymers. Another useful avenue of research is to prepare polymers with R groups which optimize liquid crystalline behavior.

In addition to the investigation of doped polymers, investigation of copolymers involving metallomacrocycles may yield systems with improved nonlinear optical activity.

ACKNOWLEDGEMENT

This material is based upon work supported by the Air Force Office of Scientific Research under Contract AFOSR F49620-85-C-0096.

REFERENCES

1. T. A. Skotheim, ed., "Handbook of Conducting Polymers", Vol. 1 and 2, Marcel Dekker, New York (1986), and references contained therein.
2. S. Kivelson, Phys. Rev. Lett. 47, 1549 (1981); Phys. Rev. B 25, 3798 (1982).
3. M. J. Rice and E. J. Mele, Phys. Rev. Lett. 49, 1455 (1982).
4. W. P. Su, J. R. Schrieffer, and A. J. Heeger, Phys. Rev. Lett. 42, 1698 (1979); Phys. Rev. B 22, 2099 (1980).
5. W. Forner, M. Seel, and J. Ladik, J. Chem. Phys. 84, 1455 (1986).
6. Z. G. Soos and S. Ramasesha, Phys. Rev. Lett. 51, 2374 (1983).
7. D. S. Boudreaux, R. R. Chance, J. L. Bredas, and R. Silvey, Phys. Rev. B 28, 6927 (1983); D. S. Boudreaux, R. R. Chance, J. F. Wolf, L. W. Shacklette, J. L. Bredas, B. Themans, J. M. Andre, and R. Silbey, J. Chem. Phys. 85, 4584 (1986); R. R. Chance, D. S. Bourdreaux, J. F. Wolf, L. W. Shacklette, R. Silbey, B. Themans, J. M. Andre, and J. L. Bredas, Syn. Met. 15:105 (1986).
8. S. Jeyadev and E. M. Conwell, Phys. Rev. B 35, 6253 (1987).
9. D. N. Rao, J. Swiatkiewicz, P. Chopra, S. K. Ghoshal, and P. N. Prasad, Appl. Phys. Lett. 48:1187 (1986).
10. P. N. Prasad, Proc. SPIE 682:120 (1986).
11. A. F. Garito, K. Y. Wong, Y. M. Cai, H. J. Man, and O. Zamani-Kharmiri, Proc. SPIE 682:2 (1986).
12. S. Etemad, G. L. Baker, and D. Jaye, Proc. SPIE 682:44 (1986).
13. G. P. Agrawal and C. Flytzanis, Chem. Phys. Lett. 44:366 (1976); G. P. Agrawal, C. Cojon, C. Flytzanis, Phys. Rev. B 17:776 (1978).

14. K. C. Rustagi and J. Ducuing, Optic Commun. 10:258 (1974).

15. L. R. Dalton, Proc. SPIE 682:77 (1986).

16. L. R. Dalton, C. Young, and P. Bryson, "Ultrastructure Processing of Ceramics, Glasses, and Composites," ed. J. D. MacKenzie and D. R. Ulrich, John Wiley and Sons, New York (1987).

17. L. R. Dalton, J. Thomson, and H. S. Nalwa, Polymer 28:543 (1987).

18. A. O. Patil, Y. Shenoue, F. Wudl, and A. J. Heeger, J. Am. Chem. Soc. 109:1858 (1987).

19. R. L. Elsenbaumer, K. Y. Jen, G. G. Miller, and L. W. Shacklette, Syn. Met. 18:277 (1987).

20. K. Kaeriyama, M. Sato, and S. Tanaka, Syn. Met. 18:233 (1987).

21. M. Sato, S. Tanaka, and K. Kaeriyama, Syn. Met. 18:229 (1987).

22. D. R. Gagnoon, C. S. Brown, J. N. Winter, and J. Barker, Polymer 28:601 (1987); and references contained therein.

23. W. J. Feast, M. J. Taylor, and J. N. Winter, Polymer 28:593 (1987).

24. J. E. Frommer and R. R. Chance, "Encyclopedia of Polymer Science and Engineering," ed. M. Grayson and J. Droschwitz, Wiley, New York (1987); Acc. Chem. Res. 19:2 (1986).

25. T. Ito, H. Shirakawa, and S. Ikedo, J. Polym. Sci., Polym. Chem. Edn. 12:11 (1975).

26. J. H. Edwards, W. J. Feast, and D. C. Bott, Polymer 25:395 (1984).

27. J. K. Stille and E. Mainen, Polym. Lett. 4:39 (1966).

28. J. K. Stille and E. Mainen, Macromolecules 1:36 (1963).

29. J. K. Stille and M. E. Freeburger, J. Polym. Sci. A-1, 6:161 (1968).

30. F. E. Arnold, J. Appl. Polym. Sci. 15:2035 (1971).

31. F. E. Arnold and R. L. Van Deusen, Macromolecules 2:497 (1969).

32. R. L. Van Deusen, Polym. Lett. 4:211 (1966).

33. R. L. Van Deusen, O. K. Goins, and A. J. Sicree, J. Polym. Sci. A-1, 6:1777 (1968).

34. L. F. Fieser and E. M. Chamberlain, J. Amer. Chem. Soc. 17:579 (1895).

35. C. L. Jackson and H. S. Grindley, Amer. J. Chem. 17:579 (1895).

36. P. Boldt, Chem. Ber. 100:1270 (1967); A. Muller, A. Raltschewa, and M. Papp, Ber. Dtsch. Chem. Ges. 75:692 (1942).

37. H. Bollmann, H. Becker, M. Corell, H. Streeck, and G. Langbein, Ann. 531:1 (1937); F. G. Oberender and J. A. Dixon, J. Org. Chem. 24:1226 (1959); H. Cho and R. G. Harvey, Tet. Lett. 1491 (1974); J. K. Stille and E. L. Mainen, Polym. Lett. 4:655 (1966); K. Kmai, M. Kurihara, L. Mathais, J. Wittmann, W. B. Alston, and J. K. Stille, Macromolecules 6:158 (1973).

38. R. B. Merrifield, J. Am. Chem. Soc. 85:2149 (1963).

39. C. L. Young, D. Whitney, A. I. Vistnes, and L. R. Dalton, Ann. Rev. Phys. Chem. 37:459 (1986).

40. L. R. Dalton, "EPR and Advanced EPR Studies of Biological Systems," CRC Press, Boca Raton (1985).

41. H. Thomann, L. R. Dalton, M. Grabowski, and T. C. Clarke, Phys. Rev. B 31:3141 (1985).

OPTICAL PROPERTIES OF POLY(p-PHENYLENE VINYLENE)

Jan Obrzut and Frank E. Karasz

Department of Polymer Science and Engineering
University of Massachusetts
Amherst, MA 01003

ABSTRACT

The optical spectra of poly(p-phenylene vinylene) thin films have been measured at 298 K and 77 K. The absorption band of the conjugated polymer backbone is located at 3.09 eV and is split into several components at 77 K. This absorption can most reasonably be described on the basis of vibrationally coupled exciton states. The evolution of the spectrum of the precursor polymer from monomer, dimer and oligomeric sequences is discussed.

INTRODUCTION

The discovery that it is possible to control the electrical conductivity of conjugated polymers such as polyacetylene, polyphenylene or polypyrrole from the insulating to the metallic range has stimulated substantial efforts to understand the relevant electronic states[1].

Much of the theoretical work on conjugated organic polymers applicable to model systems has utilized relatively simple methods such as the Hückel type approach. Using this model, the bathochromic shift of the lowest energy absorption band was calculated for poly(p-phenylene vinylene) (PPV) oligomers; this supported the extrapolation to infinite chain length[2]. A current approach suggesting the use of semiempirical one electron crystal spin orbitals[3] and the use of the one-electron valence effective Hamiltonian technique (VEH)[4], appears attractive for extension to calculations of semicrystalline polymers. In the VEH method, the band structure is obtained from the eigenvalues of the set of secular equations in which one electron orbitals are expressed as linear combinations of Bloch functions with a reflecting periodic potential in the polymer chain[5,6]. Thus the energy band structure, which appears in the presence of a periodic potential, supports the charge carrier transport model. On the other hand, the model of localized molecular states is still used in the interpretation of the electronic spectra, since it offers a better approach when the structural variety of "real" polymers is considered.

The molecular orbitals of the polymer are modified during charge transfer leading to the formation of defects which are more or less localized in the electronic structure.

The existence of charged solitons in polyacetylene[7] and results obtained for polythiophene[8], polypyrrole[9] or polyphenylene, stimulated construction of a model in which polarons and bipolarons were the predominant excitation, and in which the charge was stored in bipolarons[10].

In this work an experimental and theoretical investigation of the electronic structure of PPV is presented. The electronic spectra of the high molecular weight polymer with different conjugated sequence lengths are analyzed using INDO/S-CI[11], VEH[6] and vibrationally coupled exciton theoretical treatments[12].

EXPERIMENTAL

Poly(p-phenylene vinylene) was synthesized from a sulfonium salt precursor polymer[13]. The UV-Vis-NIR spectra of ca.200 nm thick polymer films were recorded using a Perkin-Elmer Lambda 9 Spectrophotometer. Calculations were performed on the CDC CYBER CY730 computer at the University of Massachusetts Computer Center.

RESULTS AND DISCUSSION

A. General Features of the PPV Spectrum and Its Interpretation in the UV Region.

Poly(p-phenylene vinylene) is obtained by a thermal elimination reaction of the precursor polymer sulfonium salt. Thus its conjugated structure evolves from individual aromatic sites through dimers, trimers, etc. and eventually to an extensively conjugated system.

The UV spectrum of the polymer film with s>n (Fig.1) consists of absorption bands at 200.2 nm, 231.0 nm, 268.0 nm, 310.2 nm, 327.1 nm and 341.0 nm. The first three of these absorption bands diminish

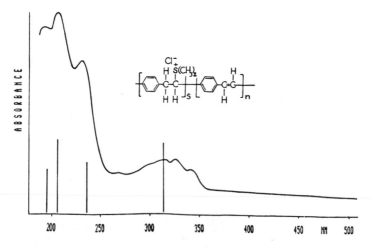

Fig. 1. Spectrum of a precursor polymer film with n:s ~ 1:2. Vertical lines denote INDO/S-CI predictions of singlet-singlet π-π^* transitions for stilbene.

progressively as the number of phenylene vinylene units, n, increases. The 200 nm absorption maximum has been attributed to a $(3p_z)^2-(3p_z)(3d)$ transition of atomic character in the sulfonium chromophores with some contribution from the benzene ring (203.5 nm). The next two absorption bands, at 231 nm and 268 nm are common features of arylsulfonium salts[14]. The four overlapping absorption bands between 290 and 360 nm in Fig. 1 arise from the conjugated aromatic system which is formed while the film is being dried at room temperature. The partial elimination to yield occasional phenylene vinylene sequences yields stilbene-like chromophores in the polymer chain.

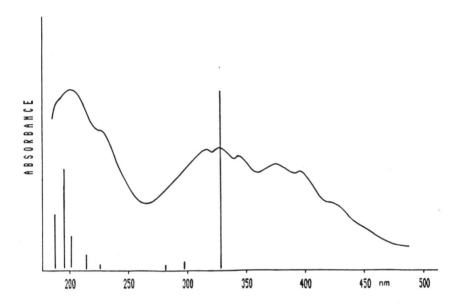

Fig. 2. Spectrum of polymer film with n:s=1:5. Vertical lines denote INDO/S-CI predictions of singlet $\pi-\pi^*$ transitions for distilbene vinylene.

As the elimination reaction progresses, the n:s ratio increases and absorptions presumably related to longer sequences of the conjugated system become visible in the spectrum of Figure 2.

The broad absorption band beginning at 270 nm changes character as the conjugated π electron system is created. Four absorption maxima in the spectrum of Fig. 1, at 310.2 nm, 317 nm, 327.1 nm and 341.9 nm have the same positions as corresponding maxima in the spectrum of Fig. 2 with increased intensities due to the larger number of excitable chromophore units.

In the spectrum shown in Fig. 2, new absorption bands also appear at fixed positions at 375.5 nm, 393.1 nm, 424.5 nm and 451.0 nm. This result is rather unexpected since traditionally the lowest $\pi-\pi^*$ transition in conjugated polymers should evolve toward the red part of the spectrum as the conjugated length increases.

A spectrum of poly(p-phenylene vinylene) is shown in Fig. 3.

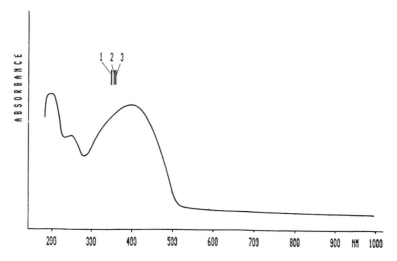

Fig. 3 Spectrum of a 200 nm thick PPV film. Vertical lines denote
the lowest π-π^* transition calculated for oligophenylene vinylenes:
1:$(oPV)_5$; 2: $(oPV)_6$; 3:extrapolated to an infinite number of units.

In the UV region the highest energy absorption band is located at
200 nm, the next one is at 244.8 nm while a broad polymer absorption
band is spread out between 270 nm and 500 nm with a maximum at 402 nm
and an absorption band edge at 511 nm (2.49 eV). The 200 nm and 244.8
nm absorption maxima are related to localized molecular states as is
shown in Fig. 2 where they are quantitatively simulated as ^1A_g-^1B_u
π-π^* electronic transitions in the distilbene vinylene model compound
(C_{2h} symmetry point group). In contrast, the experimental and theoretical results shown in Fig. 1-3 indicate that the 402 nm absorption
band which is of fundamental importance for the polymer electronic
structure could be a result of overlapping transitions from localized
energy levels rather than an effect of extended conjugation.

B. Visible Region (Polymer Band)

Electronic-Vibrational Coupling in the Exciton Model. A more
adequate explanation of the "polymer band" may be obtained by considering electronic-vibrational coupling (E-V coupling) for a molecular aggregate. The model consists of an array of non-rigid molecules
in a rigid lattice. Therefore the electronic-vibrational wave functions are products of electronic and vibrational wave functions for
the whole aggregate. The intermolecular interactions are assumed to
be purely electronic. The lowest energy ^1B_u transition for the
monomer system is composed of vibrational states, four of which are
seen in Fig. 1. At this level of elimination the n:s ratio is 1:2,
the stilbene chromohores are separated by an arylsulfonium group and
monomer vibrations are not affected by the polymeric state of aggregation. Simultaneous elimination of both sulfonium groups surrounding
the monomer creates a dimeric system which corresponds to the absorption band shown in Fig. 2 . Vibrations are excited when an electronic transition occurs in the monomer(1). Therefore the vibrational
frequency approaches the same normal modes as that of the electronically unexcited monomer(2), because the excitation is spread over the
dimeric state. The energies of the first three E-V levels obtained as
second order corrections to vibronic levels of the ^1B_u electronic

276

excited state of the monomer are 3.67 eV, 3.82 eV and 3.99 eV. In the case of the dimer they are split by 0.45 eV. The predicted absorption band shape of the dimer is shown in Fig. 4A where the vertical lines denote relative energies and intensities of Q(α) and Q(β) transitions limited to the first two vibrational components.

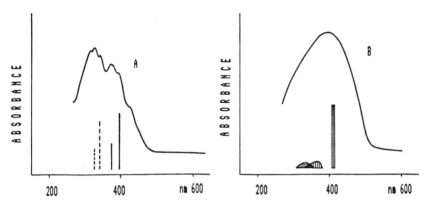

Fig. 4. A. The lowest absorption band and predicted absorption band shape for the dimer. Full vertical lines represent the Q(α) polarized vibronic absorption band, broken lines the Q(β) polarized band. B. The lowest absorption band and predicted absorption band shape for PPV. Vertical shading indicates Q(0) polarization, horizontal shading, Q($\pi\underline{r}$) polarization.

To obtain agreement with experiment, the absolute positions of the vertical lines have been adjusted by shifting each by 0.22eV to lower energies. In the eliminated material, the polymer band (Fig. 4B) does not show vibrational structure at room temperature. The poly(p-phenylene vinylene) chain may be treated as a linear molecular aggregate with two monomers per unit cell and with lattice spacing a. The sites are translationally nonequivalent. Hence the band consists of two components corresponding respectively to the transitions of the states characterized by the wave vector, $\underline{k}=0$ and $\underline{k}=\pi r$. With the phase factor chosen to assign $\underline{k}=0$ to the lower of the levels to which the transition is allowed, the expected band system when E-V functions are included is shown in Fig. 4B. The calculated maximum of the excitonic absorption is at 3.06 eV (405 nm) with splitting between components of 0.6 eV. In comparison to the experimental spectrum the calculated intensity does not contribute to the total absorption at about 370 nm.

VEH Band Structure. A different approach to interpreting the polymer band in PPV can be made using the valence effective Hamiltonian method (VEH).

The general assumption for this technique, based on a tight binding model, is that the polymer structure, reproduced by translational operations of the primitive unit cell, allows one to apply the periodic potential and the Bloch theorem for valence electrons. Calculations for a one dimensional PPV crystal with a lattice period

a_x = 6.64Å gives a 2.26 eV energy gap (Eg) and a 2.9 eV band width (Fig. 5).

The bands arise from 4 π and 15 π doubly occupied molecular orbitals in the p-phenylene vinylene unit cell and reflect a periodic potential resulting from the geometry of the system. An analysis of the LCAO coefficients shows that the highest occupied valence band derives from the four $2p_z$ orbitals of carbon atoms connected to the polymer backbone. The eight π electrons included in the unit cell and the appreciable value for Eg prevent significant conductivity without any other modification. The 2.26 eV energy gap can be compared to the absorption edge (see the low temperature spectrum of a PPV film in Fig. 6), namely to the first vibrational state at 497 nm (2.495 eV).

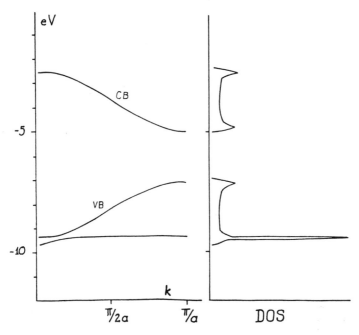

Fig. 5. Energy bands and corresponding density of states, (DOS), for a stereo-regular PPV chain. VB-valence band, CB-conducting band.

Although those results are reasonably consistent with the observed transparency of the material, the calculated band width is almost twice the experimental value and overlaps the energy region where only transitions between localized π MO levels can occur.

Polymer band at low temperature. In spite of the discrepancy, the E-V coupling theory offers a basically correct description of the "polymer band". The polymer state has few major effects upon the spectra. One is a general displacement of the band to the red compared to its location in the monomer as was discussed in detail above. Another is the partial or complete removal of degeneracy in the monomer electronic state and violation of the free monomer selection rules. Therefore the "polymer band" in Fig. 4B may include the increased number of lines over the number predicted above. Such a conclusion, particularly of overlapping exciton states perturbed by vibrations is strongly supported by the PPV spectrum taken at low temperature. In Fig. 6 nine components of a vibrational progression in the "polymer band" at 77 K are clearly visible.

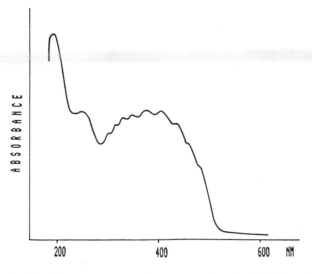

Fig. 6. The lowest energy absorption band of PPV at 77 K.

Since the phonon absorption process decreases when the temperature is lowered, the polymer band containing the electronic transitions becomes sharper.

The results and discussion presented above indicate that the exciton electronic-vibrational coupling model is the best available approach for understanding the nature of the lowest electronic excitations in poly($\underline{p}$-phenylene vinylene).

CONCLUSION

The electronic spectrum of the PPV film in the insulating form reveals three absorption bands with maxima at 6.12 eV, 5.06 eV and 3.08 eV and an absorption edge at 2.49 eV.

Applied quantum chemical techniques have shown that the 6.12 eV absorption band contains two intense 6.35 eV and 6.17 eV singlet-singlet (π-π^*) transitions, localized in the electronic structure of a short polymeric sequence. The next π-π^* transition at 5.01 eV reveals a similar character. In contrast, the 3.76 eV absorption spread over several polymer units is perturbed by electronic-vibrational coupling. The 3.06 eV energy of the lowest transition in neutral PPV calculated using the exciton electronic vibrational model supports the exciton character of the 3.08 eV polymer band.

ACKNOWLEDGEMENT

The authors thank Dr. J. Lipinski and Dr. J. L. Brèdas who provided the INDO/S-CI and VEH programs respectively. This work was supported by AFOSR Grant 86-0100.

REFERENCES

1. W. Hayes, Contemp. Phys. 26: 421 (1985).
2. H.H. Hörhold and R. Bergman, Advances in the Chemistry of Thermally Stable Polymers, (Proceedings) Warsaw 1977 p. 29-48.
3. M.J.S. Dewar, Y. Yamaguchi, J. Chem. Phys. 43: 145 (1979).
4. G. Nicolas and Ph. Durand, J. Chem. Phys. 72: 453 (1980).
5. J.M. Andre, G.R. Ladik; 1984 "Quantum Chemistry of Polymers", Wiley and Sons, New York.
6. J.L. Brèdas, R.R. Chance, R. Silbey, Phys. Rev. B 26(10) (1982).
7. A. Feldblum, J.H. Kaufman, S. Etamad and A.J. Heeger, Phy. Rev. B. 26: 815 (1982).
8. A.J. Heeger, Poly. J. 17: 201 (1985).
9. E.M. Genies and J.M. Pernaut, J. Electronanal. Chem. 191: 111 (1985).
10. J.L. Brèdas, J.C. Scott, K. Yakuski and G.B. Street, Phys. Rev. B. 30 (1984).
11. J. Lipinski, A. Nowek and H. Chojmacki, Acta Physica Polonica A53: 229 (1978).
12. E.G. McRae, Aust. J. Chem. 14: 329 (1961).
13. D.R. Gagnon, J.D. Capistran, F.E. Karasz and R.W. Lenz, ACS Polymer Preprints 25(2): 284 (1984).
14. C.J.M. Stirling, 1981 "Chemistry of the Sulphonium Group", Wiley and Sons, New York.

HOLES, ELECTRONS, POLARONS, AND BIPOLARONS AND THE THERMODYNAMICS OF

ELECTRICALLY ACTIVE DOPANTS IN CONDUCTING POLYMERS

H. Reiss and Dai-uk Kim

Department of Chemistry
University of California
Los Angeles, California 90024

ABSTRACT

This paper describes how measurements of the reversible (equilibrium) distribution of an electrically active dopant, between an external phase and a conducting polymer, can be used to investigate both electronic species and electron energy level structures in such polymers. Examples are presented, involving conventional inorganic semiconductors, and it is shown how complications anticipated with conducting polymers have occurred and been overcome in these systems. Following this, experiments on both absorption isotherms and conductivity for vapor phase iodine in both polythiophene and azite are described and analyzed for the determination of relevant species. The formations of triiodide and pentaiodide ions are indicated, and bipolarons appear to form in polythiophene.

INTRODUCTION

The intent of this paper is to draw attention to the usefulness of measurements of the equilibrium distribution of an electrically active dopant between a conducting polymer and some external phase. Such measurements, essentially thermodynamic in nature, can throw light on the electronic "species" (e.g. polarons, bipolarons, etc.) generated by such dopants, and can even provide information concerning energy levels and band structure. A simple example of this approach, having nothing to do with conducting polymers, concerns the solubility of hydrogen in palladium. It is well known that the solubility depends upon the square root of the hydrogen pressure, indicating that hydrogen dissolves as atoms rather than molecules. Studies of the solubility as a function of temperature could yield information on the dissociation energy of hydrogen.

To provide examples of this approach, however, one need not stray so far from conducting polymers as the case of hydrogen and palladium. The method has been abundantly applied, over a period of fifty years, to conventional inorganic semiconductors[1,2,3,4]. Why it has not been applied to conducting polymers is not entirely clear, but some objections to its use have been raised on the following basis:

(1) Electrically active dopants, in conducting polymers, are largely

dissolved in an irreversible manner, and are therefore not subject to analysis by thermodynamic means.

(2) The uptake of an electrically active dopant by a conducting polymer is usually attended by dramatic conformational changes[5].

(3) The levels of dopant concentration, typical of conducting polymers, are orders of magnitude larger than those in conventional semiconductors. This leads to unusual mechanisms of diffusion as well as to a host of other phenomena, unknown in conventional semiconductors.

The high concentration of dopant, mentioned in item 3 of the above list, is indeed a serious question, particularly because it can prevent the solute from behaving in a thermodynamically ideal manner such that thermodynamic activities can be replaced by concentrations. For example, in the case of hydrogen and palladium, the square root relationship depends, strictly speaking, on the thermodynamic activity of hydrogen in palladium being representable by concentration. On the other hand, this merely suggests that experiments on solubility in conducting polymers should be conducted at lower concentrations, and that methods should be developed to render measurements at such low concentrations feasible.

Insofar as questions of reversibility, conformational changes, and diffusion are concerned, all of these problems arose to a greater or lesser extent with conventional semiconductors, and the plan of this paper will be to first show that this was the case, by reference to a particularly simple system, so that the stated objections do not constitute a strong reason for excluding conducting polymers from the approach. Following this, the results of applying the method to the absorption of iodine vapor in two conducting polymers (polythiophene and azite)[6] will be described. It will be seen that these studies, incomplete as they are, already reveal many interesting phenomena, and in the case of polythiophene, give support to the existence of bipolarons as current carriers.

Finally, it should be pointed out that, especially in conducting polymers where resistivity may be dominated by "interchain hopping"[7] of electrons, a thermodynamically reversible means of estimating carrier concentration is independent of these interesting, but difficult-to-characterize, processes. Thus, especially at low concentrations, the study of reversible solubility offers the promise of measuring the carrier concentration "internal to a polymer molecule" without interference from interchain hopping, a problem endemic to irreversible measurements such as those involving conductivity.

THE METHOD APPLIED TO CONVENTIONAL SEMICONDUCTORS

A particularly simple example, in which the study of the solubility of an electrically active dopant in a semiconductor provides information on the mechanism of ionization of that dopant, is provided by the case of lithium, as a dopant for the elemental semiconductors, silicon and germanium.[3,4] Lithium is known to behave as a donor in both silicon and germanium, and provides a donor level lying about 0.01 electron volts below the bottom of the conduction band. As a result, lithium in silicon or germanium, at room temperature, is almost fully ionized, and each lithium atom provides one conduction band electron. Furthermore, it is also known that lithium, in silicon or germanium, occupies an interstitial position, so that the lithium ion, smaller than a helium atom, can diffuse through the open (diamond lattice) structures of these crystals with great rapidity. In fact, at room temperature, and within one year, lithium could saturate, by diffusion, a sheet of germanium 0.5 millimeter thick. This relatively rapid equilibration makes it ideal for the study of reversible solubility at relatively low temperatures.

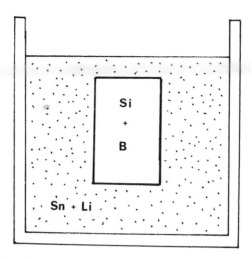

Fig. 1. Boron doped silicon immersed in
molten tin containing lithium.

A particularly simple experiment, in this respect, is illustrated in
Figure 1 which shows a crystal of silicon doped with boron (an acceptor)
immersed in a bath of molten tin containing lithium as a solute[3]. The
situation is such that the silicon crystal does not dissolve in the tin.
Furthermore, boron, unlike lithium, occupies a substitutional position in the
crystal, and is an extremely slow diffuser. Thus, at the low temperatures of
the experiment, boron is confined to the silicon. The boron acceptor levels
lie only a few hundredths of an electron volt above the top of the valence
band, and so boron, like lithium in silicon, is almost completely ionized.
The energy level diagram, when both lithium and boron are dissolved in
silicon, is illustrated in Figure 2 where the conduction band, valence band,
and the donor and acceptor levels are indicated, and where E_C, E_V, E_D, and E_A
are the energies of the conduction band edge, the valence band edge, the

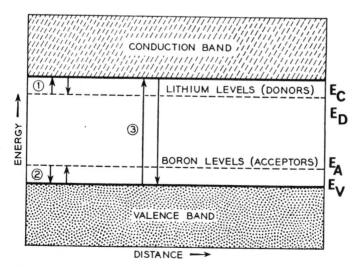

Fig. 2. Energy band diagram for silicon containing
lithium and boron.

donor level, and the acceptor level. The electronic transitions, indicated by the numbers 1, 2, and 3 are, respectively, the ionization of lithium, the ionization of boron, and the excitation of an electron across the forbidden gap.

Corresponding to the various processes in Figures 1 and 2, we can write the following set of interconnected chemical-like equilibria.

$$
\begin{array}{ccc}
Li(Sn) & \rightleftharpoons Li^+(Si) & + e^- \\
+ & & + \\
B^-(Si) & + e^+ \\
\updownarrow & & \updownarrow \\
[LiB^-] & [e^+e^-]
\end{array}
\qquad (1)
$$

In Eq. (1) the phases (Sn, Si) are indicated by the symbols in the parentheses while e^- and e^+ represent a conduction band electron and a valence band hole, respectively. The quantity e^+e^- in square brackets represents a "recombined" hole-electron pair (actually an electron in the valence band). The other quantity in square brackets denotes a coulombically bound ion pair, formed from a lithium and a boron ion. At the temperature of the experiment of Figure 1, such ion pair formation is negligible, and for the moment we can ignore it, although we shall return to its consideration (at lower temperatures) later. The horizontal equilibrium at the top of Eq. (1) indicates the dissolving of lithium in silicon, and shows the lithium to be completely ionized. A more precise version of this equilibrium would exhibit a two-stage process in which the lithium first dissolved in silicon, in the unionized state, followed by an equilibrium representing ionization. This last equilibrium would correspond to the process marked with a 1 in Figure 2. However, since lithium is almost completely ionized, only a minor error is incurred in employing the single stage process. The same is true of the second horizontal line in Eq. (1) which shows boron to be completely ionized. A more exact account would replace the second line in Eq. (1) with an equilibrium in which unionized boron atoms (in silicon) dissociate to produce the species actually shown on the second line. This equilibrium would correspond to the process marked with a 2 in Figure 2. The vertical equilibrium in Eq. (1) involving the "recombined" hole-electron pair corresponds to the transition, in Figure 2, marked with a 3. As indicated earlier, for the time being, we ignore the equilibrium involving the formation of an ion pair.

Now, each of the equilibria in Eq. (1) can be characterized by an "equilibrium constant" involving the customary products and quotients of the thermodynamic activities of the relevant species. When all these species are sufficiently dilute the activities can be replaced by concentrations, i.e. the activity coefficients approach unity. It is interesting to note that the last species to admit of a unit activity coefficient, as the dilution is increased, is usually the electron or "hole." The reason for this involves the late passage from fermi statistics[3], as dilution is increased, which, for the conduction and valence bands, occurs when the fermi level is at least several kT (where k is the Boltzmann constant and T is temperature) from either band edge. Indeed when fermi statistics must be used, the activity coefficients are of an interesting variety[8], expressing the interaction implicit in the Pauli exclusion principle. However, we shall restrict attention to situations in which the various species are present in low concentration so that classical statistics applies, and all activities may be replaced by concentrations.

For chemists, the set of equilibria, appearing in Eq. (1) are a familiar arrangement. For example, if e^- is regarded as the analog of the hydroxyl ion in aqueous solution, and e^+ the analog of the hydrogen ion, then $[e^+e^-]$ becomes the analog of the "weakly dissociated" compound, water. Furthermore the ionizable donor is the analog of a base while the acceptor is the analog of an acid. It is then immediately clear that the solubility of the donor, lithium, in silicon should be increased by the presence of the acceptor, boron, i.e. an acid should dissolve a base.

In order to express this conclusion in quantitative terms, it is necessary to write the equilibrium constant expressions (in terms of concentrations) for each of the equilibria in Eq. (1). To these expressions we add the requirement of electroneutrality, namely, that the sum of the concentrations of e^- and B^- must equal the sum of the concentrations of Li^+ and e^+. The simultaneous solution of these various equations allows one to express the solubility (concentration) of lithium in silicon as a function of the boron concentration. The relevant expression is

$$D = \frac{A}{1+(1+(2n_i/D_0)^2)^{1/2}} + [\{\frac{A}{1+(2n_i/D_0)^2)^{1/2}}\}^2 + D_0^2]^{1/2} \qquad (2)$$

In this equation D represents the concentration of lithium and A the concentration of boron, while D_0 represents the concentration (solubility) of lithium when the silicon contains no boron. n_i is the so called intrinsic concentration of carriers (in silicon). In fact the equilibrium constant for the vertical equilibrium, in Eq. (1), involving holes and electrons is simply n_i^2, and is analogous to K_W, the equilibrium constant for the dissociation of water. D_0 makes its appearance in Eq. (2) because of its relation to the equilibrium constant K for the equilibrium in the first line of Eq. (1). Indeed K is given by

$$K = (D_0^2/2) + \{(D_0^4/4) + D_0^2 n_i^2\}^{1/2} \qquad (3)$$

How good is the relationship in Eq. (2)? Figure 3, which compares theory and experiment, provides the answer. In the figure the solubility of

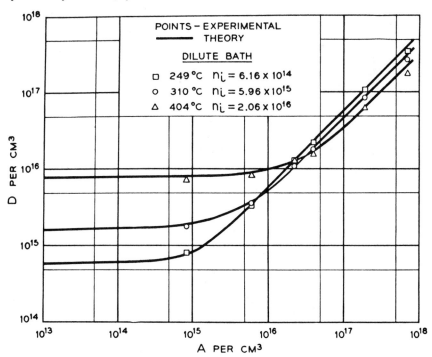

Fig. 3. Isotherms showing the solubility of lithium D, in silicon as a function of boron doping A, for an external phase of tin containing 0.18 percent lithium.

lithium as a function of boron concentration is illustrated for three temperatures, 249°C, 310°C, and 404°C, respectively[3]. The corresponding values of n_i, measurable by separate means, are listed, in the figure, next to the temperatures. These data are, of course, for a particular fixed concentration of lithium in the molten tin phase. Concentrations are expressed in atoms per cubic centimeter. The squares, circles, and triangles represent experimental data while the curves are plots of Eq. (2). Notice that Figure 3 is a log-log plot so that very considerable increases in solubility are caused by the presence of boron. Furthermore, the agreement between experiment and theory is excellent.

Now, in this study, both n_i and K are measured beforehand (the latter from Eq. (3)). Furthermore the agreement between theory and experiment depends sensitively on the theory having been referred to the correct mechanisms of equilibrium and ionization, i.e. Eq. (1). However, if the mechanisms of ionization and the energy level spectrum or density of states (which enters into the equilibrium constant) had not been known, they could have been inferred by fitting the theory to the experimental data. The mechanisms of equilibrium and ionization could be confirmed by the fit, the theory having been based on an assumed mechanism, and the energy level diagram could be determined from the dependence (required for a fit to the experimental data) of the equilibrium constants on temperature. For example, the ratio of the band gap energy to the temperature is contained exponentially in n_i^2, the equilibrium constant for the vertical process in Eq. (1).

This simple example has been presented in order to illustrate how the measurement of solubility, in conducting polymer phases, might be used to acquire information concerning electronic species, energy diagrams, and mechanisms of equilibration and ionization. As indicated earlier, the method is more powerful when dopant concentrations are low enough to allow concentrations to replace thermodynamic activities. This is a goal to be pursued, but, as we shall see, valuable information appears to be available, even at higher concentrations.

We now turn to some of the cautionary notes, sounded in Section I, emphasizing the differences between conducting polymers and inorganic semiconductors. First, there is the question concerning the nonreversibility of the absorption of dopants by conducting polymers. In response to this concern, it should be pointed out that some dopants may dissolve in polymers reversibly, and that, on the other side, many electrically active dopants in inorganic semiconductors are not dissolved in a fully reversible manner. In this respect oxygen is a good example since it is electrically active, but can lead to the highly irreversible internal oxidation of silicon[9,10].

The next cautionary note deals with the very pronounced conformational changes which accompany the doping of polymers with an electrically active species. Conformational or structural changes are frequent occurrences when inorganic semiconductors are electroactively doped. A particularly clear example is offered by the following.

Consider crystalline ferrous oxide in equilibrium with gaseous oxygen. Ferrous oxide, FeO, is notably nonstoichiometric. In fact oxygen in the crystal may exceed iron by as much as 10%. The lattice consists of ferrous ions and oxide ions, and the excess of oxygen over iron has been shown to be due to the existence of ferrous ion vacancies. The following equilibrium has been postulated[11]

$$\frac{1}{2}O_2(gas) \rightleftharpoons O^{2-}(FeO) + \square_{Fe} + 2e^+ \qquad (4)$$

in which the notation in the parenthesis again indicates the phase and the symbol $\Box_{Fe}$ indicates a ferrous ion vacancy. Again, e^+ denotes a hole. The stoichiometry of the process described in Eq. (4) demands that the concentration p of holes be twice the concentration C_v of vacancies. Denoting the pressure of oxygen as P_{O_2} we may write for the equilibrium constant corresponding to in Eq. (4)

$$\frac{p^2 C_v}{p_{O_2}^{1/2}} = K \tag{5}$$

Introducing the requirement that p be twice C_v into this equation, and solving for small p gives

$$p = (2K)^{1/3} p_{O_2}^{1/6} \tag{6}$$

Now, if the conductivity of ferrous oxide is proportional to the hole concentration p, then, according to Eq. (6), it should vary as the 1/6 power of the oxygen pressure. Figure 4 exhibits experimental plots[11] of the logarithm of that conductivity versus the logarithm of oxygen pressure for temperatures of 950° C and 1000° C, respectively. The slopes of these curves are each 1/6, and therefore provide compelling evidence for the validity of the mechanism assumed in Eq. (4).

It should be noted that the introduction of vacancies, and the accompanying nonstoichiometry, corresponds to a structural or "conformational" change. Nonetheless it can be dealt with, even though the non-

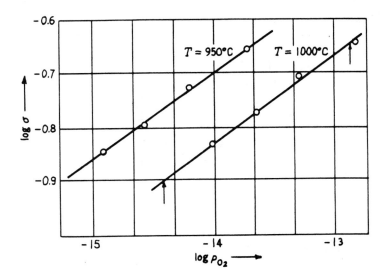

Fig. 4. Dependence of Electrical Conductivity of FeO upon oxygen pressure.

stoichiometry is as large as 10%. The presence of conformational change does not forbid the application of methods based on solubility, as long as reversibility can be achieved.

Next we address the cautionary note dealing with the possible complex nature of diffusion in conducting polymers. Actually, very complex cases of diffusion also occur in inorganic semiconductors. For example, there are well documented situations in which "uphill" diffusion takes place due to the presence of electrical fields resulting from nonuniform distributions of electrically active dopants[12]. On the other side, nonreversible diffusion in conducting polymers can, in fact, be reversible, and only give the appearance of nonreversibility. A model situation of this type is discussed in a recent paper[13].

A particularly simple case of nonsimple diffusion involves the ion pairing process appearing in Eq. (1). If the acceptor in the pair is substitutional, and cannot diffuse, then the mobile interstitial donor (lithium) is effectively "trapped" when it is paired, and the diffusion coefficient of lithium should be reduced in the presence of immobile acceptor. Figure 5, which contains a plot of the logarithm of the diffusion

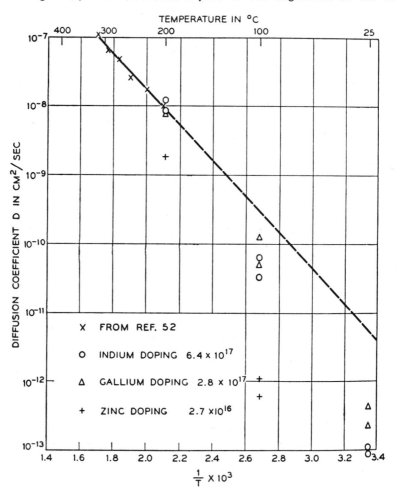

Fig. 5. Plot of diffusivity of lithium in undoped germanium as a function of temperature--also showing points for apparant diffusivities of lithium in variously doped specimens.

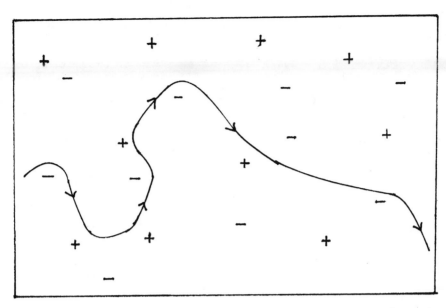

Fig. 6. Circuitous path of a positive hole through an array of unpaired
positive and negative donor and acceptor ions in a semiconductor.

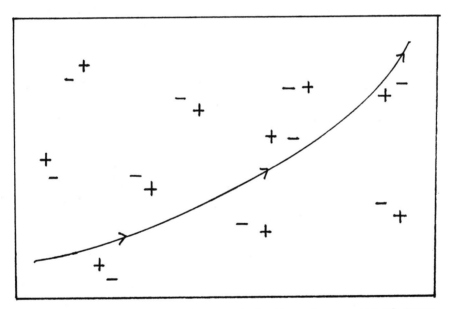

Fig. 7.   Direct path of a positive hole through an array of donor-
acceptor ion pairs in a semiconductor.

coefficient of lithium in silicon versus the reciprocal of temperature,
demonstrates this phenomenon. The straight line in the figure shows the
variation of the diffusion coefficient with temperature when silicon contains
no acceptor. The circles, triangles, and crosses are experimental values of
the diffusion coefficient in the presence of the acceptors indium, gallium,
and zinc at the concentrations (atoms per cubic centimeter) indicated in the
box at the lower left hand corner. It should be mentioned that zinc is a
doubly ionizable acceptor with energy levels lying respectively at 0.03 and
0.09 electron volts above the edge of the valence band. A doubly charged

acceptor is obviously more efficient in the trapping of lithium ions than a
singly charged one. As the data show, at 200° C, the diffusion coefficient
in the presence of both indium and gallium is unaffected, but in the presence
of zinc it is reduced by about a factor of 5. This implies that, at this
temperature, essentially no ion pairs are formed by singly charged acceptors,
but that a considerable number are formed by the doubly charged acceptor. As
the temperature is lowered, e.g. to 100° C, it will be seen that even the
singly charged acceptors reduce the diffusion coefficient below that for pure
silicon, and, for the doubly charged acceptor, the reduction is of the order
of several hundred fold. Thus, as the temperature is lowered to 100° C, even
the singly charged acceptors form ion pairs. The process is further
augmented at still lower temperatures, e.g. at 25° C.

There are many ways to demonstrate the presence of ion pairs and to
measure the degree of pairing. One of these involves the effect of ion
pairing on the mobility of a current carrier, e.g. a hole, in the
semiconductor. Figures 6 and 7 illustrate this effect. Figure 6 is a
schematic of the path of such a hole as it proceeds through a silicon crystal
in which the donor and acceptor ions are not paired. Due to the strong
coulomb fields experienced by the hole, it follows an extremely tortuous path
so that its mobility is obviously reduced. However, in a sample in which
pairing has occurred, the hole is only subject to the much weaker dipole
fields of the pairs. In fact (due to the hole's low effective mass) the
thermal deBroglie wave length of the hole is much larger than the spacial
extent of a single ion pair, and so it barely notices the presence of the
pairs, and follows a rather direct path through the crystal.

If one starts with a crystal containing only acceptor ions, it is to be
expected, due to the coulomb forces, that the path of the hole will be
tortuous and its mobility small. If lithium is then diffused into the
crystal, and ion pairs are formed, the mobility should be increased. The
results[3] of just such an experiment are exhibited in Figure 8 in which the

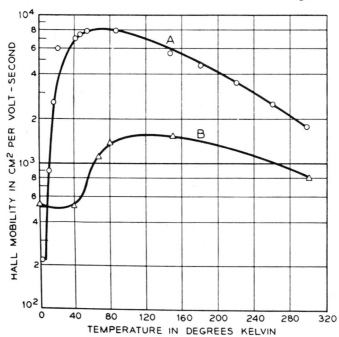

Fig. 8. Plot of Hall mobility as a function of temperature for germanium
containing 3 X 10^{17} cm^{-3} gallium. Curve A is for a sample also containing
2.8 X 10^{17} cm^{-3} lithium.

Hall mobility of the hole is plotted versus temperature. Curve B corresponds to a sample containing only the acceptor, gallium, at a concentration of about $3 \times 10^{17} cm^{-3}$. Curve A shows what happens when lithium is diffused into the sample to a concentration of approximately $2.8 \times 10^{17} cm^{-3}$. Ion pairs form, and the mobility is very substantially increased. As the temperature is increased, curve A converges on curve B since the ion pairs dissociate. The two curves cross at the temperature where the pairs are 50% dissociated.

Another means of confirming the presence of ion pairs involves the measurement of the rate of pairing. In a typical experiment[3], a silicon crystal containing about $10^{16} cm^{-3}$ pairs, consisting of lithium and gallium ions, is heated to slightly above room temperature where most of the pairs are dissociated. Then the crystal is quickly plunged into liquid nitrogen so as to quench the pairs in their dissociated states. Following this, the temperature is raised to 195° K (dry ice temperature) where the lithium ions are mobile enough to engage in pairing. As pairing proceeds, the mobility of the excess holes increases, and since the concentration of holes remains constant, the conductivity increases. Pairing is then followed by measuring conductivity σ as a function of time, t. According to theory, conductivity should exhibit the following dependence on time,

$$\ln(\sigma_\infty - \sigma) = \alpha + (t/\tau) \qquad (7)$$

In this equation τ is the relaxation time for the pairing, given by[3]

$$\tau = \frac{\kappa k T}{4\pi q^2 N D} \qquad (8)$$

in which κ is the dielectric constant of the medium, q is the charge on the ion, N represents the concentration of donor or acceptor ions (almost equal concentrations), and D is the diffusion coefficient of lithium in pure silicon. Returning to Eq. (7) α is a constant and σ_∞ is the conductivity at infinite time (when the maximum pairing has been achieved).

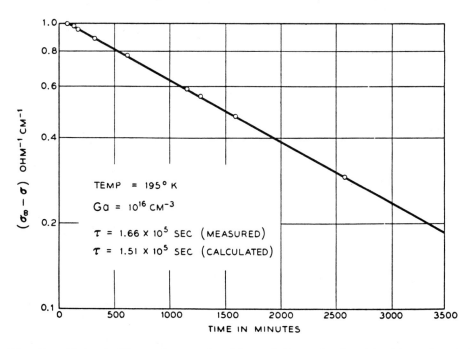

Fig. 9. Plot of $(T_\infty - \sigma)$ as a function of time showing kinetics of pairing.

Figure 9 illustrates the extent of agreement, between theory and experiment, which can be achieved. In it, the left member of Eq. (7) is plotted versus time. The points represent experimental data, and fit a straight line as Eq. (7) requires. Furthermore the slope of the curve yields an experimental value for the reciprocal of the relaxation time τ, and as can be seen from the inset at the lower left of the figure, agreement between the measured and calculated relaxation times is excellent.

In a certain sense, ion pairs resemble charged polarons engaging in variable range hopping[14]. In order for such hopping to occur, the positive hole in the polaron must dissociate from the negative (acceptor) ion which pins it. Similarly, the lithium ion, in an ion pair, must escape from the stationary acceptor ion in order for dissociation to occur. A theory has been developed for the average mobility (its contribution to the overall current) of the lithium ion as a function of its distance from the acceptor counter ion[15].

ABSORPTION OF IODINE VAPOR IN POLYTHIOPHENE AND AZITE

The question to be addressed now is--can we find a "lithium for conducting polymers"? Preliminary evidence indicates that iodine may serve this purpose. This evidence is furnished by a recent study[16] on the absorption of iodine vapor in polythiophene, and by the additional studies reported below, for the first time, on the conductivity changes which accompany such absorption. Further evidence, of a tentative nature, is also furnished by a study of iodine absorption in azite (a new compound synthesized by O. L. Chapman and his coworkers) also reported below, for the first time.

In reference (16) a quartz microbalance was employed for the investigation of the absorption of iodine vapor in polythiophene. The entire balance was enclosed in a glass housing which could be filled with iodine vapor at fixed temperature and pressure. Iodine pressure was measured by spectral means, utilizing iodine's strong absorption in the visible. Temperature was controlled by inserting the entire glass housing into an oven. Films or powders of polythiophene were suspended from the balance, and the uptake of iodine was determined by weighing. Typical results are illustrated in Figure 10 which displays plots of uptake (micromoles of I_2) in a film versus pressure of the iodine vapor (torr), at 40, 50, 60, 70, and 80° C. The film was prepared by electrodeposition of polythiophene on a layer of gold having a thickness of 75Å. The gold was supported on a glass cover slip.

The film was electroreduced to the neutral state after deposition; its thickness was of the order of 2.5×10^{-5} cm and it contained 0.24 µmol of thiophene monomer.

The absorption of iodine by such films was found to indeed be reversible. The 40° C isotherm in the figure illustrates this point. The open circles correspond to a first determination of the isotherm. After this, iodine was removed from the film by evacuating the glass housing. After a period of about 30 minutes the film returned to its original weight, indicating the full removal of iodine. Then the isotherms at 50, 60, 70, and 80°C were determined, the iodine being removed from the film between each determination. After the 80° C measurement, iodine was again removed by evacuation, and the 40° C isotherm redetermined. The squares illustrate the results of this measurement, and clearly show the reproduction of the original isotherm. This process could be repeated many times. Similar studies were conducted with bromine and chlorine[17]. In these cases the uptake was not fully reversible, and irreversibility increased in the direction of lower atomic weights.

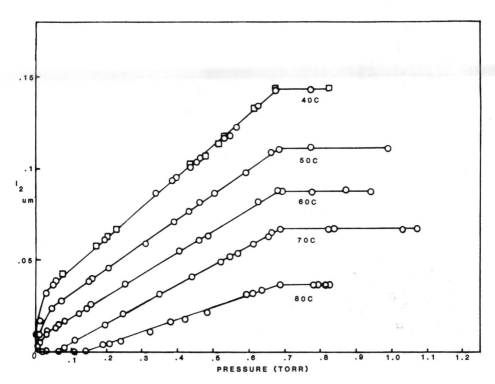

Fig. 10. Absorption isotherms for iodine in a polythiophene
film (0.24 μmol) at five temperatures. Uptake in
μmole and iodine pressure in torr. Reversibility is
indicated by separate runs (circles and squares).

The isotherms of Figure 10 exhibit a number of remarkable features. At
low pressures they are concave downward, at lower temperatures, and concave
upward at higher temperatures. Furthermore, at higher iodine pressure
(over long ranges and at all studied temperatures) iodine uptake appears to
depend linearly on iodine pressure. The most remarkable feature is the
sharp discontinuity to zero slope (at all temperatures), indicating some
sort of saturation. Curves of this shape cannot be explained by a
saturation phenomenon of the sort that occurs in, say, simple Langmuir-type
absorption. Langmuir isotherms are continuous to saturation and curved,
exhibiting linear behavior only at the very lowest pressures, and the level
of saturation is independent of temperature. In contrast, in the
I_2-polythiophene isotherms the saturation level <u>decreases</u> with an increase
in temperature.

The most striking aspect of the discontinuities is the fact that they
all occur at a pressure of about 0.68 torr, <u>independent of temperature</u>.
This suggested a <u>mechanical</u> basis for the phenomenon, and so further
measurements were undertaken in which an inert gas (helium) was mixed into
the iodine vapor. Two sets of experiments of this type were performed, one
with the helium pressure maintained at 0.375 torr and the other with helium
pressure at 0.60 torr. The results of these investigations are exhibited
in Figure 11, superimposed upon the isotherms of Figure 10. The startling
fact is that the isotherms, in the presence of helium, are identical with
those obtained in the absence of helium <u>until the total gas pressure is
about 0.68 torr</u>. This is true, independent of temperature. For example,

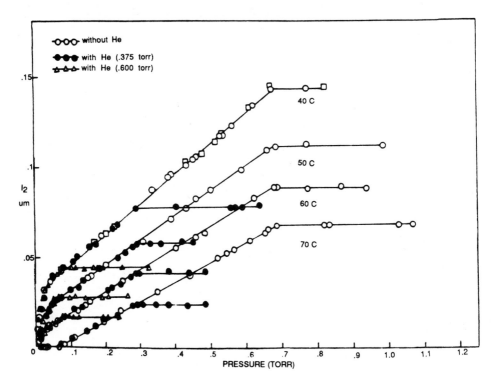

Fig. 11. Absorption isotherms for iodine in the same polythiophene film used for Figure 10, but in the presence of partial pressures of helium amounting to 0.375 and 0.600 torr, indicated by filled circles and triangles, respectively. Isotherms at only four temperatures are plotted, and the open circles and squares are again the data of Figure 10. Note that the discontinuities in slope all occur at about the same total pressure of the ambient gas.

with 0.375 torr of helium, the discontinuities in the slopes of the curves occur in the neighborhood of 0.29 torr of iodine pressure. With 0.60 torr of helium, the discontinuities occur in the neighborhood of 0.08 torr of iodine pressure.

In order to test the generality of the phenomena, absorption isotherms were also measured, using polythiophene in the form of powder. The polymers in the powder were synthesized in the laboratory of Professor F. Wudl at the University of California, Santa Barbara, and contain, on the average, 44 thiophene (monomer) units. Absorption isotherms, obtained with powder of the same weight as the film are exhibited in Figure 12, and are plotted over the isotherms appearing in Figure 10. It is apparent that the isotherms for the powder are identical with those obtained with the film, until the pressure of the discontinuity is reached. Then the powder isotherms do not exhibit the discontinuity, and continue to higher pressures with the same constant slopes.

Thus, the discontinuity is a property of the film alone, and needs to be explained.

At this point, no truly satisfactory explanation is available. In reference (16) an explanation was attempted, based on the assumed existence

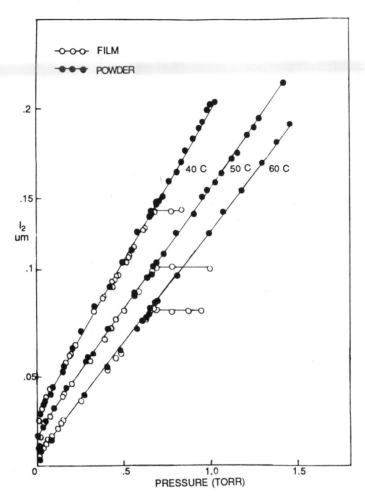

Fig. 12. Absorption isotherms for a powder sample containing 0.24 µmol
of polythiophene. Filled circles correspond to powder. The
open circles are the 30, 40, and 50°C isotherms of Figure 10.
The powder isotherms do not show the discontinuities in slope,
but, at pressures lower than that corresponding to the
discontinuity, are identical with the film isotherms.

of a charged depletion layer in the doped polymer, a Schottky diode having
been assumed to form at the interface between the doped polymer and the
layer of gold. The field in the resulting double layer was assumed to
assist the gas pressure in compressing the film in such a way that an
electromechanical instability developed when the internal forces of the
film, resisting compression, could no longer balance the combined effects
of pressure and double layer electric field. Although it was even possible
to formulate a quantitative theory for this model, it remained impossible
to explain how the discontinuity could be independent of both temperature
and level of doping. Further investigation of the discontinuity is
warranted; it may have device applications.

The range of doping in the straight line portion of the isotherms is
very high, and phenomena of the type discussed in the previous section (in

connection with lithium) based on equilibria involving electronic species would be hard to interpret at such high concentrations. The truly interesting portions of the isotherms are the curved (concave downward and concave upward) portions which occur at low pressures. Unfortunately the precision with which iodine pressure could be measured by the technique employed in these experiments was no greater than 0.01 torr. Therefore, experiments in this most interesting low pressure range must await an improvement of technique (e.g. the use of iodine fluorescence).

Recently, we measured the conductivity of the film involved in Figures 10-12, as a function of iodine pressure. These measurements were performed at 50° C using a four point probe. The reliability of the measurements was impaired by the fact that the films were by no means perfectly uniform (as indicated by scanning electron microscope studies), and also because of the underlying, thin (but nevertheless present) layer of gold. Figure 13 shows the results. Again, helium was mixed into the iodine vapor at two partial pressures, 0.125 torr and 0.375 torr. In the figure the logarithm of the conductivity is plotted versus the logarithm of the iodine pressure. Once

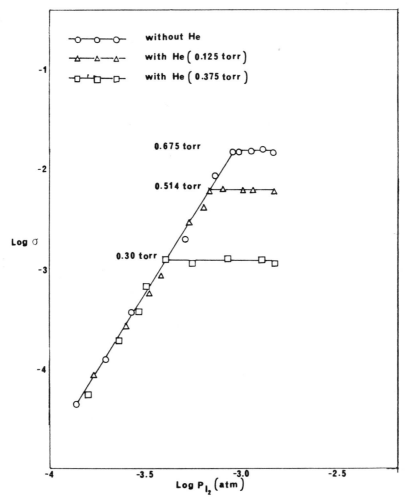

Fig. 13. Plot of log conductivity versus log iodine pressure for the film of Figures 10, 11, and 12. Note that the slope has the value 2.6.

more, we see the remarkable discontinuities in slope which always occur at approximately the same <u>total</u> pressure of the ambient gas mixture. However, <u>here</u> it is <u>conductivity</u> which is measured, not <u>iodine uptake</u>. The slope of the log-log plot is approximately 2.6. This implies that the conductivity varies as the 2.6 power of the iodine pressure.

The situation is therefore somewhat reminiscent of that found in connection with the conductivty of ferrous oxide, depicted in Figure 4. There the conductivity varied as the 1/6 power of the oxygen pressure. In that case, we were able to explain the variation in terms of a physically acceptable model. Can we do the same for iodine in polythiophene, and therefore throw some light on the electronic species involved? The present case is somewhat more complicated by the fact that not only does the conductivity vary approximately as the 2.6 power of the iodine pressure, but also, according to Figures 10-12, the total uptake of iodine varies approximately linearly with pressure. Any postulated model must account simultaneously for these two behaviors.

Furthermore, the level of doping is so high that, if conductivity is "domain limited" as appears to be the case in crystalline polyacetylene[18], the connectivity of the domains almost certainly corresponds to that beyond the critical point for percolation. Bearing this in mind, we assume the existence of the equilibria described by the following two equations.

$$I_2(gas) \rightleftharpoons I_2(PT) \tag{9}$$

$$5I_2(PT) + \sigma \rightleftharpoons I_5^- + I_5^- \cdot (e^+)_2 \tag{10}$$

These processes, and the symbols used to represent them, require some discussion. Eq. (9) represents the reversible distribution of I_2 between the vapor phase (gas) and the polythiophene film (PT). $I_2(PT)$ denotes I_2 molecules in the film. Assuming the validity, or near validity, of Henry's law we may write

$$[I_2] = \alpha P \tag{11}$$

where the square bracket stands for concentration, α is the Henry's law constant, and P is the pressure of the iodine vapor.

In Eq. (10) σ represents an "effective" site on the polymer molecule at which the species on the right of the equation can be formed. We will assume that the total iodine in the species represents only a few percent of all of the iodine dissolved in the film, most of the iodine remaining in the form $I_2(PT)$. Under this circumstance the concentration of σ can be approximated as constant, and we may write an equilibrium constant relation, for the process in Eq. (10), having the following form

$$\frac{[I_5^-] [I_5^- \cdot (e^+)_2]}{[I_2]^5} = K_B \tag{12}$$

in which the square brackets again stand for concentration, and all of the species involved in the process are situated within the film. The species, on the right of Eq. (10), whose concentrations appear in the numerator of Eq. (12), consist of a pentaiodide ion (a known stable species) and a bipolaron $(e^+)_2$ "pinned" to a pentaiodide ion.

These species are rationalized as follows. The bipolaron is reasonable on the usual basis that an isolated polaron would require a long run of polymer chain having a high energy configuration[19]. Polyacetylene is the exception since the polymer chain on each side of the polaron (or soliton) will represent one of two energetically degenerate configurations. As a result, except for polyacetylene, two polarons want to be as close as possible, and therefore form a bipolaron. However, the bipolaron need not be pinned by two pentaiodide ions equally close to it. The obvious reason for this lies in the coulomb repulsion between the two negative ions. The final situation is then a balance between the need for the two polarons to be close and the two anions to be far from one another.

On the other hand, even though the species on the right of Eq. (10) constitute only a small fraction of the dissolved iodine, the overall level of doping is sufficiently high, so that that fraction represents a large number of pentaiodide ions in an absolute sense. The bipolaron $(e^+)_2$ may then move (much as the electrons in a metal move among the core positive ions in the metal) with relative ease, and constitute the carriers of electric current. To a first approximation then, the conductivity of the film should be proportional to the concentration of bipolarons. This last assumption is compromised somewhat by the fact that, as iodine pressure is varied, the concentration of pentaiodide ions will also vary so that the distance between ions will change, and the mobility of the bipolaron may not be constant. On the other hand as mentioned earlier, at the high concentrations with which we are concerned, the critical percolation density will probably have been exceeded, and the mobility should not be too sensitive to concentration.

Nevertheless, if we assume, in accordance with the previous argument, that the conductivity is proportional to the bipolaron concentration we may use Eqs. (11) and (12) to determine the dependence of that concentration on iodine pressure, and, therefore, of the dependence of the conductivity on pressure. However we need one last relation which follows from the stoichiometry of Eq. (10). This is

$$[(e^+)_2] = [I_5^- \cdot (e^+)_2] = [I_5^-] \tag{13}$$

Solving Eqs. (11), (12), and (13) we find

$$[(e^+)_2] = (K_B \alpha^5)^{1/2} P^{5/2} = \gamma P^{2.5} \tag{14}$$

in which γ is a constant. Thus we see that, according to this model, the concentration of bipolarons should depend on iodine pressure to the 2.5 power, and therefore the conductivity should depend on pressure to the same power. The exponent 2.5 in Eq. (14) is attractively close to the exponent 2.6 in the observed pressure dependence of Figure 13.

Thus, within the limitations of the tentative nature of the argument, we have learned something about the electronic species from a reversible measurement of solubility, and the approach has been similar to that used effectively in connection with conventional, inorganic semiconductors.

However, we must also explain the linear dependence of iodine uptake on iodine pressure. The total uptake (in units of I_2) is given by

$$[I_2]_{Tot} = [I_2] + \frac{5}{2}[I_5^-] + \frac{5}{2}[I_5^- \cdot (e^+)_2]$$

$$= [I_2] + 5[(e^+)_2] \tag{15}$$

$$= \alpha P + 5(K_B \alpha^5)^{1/2} P^{5/2}$$

where we have used Eqs. (11), (13), and (14). The quantity $[I_2]_{Tot}$, i.e. the total uptake, will appear to be approximately linear in P if the second term on the extreme right of Eq. (15) is sufficiently small compared to the

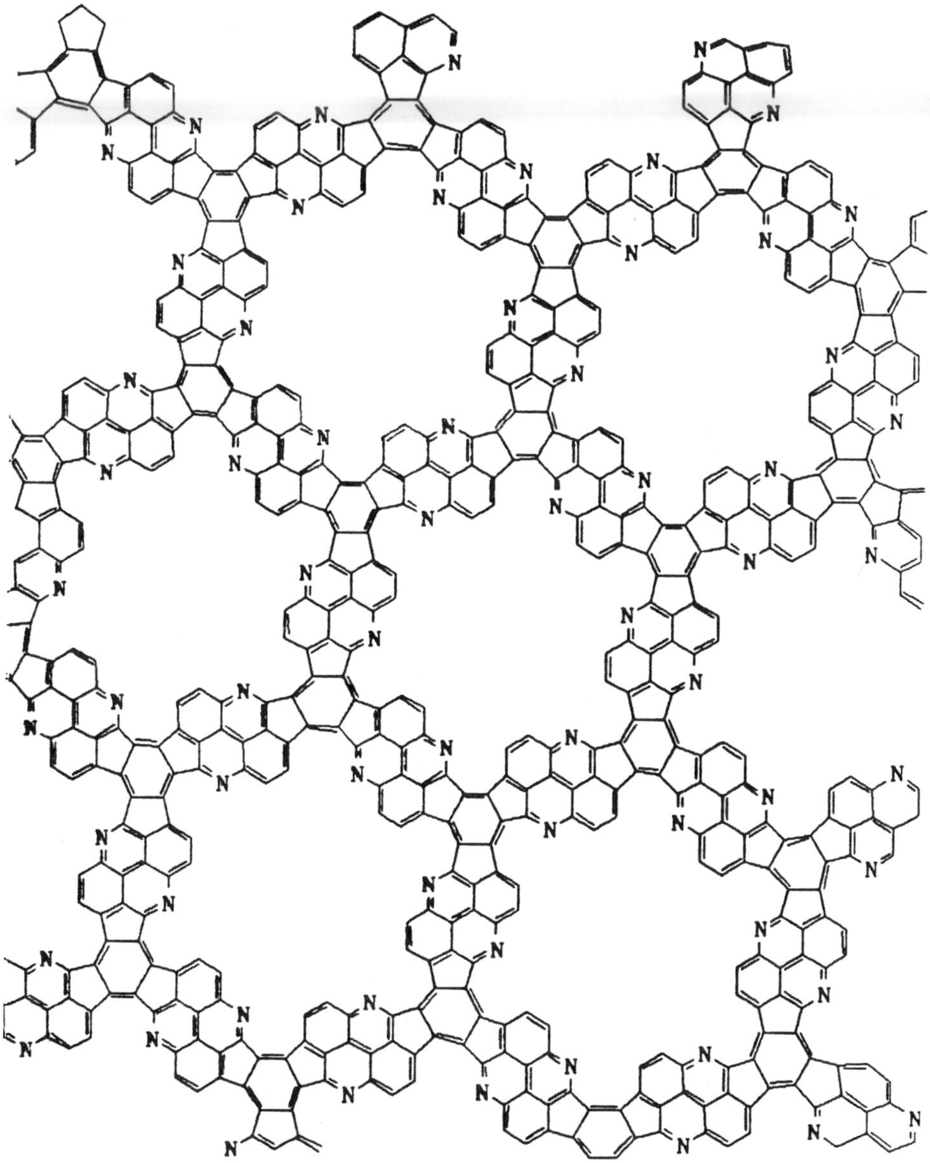

Fig. 14. Structure of Azite according to O. L. Chapman.

first term. But we have already chosen K_B in Eq. (12) to be small enough so that this is the case, i.e. K_B has been chosen small enough so that only a few percent of the total absorbed iodine is in the form of the species on the right of Eq. (10). Thus, within the precision of measurement, we have explained the general features of both the uptake and conductivity curves. Note that we have not explained the concave downward and concave upward curvatures at low pressures in the uptake curves, nor have we explained the remarkable discontinuities in slope which appear in Figures 10-12.

Recently Professor O. L. Chapman and his coworkers synthesized the compound whose structure is exhibited in Figure 14. They have named this

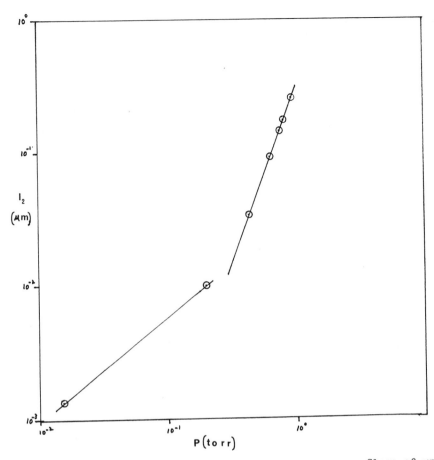

Fig. 15. Plot of I_2 uptake in Azite versus I_2 pressure. Slope of upper portion of curve is 2.6. Slope of lower portion is 1.56.

compound "azite." At their request we measured the absorption isotherm for iodine vapor in azite. Again, absorption proved to be reversible. Figure 15 exhibits the experimental data as a plot of logarithm of iodine uptake as a function logarithm of iodine pressure, at 50° C. This work is very preliminary, but the plot is observed to consist of two linear segments, one with a slope of 3/2 and the other with a slope of 5/2. Without going into detail, it should be obvious, by now, that this suggests that iodine absorbed, at the lower pressures, in the form of I_3^-, and at higher pressures as I_5^-, i.e. as triiodide and pentaiodide, respectively. This is to be contrasted with the isotherm for polythiophene where, according to the previous analysis the uptake is linear with iodine pressure, suggesting that the uptake is mainly in the form of I_2. Conductivity measurements are not yet available for azite containing iodine.

In closing this paper it should be remarked that conducting polymers can apparently be doped electrochemically in a reversible manner[20]. If this is the case, analysis leading to the identification of "species," similar to those described in the present paper should be possible. In the case of electrochemical doping, the thermodynamic activity of the doping agent in the electrolyte solution would take the place of the iodine pressure in the present paper. Furthermore, the situation would be even more favorable than in the present paper, since measurements involving much lower dopant concentrations should be possible. In addition, the electrode potential involved in the doping process is also measurable, and this represents another degree of freedom, capable of supplying information on species. Theoretical analysis of model systems undergoing electrochemical doping will be presented in a forthcoming paper[21].

ACKNOWLEDGEMENT

This work was supported by AFOSR grant #F49620-86-C-0060.

REFERENCES

1. F. A. Kröger, "The Chemistry of Imperfect Crystals," North Holland Publishing Co., Amsterdam, 1964.
2. H. Schmalzried, "Solid State Reactions," Academic Press, Inc., New York, 1974.
3. H. Reiss, C. S. Fuller, and F. J. Morin, Bell Syst. Tech. Journ. 35, 535 (1956).
4. H. Reiss, in Proceedings of the Robert A. Welch Foundation Conferences on Chemical Research XIV "Solid State Chemistry," p. 145, (1970).
5. B. Francois, M. Bernanrd, and J. J. Andre, J. Chem. Phys. 75, 4142 (1981).
6. Private communication from Professor O. L. Chapman, Department of Chemistry and Biochemistry, University of California, Los Angeles.
7. R. R. Chance, D. S. Boudreaux, H. Eckhardt, R. L. Elsenbaumer, J. L. Bredas, and R. Silbey, in "Quantum Chemistry of Polymers-Solid State Aspects" ed. by J. Ladik et al., D. Reidel Publishing Co. 1984.
8. A. J. Rosenberg, J. Chem. Phys. 33, 665 (1960).
9. W. Kaiser, H. L. Frisch, and H. Reiss, Phys. Rev. 112, 1546 (1958).
10. J. R. Patel, K. A. Jackson, and H. Reiss, J. Appl. Phys. 48, 5279 (1977).
11. K. Hauffe and H. Pfeiffer, Z. Metallk. 44, 27 (1953).
12. H. Reiss and C. S. Fuller in "Semiconductor" ed. by N. B. Hannay, Ch. 6, Reinhold Publishing Corp. New York, 1959.
13. H. Reiss and W. D. Murphy, J. Phys. Chem. 89, 2596 (1985).
14. A. J. Epstein, H. Rommelmann, R. Bigelow, H. W. Gibson, D. M. Hoffman, and D. B. Tanner, Phys. Rev. Lett. 50, 1866 (1981).
15. H. Reiss, J. Chem. Phys. 25, 4087 (1956).
16. H. Reiss and Dai-uk Kim, J. Phys. Chem. 90 1973 (1986).
17. Dai-uk Kim, unpublished results.
18. L. W. Shacklette and J. E. Toth, Phys. Rev. B32, 5892 (1985).
19. R. R. Chance, D. S. Boudreaux, J. L. Bredas, and R. Silbey, in "Handbook on Conducting Polymers," ed. by T. Skotheim, Marcel Dekker, New York.
20. P. M. McManus, Sze Cheng Yang, and Richard J. Cushman, J. Chem. Soc., Chem. Commun. 1556, (1985).
21. Submitted to the Journal of Physical Chemistry.

TRANSIENT CURRENT SPECTROSCOPY IN POLYMERIC FILMS

E. Muller-Horsche and D. Haarer

Physikalisches Institut and Bayreuther Institut
fur Forschung
Universitat Bayreuth
Postfach 3008, D-8580 Bayreuth, FRG

H. Scher

Standard Oil, Research and Development
4440 Warrensville Heights, Ohio 44128

ABSTRACT

We briefly review dispersive transport and treat in more detail the multiple-trapping mechanism. We consider the problem of transient current spectroscopy: Given the experimental current I(t) how does one extract a distribution of rates? Extensive new measurements were made in a well studied polymeric material.

Photocurrents in polyvinylcarbarzole were studied over 10 decades in time. An analytical approach is proposed for extracting rate distributions from the measured photocurrents. We find that the trapping-rate distribution is not exponential but "flat" and that it shows a cutoff at low rates. This distribution gives rise to the novel feature of a gradual transition from dispersive to nondispersive charge transport during a single transit. The temperature and field dependence of the cutoff rate r_c was studied in the framework of a Poole-Frenkel model.

INTRODUCTION

It is well documented that photocurrents in polymer systems like PVK show dispersive behavior. The general theory of dispersive transport was developed by Scher and Montroll (SM).[1] In the theory, the key physical idea is that the microscopic processes which control carrier transport are governed by a distribution of event times which is broad over the time range of experimental observation. SM used the formalism of continuous-time random walk (CTRW) to describe carrier motion with an event-time distribution $\psi(t)$; the event time could either be due to, e.g., hopping between localized states or release from a trap. SM showed that for a $\psi(t)$ with a slowly varying algebraic time dependence, $\psi(t) \sim t^{-(1+\alpha)}$, with a disorder parameter α, in the range $0 < \alpha < 1$, one could account for all of the key features of dispersive transport as shown in Fig. 1.

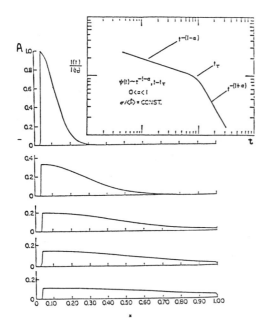

Figure 1. The amplitude
of a charge packet vs.
displacement in the film
(in units of sample
thickness). Shown are the
packet dispersion at 5
time intervals. The inset
is a log-log plot of the
normalized $I(t)$ vs. t.

Noolandi[2] and Schmidlin[3] showed that the conventional set
of coupled kinetic equations, describing carrier motion with
repeated trapping (w_i) and thermal release (r_i), can be cast
into the framework of the CTRW and is therefore equivalent.
They demonstrated the exact form of the $\psi(t)$ for multiple
trapping (MT) in terms of $\{w_i, r_i\}$; therefore, MT is a subset of
the microscopic transport processes describable by the $\psi(t)$.

Determination of Rate Distribution from I(t)

The method, which will be presented below, is adapted to
the needs of the experimentalist; in particular, it does not
make assumptions about rate distributions. It can be
demonstrated that the rate distributions can be extracted from
the experimental data. There is a price, however, which we
have to pay for making no restricting assumptions about the
trap distributions. We have to perform experiments, which
cover many orders of magnitude in time or rate. In our case
we performed experiments over 10 decades and derived rate
distributions over 5-6 decades. This "loss in information" is
due to comparatively broad distribution functions, which enter
into our Laplace analysis. We do not want to rule out that
there is a more sophisticated approach, which has a higher
intrinsic resolution. Our main goal is, at the present time,
to raise the issue of rate distributions, discuss our results
versus the usefulness of effective mobilities μ_{eff} and to
present our data for further theoretical considerations.

$$\frac{\partial p(z,t)}{\partial t} = q(z,t) + \Sigma p_i(z,t) r_i - p(z,t) \sum_i w_i - \frac{\partial}{\partial z} p(z,t) \mu E(z,t),$$

$$(1)$$

$$\frac{\partial p_i}{\partial t} = p w_i - p_i r_i \quad.$$

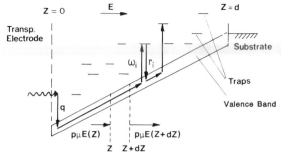

Figure. 2. Multiple trapping parameters (for holes); q is the carrier production term, w_i the trapping rate, r_i the detrapping rate, E the electric field, d the sample thickness, p the carrier density, and z is the integration coordinate.

Figure 2 and Eq. (1) describe the simple MT model. The trapping rates w_i and detrapping rates r_i populate and depopulate the charge densities $p(z,t)$ and $p_i(z,t)$ in the conduction band and in the traps, respectively. The function $q(z,t)$ stands for the rate of charge carrier production. $\partial(p\mu E)/\partial z$ describes the process of charge accumulation by a gradient in the local flux. In the following we assume that space-charge effects can be neglected and, hence, we assume a constant electric field across the sample (see Fig. 2).

Assuming pulsed excitation in our sample with light of an infinitely short penetration depth we calculate the Laplace transform of the total current I(t)

$$\tilde{I}(r) = N \frac{1-\exp[-a(r)t_0]}{a(r)t_0} \tag{2}$$

$t_0 = d/\mu E$ stands for the microscopic transit time, i.e., the transit time of a fictive charge carrier which remains in the band as a free carrier as it traverses the sample of thickness d under the influence of a field E. The quantity $a(r)$ contains the rate variable r in Laplace space as well as the trapping and detrapping rates w_i and r_i,

$$a(r) = r \left[1 + \sum_i \frac{w_i}{r+r_i} \right] \quad , \tag{3}$$

it is proportional to $\tilde{\psi}^{-1}(r)$, where $\tilde{\psi}$ is the Laplace transform of $\psi(t)$.

Note, that $a(r)$ [Eq. (3)] contains all the model parameters r^* and w_i, which we like to determine. To do so, we introduce the logarithms $r^* = \ln r$ and r_i and rewrite Eq. (3),

$$a(r^*)t_0 = e^{r^*} t_0 + \sum_i M_i \frac{1}{1+e^{r_i^*-r^*}} \tag{4}$$

305

We have also introduced the quantity $M_i \equiv \omega_i t_0$ (number of trapping events into the ith level during the microscopic transit time t_0). Differentiation of Eq. (4) with respect to r^* yields

$$F(r^*) := d(at_0)^*/dr^* - e^{r^*} t_0 = \sum_i [M_i f(r_i^* - r^*)]. \qquad (5)$$

This represents a superposition of the bell-shaped curves $M_i f(r_i^* - r^*)$ centered around the detrapping rates r_i and weighted with M_i. Note that $f(x)$ is normalized to 1. Its shape is depicted in Fig. 3.

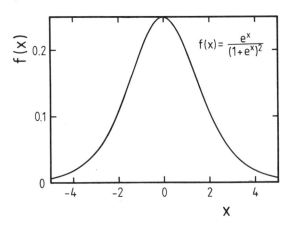

$$f(x) = \frac{e^x}{(1+e^x)^2}$$

Figure 3. f(x) function limiting our resolution in the Laplace transform. The function is normalized to 1. f(x), plotted on a log-log scale, will appear like a rounded-off triangle with slopes 1 on either side (see Fig. 5).

The functional dependence of the distribution $F(r^*)$, as defined in Eq. (5), on the quantities M_i and r_i suggests the following generalization to a continuum of traps:

$$F(r^*) = \int M(s^*) f(s^* - r^*) ds^* , \qquad (6)$$

with the following definition of $M(r^*)$: $M(r^*) dr^*$ gives the number of trapping events during the time t_0 into traps having detrapping rates between r^* and $r^* + dr^*$. It is $M(r^*)$ that gives us information about the trap density through the relation $M(r^*) = t_0 b_t n(r^*)$, where b_t is the trap capture coefficient and $n(r^*) dr^*$ the trap density in the interval r^*, $r^* + dr^*$. One can relate $n(r^*)$ to the density of states $\rho(\epsilon)$, once $r(\epsilon)$ is specified (see below).

The distribution function $F(r^*)$ can be obtained from our experimental photocurrents, which yield $a(r)t_0$ [via Eq. (2)] and the definition in Eq. (5). The right-hand side of Eq. (6) corresponds to a rate distribution $M(r^*)$, which is broadened by the function f. A deconvolution of $F(r^*)$ is not very well defined because of the width of f and because of boundary problems. Therefore we treat $F(r^*)$ as a broadened distribution of $M(r^*)$ and try to get information from our experiments by measuring the photocurrents over many orders of magnitudes in time (e.g., 10 decades in the experimental results presented in a later section). This corresponds to a large range of available $F(r^*)$ data on the r^* scale and the larger this interval, the smaller the relative width of f. This width of $f(r^*)$ is a clearer statement of the resolution limit of using I(t) as a spectroscopy tool than has been discussed in the literature.

Experimental Results and Discussion

As samples we used commercial PVK (102, Luvican, trade name BASF) which was purified by repeated precipitation from solution and which contained less than 0.05% monomer carbazole. The films were cast from THF solution. To maintain a well-defined sample thickness, we used a technique in which a spatula could be moved parallel to a precision surface with an accuracy of <1μ. The substrate onto which the films were spread was mylar coated with a thin Al layer. A typical sample thickness was 10μm. As the second electrode we used a thin layer of semitransparent, evaporated aluminum.

As a light source we used a pulsed nitrogen laser with a pulse energy of about 5 mJ at the wavelength 337 nm. The pulse width was 10 ns. With the above excitation conditions, the penetration depth of the light was <0.2 μm, i.e. about 50 times less than the sample thickness. This provided adequate conditions for surface excitation, i.e., for justifying a δ-function-like charge carrier distribution for our calculations at t=0.

One of the major challenges of the experiment was to measure currents in the time domain of 1 ns to 10 s with a maximum sensitivity of 10^{-10} Å. There is no single transient recorder which would provide a large enough dynamic range, therefore we subdivided each curve into five segments which were measured in subsequent experiments. The fast decay was monitored with a Tektronix 7912 digitizer (risetime 1 ns) and the slow part of the decay was monitored with a logarithmic time sweep (for details see Ref. 4). In the "long-time regime" we also used an amplifier with logarithmic conversion. The various sections of the decay curve were digitized; the linear sections of the decay curve were converted to log scales. The complete decay curve was subsequently displayed on a screen as depicted in Fig. 4. The figure also shows the limits of the various individual decay sections.

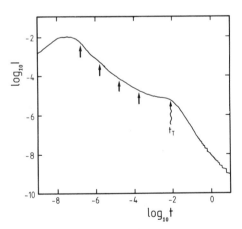

Figure 4. Typical experimental photocurrent; the arrows mark the limits of five individual measurements, which are joined to a single curve (see text). The wavy arrow marks the transit time t_T; there the influence of the back electrodes leads to a dropoff in current.

Using the above-described setup, we were able to measure the photocurrent in our PVK sample over ten decades in time and over more than eight decades in intensity. Figures 5 and 6 show our results as a function of sample temperature and electric field strength.

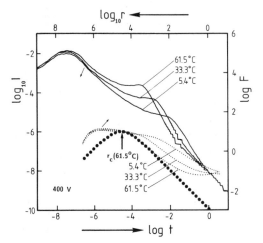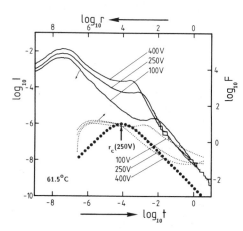

Figure. 5. Solid lines:
Photocurrent of PVK (10 μm
thick) at constant voltage
(400 V) and variable
temperature. The
experimental data cover ten
decades (left and lower
axis). Dotted lines: Rate
distributions as calculated
by our Laplace transform
(right and upper axis).
Note, that for short times
(fast rates) the validity of
our evaluation scheme breaks
down in the μs region (see
text). The arrows point
into the directions of the
axes. The dotted line
(heavy dots) gives the
contribution of one rate r_c
(61.5°C 400 V).

Figure 6. Solid lines:
Photocurrent of PVK (10 μm
thick) at constant
temperature (61.5°C) and
variable field strength.
Dotted lines: Rate
distribution as
calculated by our Laplace
transform. The dotted
line (heavy dots) gives
the contribution of one
$rate_c$ (61.6°C, 250 V).

Based on our experimental data we come to the conclusion
that the distribution of trapping rates is flat over the
parameter range $r < 10^{-6} s^{-1}$ which we can evaluate experimentally.
At a critical minimal rate r_c the distribution of rates falls
off with the resolution of our Laplace transform analysis.

The large time range of the experiment has enabled us to
encompass these two distinct features of the trapping rate
distribution. These features cause a transition in I(t) from
a highly dispersive transport to an essentially non-dispersive
one. This type of transition has been observed in both a-Se
and a-Si:H, however, as a function of T. The transition in
I(t) observed in Fig. 4 occurs within a single transite time.
The form for I(t) for $t < t_T$ for dispersive transport is

$$I(t) \propto t^{-(1-\alpha)} \,, \quad t < t_T \;.$$ (7)

We can characterize the transition in Fig. 4 as one in which $a \simeq 0.0$ for $-7.1 < \log t < -5.25$ and $a \to 1.0$ for $\log t < -2.4$. The change of slopes is gradual--nearly three decades.

We will show that a characterization of $I(t)$ with a time dependent $a(\log t)$ can lead directly to an approximate form for $\rho(\epsilon)$, which is in agreement with our more accurate determination of $F(r)$.

The MT model is a particularly simple mechanism for generating an algebraic $\psi(t)$ as it involves the independent sampling of a single random variable ϵ (the trap energy) with a general distribution $\rho(\epsilon)$. An excellent approximation to $\psi(t)$ for MT, in the repeated trapping and release time range, i.e., $r\tau << 1$, $\tau^{-1} \equiv \Sigma_i w_i$, is

$$\psi(t) = \sum_i \tau w_i r_i e^{-r_i t} \tag{8}$$

the weighted sum of the probability per unit time to be released from the ith level. One can immediately derive $a = T/T_0$, as was first done using Eq. (8) (Ref. 4) for $\rho(\epsilon) \alpha \exp(-\epsilon/kT_0$ and also derive the form of $\psi(t)$ for a finite width $\rho(\epsilon)$ with a minimum r_c cutoff [Eqs. (43) and (44) of Ref. 5]. An interesting interpretation has been made of the form of $\psi(t)$ in Eq. (8). Shlesinger[1] has introduced the notion of fractal time in analogy with a spatial RW occurring on all length scales. Shlesinger distinguishes each level or hierarchy with a multiplicative (power-law) weighting,

$$\psi(t) = \frac{1-a}{a} \sum_{n=1}^{\infty} a^n b^n \exp(-b^n t) \tag{9}$$

and for $b < a < 1$ one has $\psi(t) \alpha t^{-1-a}$, with

$$a = \frac{\ln a}{\ln b} \quad . \tag{10}$$

We can generate the continuum version of Eq. (10) for the MT model by rewriting

$$a = \frac{T}{T_0} = \frac{-\epsilon/kT_0}{-\epsilon/kT} = \frac{\ln[\rho(\epsilon)/\rho_0]}{\ln[r(\epsilon)/\nu]} \quad , \tag{11}$$

where $r(\epsilon) = \nu \exp(-\epsilon/kT)$ is the release rate at the effective energy ϵ, which can include field-induced barrier lowering. For a general $\rho(\epsilon)$ we conjecture that

$$a(\ln t) = \frac{\ln[\rho(\epsilon_d)/\rho_0]}{\ln[r(\epsilon_d)/\nu]} \quad , \tag{12}$$

where

$$\epsilon_d = kT\ln(\nu t), \tag{13}$$

309

the time-dependent demarcation energy. In Eq. (12) we have a direct relation between the slope parameter of I(t) and the density of states. We have calculated $\rho(\epsilon)$ using Eq. (12) and the time-dependent tangent to the [logI(t)]-(logt) curve in Fig. 4. The expression above has been used for r(ϵ). The general feature of the density of states are as follows: (1) $\rho(\epsilon)$ is flat ($T_0 \gg T$) for $\epsilon_0 < 0.61$ eV; (2) a slow rollover occurs for $0.61 < \epsilon_0 < 0.64$ eV with an average $T_0 = 80$ meV; (3) between 0.65 and 0.69 eV the average $T_0 = 50$ meV; (4) at $\epsilon_0 > 0.69$ eV there is a very sharp cutoff of $\rho(\epsilon)$, with the effective $T_0 \ll T$ (ϵ_0 denotes the field-independent part of ϵ). These features, derived from Eqs. (12) and (13), are in very good agreement with the more accurately determined F(r) of Fig. 5, and the value of $\epsilon_0 = 0.69$ eV for r_c. The gradual change from highly dispersive to nondispersive transport corresponds to a quasiexponential range of $\rho(\epsilon)$ ($82 \gtrsim T_0 \gtrsim 47$ meV) with an energy range nearly 0.1 eV wide. The validity of Eqs. (12) and (13) has the same range as the simpler, more intuitive treatment of MT in Ref. (6). The approach of Ref. 6 would yield similar results for $\rho(\epsilon)$, derived from our I(t) data. However, Schiff (Ref. 7) has cautioned that there are $\rho(\epsilon)$, e.g., nonmonotonic ones, that would lead to I(t) features, using the method of Ref. 6, in disagreement with experiment. one purpose of the present paper is to present a general procedure to extract a rate spectrum from I(t) data and show the intrinsic resolution of this deconvolution.

The nearly nondispersive transport usually ascribed to PVK occurs over a time range less than 2 decades prior to t_T. If one restricts measurement of I(t) to this time range than one would conclude that there is a very narrow range of trap energies instead of the broad one with a cutoff determined in the present study. The release rate r_c at the cutoff, dominates μ_{eff}, and may be a better characterization of the transport properties of PVK.

Let us show this by evaluating

$$\mu_{eff} = (d^2/t_T)U, \qquad (14)$$

d is the sample thickness and t_T is the effective transit time, as taken from the experimental decay curves (see Fig. 4). The results are shown in an Arrhenius plot in Fig. 7. Note, that in a simple model, the lines should intersect at 1/T=0 and yield the "band mobility," i.e., the mobility in the absence of trapping. If we adopt the empirical procedure of introducing an effective temperature T', with

$$1/T' = 1/T - 1/T^* \qquad (15)$$

we obtain an intersection point at the temperature $T^* = 540$ K and a value for the effective band mobility of $\mu_{eff} = 0.02$ (cm^2/Vs).

We can express the effective mobility μ_{eff} as a function of the electric field strength by assuming a field-dependent barrier height and get

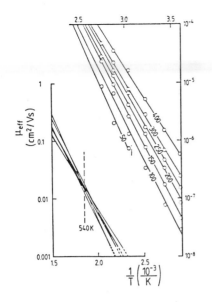

Figure. 7. Effective
mobility as measured by
the transite time,
plotted vs. 1/T. Note,
that the straight lines
intersect at 540K. (For
the sake of saving
space, the figure is
subdivided into two
sections.) The numbers
label the applied
voltages.

$$\mu_{eff} = \mu_o \exp[-\epsilon(E)/kT'], \tag{16}$$

where $\epsilon(E)$ is the field-dependent activation energy.[8] If we
assume that a Poole-Frenkel mechanism is responsible for
lowering the barrier height, then we get

$$\epsilon(E) = \epsilon_o - \beta \sqrt{E} \tag{17}$$

Here ϵ_o is the barrier height in the absence of a field and
β is a nonadjustable parameter of 4.4×10^{-4} e $\sqrt{V}$ - cm .

Our experimental data yield $\epsilon_o = 0.64$ eV and $\beta = 2.4 \times 10^{-4}$ e $\sqrt{V}$ - cm, i.e., the β value is about a-factor-of-2 off from the
theoretical value.

At this point we can summarize our results with the
conclusion, that we get moderate agreement with semiempirical
theories, describing the field and temperature dependence of
the effective mobility. Above all, we do not see a convincing
argument for maintaining the concept of an effective
temperature T' to accommodate our data on a 1/T' plot.

Instead, we return to our rate distribution and examine the
critical lowest rate r_c. This rate is marked on Fig. 5 and 6
for the 61.5°C curve. The falloff of the rate distribution is
given by the falloff of our resolution function (slope 1 in a
log-log plot). Second, the lowest rate r_c shows a typical
temperature and field dependence; it decreases for lower
temperatures and increases for higher fields. This
interesting behavior encouraged us to evaluate the observed
minimal rates r_c on a 1/T plot. We investigated the following
ansatz:

$$\ln r_c = \ln \nu - \epsilon(E)/kT. \tag{18}$$

Figure 8 shows, that if we plot the data in the above manner,
we get straight lines which intersect at 1/T=0. At this
intersection point, we obtain a frequency ν of 2×10^{12} s^{-1}, which
we can interpret as an "attempt-to-escape" frequency. This
frequency is close to values, which have been discussed in the
literature.

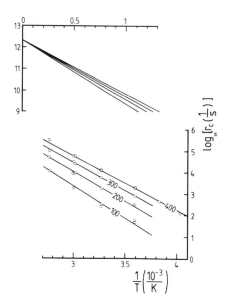

Figure 8. Slowest rate r_c, as obtained by our rate analysis, plotted vs. 1/T. Note, that the straight lines intersect at 1/T=0. (For the sake of saving space the figure is subdivided into two sections.) The numbers label the applied voltages.

Going a step further in this interpretation, we can check the validity of a Poole-Frenkel interpretation of the measured field dependence. Figure 9 shows the dependency of the rates r_c on the square root of the applied voltage. From the figure we get a straight line with an ϵ_o of 0.69 eV and with a β value of 3.4 x 10^{-4} e √ V – cm. This value is rather close to the theoretical value of 4.4 x 10^{-4} e √ V – cm (see above). In a forthcoming publication[9] we will show that the observed temperature and field dependence of the effective mobility can be modeled under the assumption that Eqs. (17) and (18) hold for all traps, so that the introduction of an effective temperature T' (15) can be avoided.

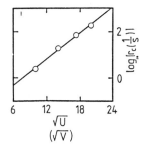

Figure 9. Poole-Frenkel Plot of the slowest rate r_c (see text).

The field of dependence in Eq. (17) can also be understood as Coulomb effect in the charge transfer step between localized molecular states. If the microscopic mobility is due to hopping, the hole can be trapped in a lower ionization energy state, forming a cation and detrapped when an electron transfers up from a neutral molecule, leaving behind a cation.

References

*   Extracts from a paper by the authors, Physical Review B35, 1273 (1987).

1. H. Scher and E. W. Montroll, Phys. Rev. B12, 2455 (1975).

2. J. Noolandi, Phys. Rev. B16, 4466 (1977).

3. F. W. Schmidlin, Phys. Rev. B26, 2362 (1977).
4. E. Muller-Horsche, Ph.D. thesis, Universitat Bayreuth, 1985 (unpublished).
5. H. Scher, in Proceedings of the Seventh International Conference on Amorphous and Liquid Semiconductors, Edinburgh, 1977 (unpublished), p. 209; G. Pfister and H. Scher, Adv. Phys. 27, 747 (1978) (referred to in text for sections and equations).
6. T. Tiedje and A. Rose, Solid State Commun. 37, 49 (1981); J. Orenstein, M. Kastner, and V. Vaninov, Philos. Masg. B 46, 23 (1982).
7. E. A. Schiff, in *Tetrahedrally-Bonded Amorphous Semiconductors*, edited by D. Adler and H. Fritzsche (Plenum, New York, 1985), p. 357.
8. W.D. Gill, J. Appl. Phys. 43, 5033 (1972).
9. H. Domes, D. Haarer, and E. Muller-Horsche (unpublished).

ELECTROCHEMISTRY OF POLYANILINE: CONSIDERATION OF A DIMER MODEL

Walter W. Focke[†] and Gary E. Wnek[*]

Department of Materials Science an Engineering
Massachusetts Institute of Technology
Cambridge, M.A.

INTRODUCTION

Polyaniline (PAn) is the electroactive polymer obtained by the oxidation of aniline in acidic media [1]. In the free base form the polymer structure is that of a poly(p-phenyleneamineimine) [2] with repeat units:

and

The oxidation state of polyaniline depends on the relative proportions of the amine and quinone-diimine units in the free base form. The emeraldine oxidation state corresponds to an equal number of amine and imine nitrogens in the polymer backbone.

Chiang and MacDiarmid [3] have drawn attention to the novel Brönsted acid doping of this polymer. Huang and MacDiarmid [4] suggested that the acid-base and electrochemical behavior of polyaniline might be modeled on the basis of dimeric repeat units. They derived a bipolaron model for the emeraldine oxidation state since their analysis neglects the formation of radical cations. Bipolarons in polyaniline correspond to fully protonated quinone-diimine units. In this paper we extend the analysis of Huang and MacDiarmid [4] to include radical cations. The resulting polaron model should be valid for symmetric dimeric oligomers e.g. N,N'-diphenyl-p-phenylenediamine. The validity of this model for the polymer is investigated.

JUSTIFICATION OF THE DIMER MODEL

To gain a better understanding of the structure and behavior of polyaniline we have compared its UV spectra and cyclic voltammetry data to those of the phenyl-capped dimeric and tetrameric oligomers. The phenyl-capped oligomers were chosen because of their symmetry (which reduces end effects) as well as for their greater chemical stability. Oligomers that have amino endgroups are hydrolytically unstable upon oxidation.

---

† Present Address: NIMR, CSIR, Box 395, Pretoria 0001, South Africa.

* Present Address: Department of Chemistry, Rensselaer Polytechnic Institute, Troy, New York 12180-3590.

Figure 1 shows UV spectra obtained for solutions of 50% oxidized material (emeraldine oxidation state) in DMSO. The spectra for the dimer are actually for an equimolar mixture of N,N'-diphenyl-p-phenylenediamine and N,N'-diphenyl-p-phenylenediimine while those for the polymer are for chemically synthesized emeraldine. In the free base forms the primary feature is that of a broad peak at 450 nm, 594 nm and 621 nm respectively for the dimer, tetramer and polymer. Upon acidification this peak, characteristic of the free base form, disappears and in its place two new peaks grow at approximately 400 - 420 nm and 700 - 900 nm depending on the chain length. The basic similarities in the spectra of oligomers and polymer in the free base as well as protonated forms is suggestive for similar molecular structures being involved [5,6].

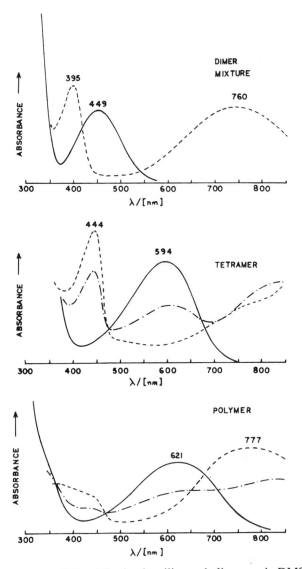

Figure 1. UV spectra for 50% oxidized polyaniline and oligomers in DMSO. Free base form: ———, on addition of a small excess of concentrated HCl: - - - - -, on addition of a small amount of HCl: — · —.

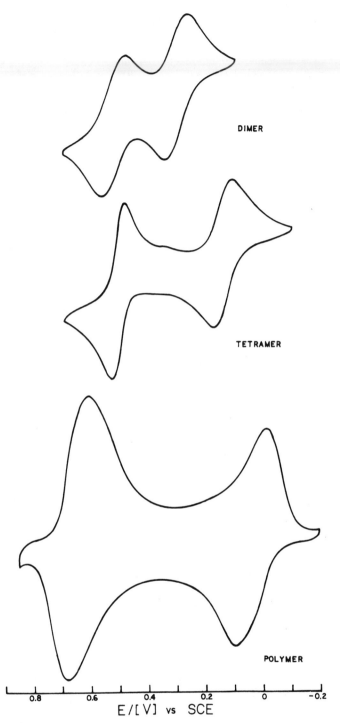

DIMER

TETRAMER

POLYMER

E/[V] vs SCE

Figure 2. Cyclic voltammograms for electrochemically grown polyaniline films and solutions of oligomers in 80% acetic acid containing 0.1 N TEAP. Scan rate 20 mV/s. Almost identical CV's are obtained if the TEAP is replaced by 0.1 N CF$_3$COOH.

Figure 2 shows cyclic voltammograms obtained in 80% acetic acid containing 0.1 N TEAP. All three voltammograms show only two sets of redox peaks. The separation between the two sets of redox peaks increases with increasing chain length. This means that it is easier to oxidize the longer molecules to an intermediate oxidation state although the stability of this state also extends to higher potentials for the longer chains. In the case of the dimer it is well established that the intermediate state corresponds to a radical cation [5-8]. The surprising similarity between all three voltammograms suggests that radical cation intermediates might also be involved in the tetramer and polymer. The fact that, for the tetramer, two successive two-electron transfers (rather than four successive one-electron transfers) are observed implies that radical ion delocalization does not extend over the whole molecule. This observation also suggests that the tetramer behaves more like two independent but coupled dimers. In conclusion, both UV spectra [5] and CV data show that the behaviour of the higher oligomers parallels that of the dimer. This provides credence for the suggestion of Huang and MacDiarmid [4] that the dimer is the "lowest common denominator".

DIMER MODEL

We now attempt to desribe the electrochemical behaviour of polyaniline in terms of the simplest model valid for the dimer. In Figure 3 the idealized repeat units for polyaniline are shown in their unprotonated forms. The model neglects the possibility of isolated amine repeat units but this is expected to be of minor importance for low oxidation states. The interconversion of the various species is shown in the reaction scheme of Figure 4.

The pH dependence of the halfwave potentials corresponding to the two redox processes observed in polyaniline were investigated by Huang, Humphrey and MacDiarmid [9]. The halfwave potential for the first redox process is independant of pH in the range $1 < pH < 4$. This behaviour is consistent with the following half-reaction:

$$R^{+\bullet} + e^- = A \tag{I}$$

$$E = E^0_A + \frac{RT}{F} ln \frac{[R^{+\bullet}]}{[A]} \tag{1}$$

The halfwave potential for the second redox process decreases at a rate of approximately 120 mV/pH [9]. This behaviour is consistent with the half-reaction:

$$Q + 2H^+ + e^- = R^{+\bullet} \tag{II}$$

$$E = E^0_R + \frac{RT}{F} ln \frac{[Q][H^+]^2}{[R^{+\bullet}]} \tag{2}$$

The overall half-reaction is obtained by combining reactions I and II :

$$Q + 2H^+ + 2e^- = A \tag{III}$$

with corresponding Nernst equation:

$$E = E^0_Q + \frac{RT}{2F} ln \frac{[Q][H^+]^2}{[A]} \tag{3}$$

where $\qquad E^0_Q = (E^0_A + E^0_R)/2$ \hfill (4)

318

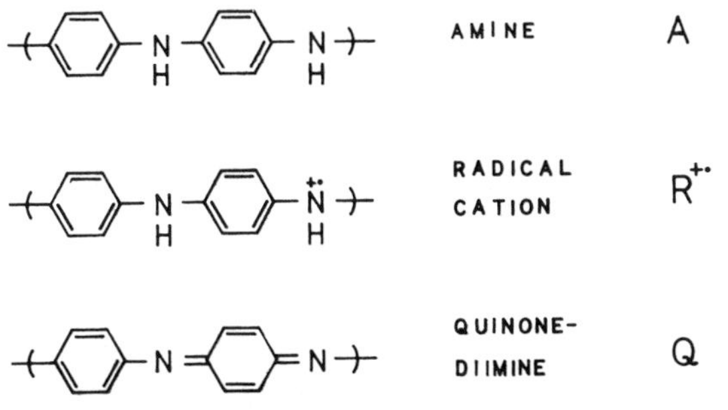

Figure 3. Idealized repeat units for the dimer model of polyaniline. Note that the species A and Q can exist in various protonated forms. Since radical cation stability requires resonance the protonation of $R^{+\bullet}$ is not considered.

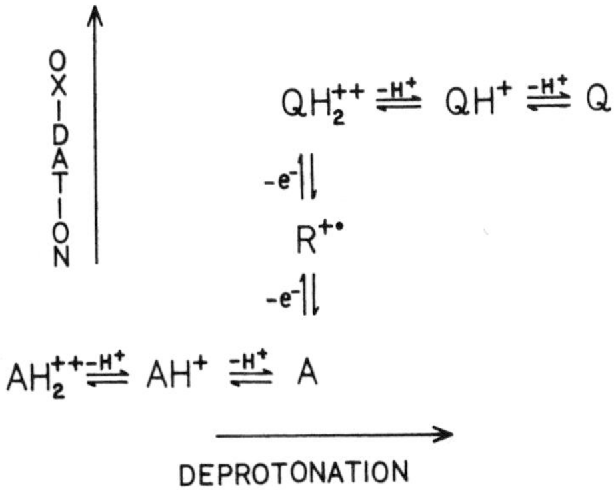

Figure 4. Reaction scheme showing the interconversion of the various species.

At equilibrium equations (1) and (2) are both valid for predicting the open circuit potential. Equating them leads to the polaron equilibrium constant:

$$K_e = \frac{[R^{+\cdot}]^2}{[A][Q][H^+]^2}$$

$$= \exp[F(E_R^0 - E_A^0)/RT] \tag{5}$$

which corresponds to the conproportionation reaction:

$$A + Q + 2H^+ = 2\,R^{+\cdot} \tag{IV}$$

Equilibrium constants for the protonation reactions are defined as follows:

$$A + H^+ = AH^+ \qquad\qquad K_{R1} = \frac{[AH^+]}{[A][H^+]} \tag{6a}$$

$$AH^+ + H^+ = AH_2^{++} \qquad\qquad K_{R2} = \frac{[AH_2^{++}]}{[AH^+][H^+]} \tag{6b}$$

$$Q + H^+ = QH^+ \qquad\qquad K_{Q1} = \frac{[QH^+]}{[Q][H^+]} \tag{6c}$$

$$QH^+ + H^+ = QH_2^{++} \qquad\qquad K_{Q1} = \frac{[QH_2^{++}]}{[QH^+][H^+]} \tag{6d}$$

It is convenient to define the oxidation state of the polymer as the ratio of quinone-diimine units to diamine units in the corresponding free base form:

$$\alpha = [Q]_0/[A]_0 \tag{7}$$

With this definition the emeraldine oxidation state corresponds to $\alpha = 1$. Using the equilibrium constants (6) the overall Nernst equation, equation (3), can be written as a function of α:

$$E = E_Q^0 + \frac{RT}{F} \ln\left\{ [H^+][(\,\gamma^2(1-\alpha)^2 + 4\alpha\phi\psi)^{1/2} - \gamma(1-\alpha)]/2\phi \right\} \tag{8}$$

where

$$\psi = 1 + K_{R1}[H^+] + K_{R1}K_{R2}[H^+]^2 \tag{9}$$

$$\phi = 1 + K_{Q1}[H^+] + K_{Q1}K_{Q2}[H^+]^2 \tag{10}$$

$$\gamma = K_e^{1/2}[H^+]/2 \tag{11}$$

Equation (8) defines the redox potential of polyaniline in terms of the oxidation state of the polymer, the hydrogen ion concentration $[H^+]$ (i.e. pH) and the equilibrium constants defined in equation (6). For the emeraldine oxidation state ($\alpha = 1$) this equation assumes a particularly simple form:

$$E = E^0_Q + \frac{RT}{F} \ln \left\{ \frac{[H^+]\{ 1 + K_{R1}[H+] + K_{R1}K_{R2}[H^+]^2\}}{1 + K_{Q1}[H^+] + K_{Q1}K_{Q2}[H^+]^2} \right\} \qquad (12)$$

The parameters in equation (12) were estimated in part by fitting to data obtained on emeraldine (see the Table for details). The trends predicted by equation (8) for other oxidation states, are shown, together with experimental data obtained by potentiometric titration, in Figure 5. In these experiments, the open circuit potential of electrochemically synthesized polyaniline, in a preset oxidation state, was followed as a function of solution pH. The latter was changed continuously at a rate of ca. 1.5 pH units/hour by titration with buffer solutions [10]. Figure 5 shows that it was indeed possible to fit data for the emeraldine oxidation state to equation (12). The agreement with other oxidation states is, however, only qualitative.

TABLE: ESTIMATED PARAMETER VALUES FOR DIMER MODEL

Parameter	Value	Estimated from:
E_A^0	0.1 V (SCE)	Cyclic.voltammetry data of Huang et al. [9]
E_R^0	0.8 V (SCE)	
E_Q^0	0.45 V (SCE	Equation (4)
K_e	7×10^{11} M^{-2}	Equation (5)
K_{R2}	~ 1	Cyclic voltammetry [9]
K_{R1}	0.01	By analogy to $\Delta pK = 2$ observed for the dimer: N,N'-diphenyl-p-phenylenediamine [14]
K_{Q2}	21	From a least squares fit of equation (12) to data obtained for the emeraldine oxidation state.
K_{Q1}	109	

It is of interest that the polaron model, equation (12), is identical to the equation derived for bipolarons by Huang and MacDiarmid [4]. This agreement applies only to the potential-pH relationship for the emeraldine oxidation state. For instance, the two models predict very different proton loadings. Both models are essentially based on equation (3) and both predict the same ratio of [Q]/[A] for emeraldine at a given pH. However, in the case of the polaron model the concentrations of quinone-diimine units (Q) and diamine units (A) will be lower owing to the formation of polarons according to reaction (IV). This reaction, by the same process, also binds additional protons to the polymer chain. This difference in the predicted degree of protonation can be used to determine which model is more appropriate.

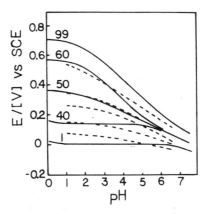

Figure 5. Comparison of predictions of the dimer model with potentiometric titration data. The solid lines are the predictions for the indicated percentage of diimine units in the polymer backbone. The broken lines were determined experimentally. The actual oxidation states corresponding to the experimental curves could not be determined with certainty due to the observed hysteresis in the ECPS data in Figure 7.

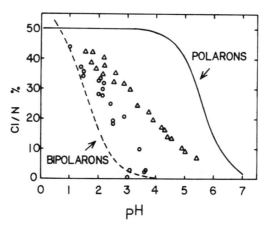

Figure 6. Comparison of experimentally determined degree of protonation of emeraldine with predictions by the polaron and bipolaron models. The degree of protonation is conveniently expressed by the mole ratio of chlorine to nitrogen which was determined by elemental analysis for the data of MacDiarmid et al. [8] (o) and calculated from measured pH changes for the present data (Δ)

In Figure 6 the predicted and experimentally determined proton loadings on emeraldine are compared. The present experimental data were obtained as follows: A given mass of dry emeraldine base was equilibrated with a known volume of 0.1 N HCl by stirring in a sealed bottle for 12 days. The final equilibrium pH was then measured and the proton uptake of the polymer calculated from this value. Also shown are data by MacDiarmid et al. [11] who equilibrated emeraldine base samples with HCl solutions of various pH. Their samples were then dried on the vacuum line and the degree of protonation determined by elemental analysis as the mole ratio of chlorine to nitrogen.

Figure 6 shows that the experimental data fall roughly halfway between the predictions for the two models. The data of MacDiarmid et al. [11] indicate lower degrees of protonation than the present results and as a result lie closer to the bipolaron curve. The discrepancy between the present data and that of MacDiarmid et al. [11] may derive from a loss of HCl from the wet polymer on vacuum drying. Even the present data indicate proton loadings considerably below the predicted values. This indicates a basic deficiency in the dimer model. It is well known [12] that the basicity or acidity of structural groups changes when they are linked together to form a polymer. This effect is due to the modulating influence of the diffuse electrostatic field induced by the charged groups held on the same polymer chain. This results in a dependence of the pK_B of unoccupied basic sites on the fraction of protonated sites. This effect could explain the low degree of protonation observed for emeraldine.

In Figure 7 the predicted potential vs injected charge curve for pH = 0 is compared to data obtained by electrochemical potential spectroscopy (ECPS) [13]. The hysteresis in the experimental curves in Figure 7 prevent the assignment of specific oxidation states to the experimental curves in Figure 5. However, the absence of the predicted "jump" at the emeraldine stage of oxidation is conspicuous. We conjecture that this implies that protonated emeraldine is not simply a pure radical cation state but may in fact contain an admixture of bipolarons. Although polarons are the energetically favored species, an array of polarons represents a state of low entropy. A decrease in total free energy may result by the conversion of a fraction of the polarons into bipolarons owing to the increase in entropy.

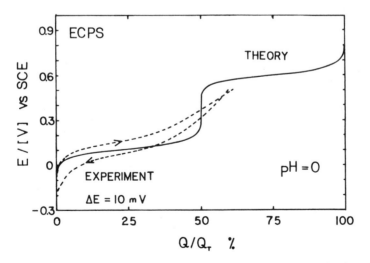

Figure 7. Comparison of electrochemical potential spectroscopy data obtained at pH = 0 with the predictions of the dimer model . ECPS [13] parameters: ΔE_{step} = 10 mV, I_{min} = 1 μA and Q_T = 200 mC.

CONCLUSIONS

Comparisons of predictions with experimental results indicate that the dimer model, although providing qualitatively correct trends, fails to describe the electrochemical behavior of polyaniline quatitatively. This failure of the present dimer model may be caused by the neglect of nearest neighbor interactions on the polymer chain. These interactions are expected to lower the effective basicity of the unprotonated imine nitrogens as the proton loading of the chain increases. In addition, the interactions may lead to a perturbation of the polaron-bipolaron equilibrium favoring an increase in the fraction of bipolarons on the chain.

ACKNOWLEDGEMENTS

This work was supported in part by a grant from the Office of Naval Research (to G. E. W.) and a fellowship from CSIR (to W. W. F.).

REFERENCES

1. R. de Surville, M. Jozefowicz, L. T. Yu, J. Perichon and R. Buvet, Electrochemical Chains using Protolytic Organic Semiconductors, Electrochim. Acta 13:1451 (1968).
2. D. Vachon, R.O. Angus, F.L. Lu, M. Nowak, Z. X. Liu, H. Schaffer, F. Wudl and A. J. Heeger, Polyaniline is Poly(p-phenylamineimine): Proof of Structure by Synthesis, Synth. Met. 18:297 (1987).
3. J. C. Chiang and A. G. MacDiarmid, "Polyaniline": Protonic Acid doping of the Emeraldine form to the Metallic Regime, Synth. Met. 13:193 (1986).
4. W. S. Huang and A. G. MacDiarmid, Polyaniline: Non-oxidative doping of the emeraldine base form to the metallic regime, In Press., J. Chem. Soc. Chem. Com.
5. P. M. McManus, S. C. Yang and R. J. Cushman, Electrochemical Doping of Polyaniline: Effects of Conductivity and Optical Spectra, J. Chem. Soc. Chem. Commun. 1556 (1985).
6. G. E. Wnek, A Proposal for the Mechanism of Conduction in Polyaniline, Synth. Met. 15:213 (1986).
7. H. Linschitz, J. Rennert and T. M. Korn, Symmetrical Semiquinone Formation by Reversible Photo-oxidation and Photo-reduction, J. Am. Chem. Soc. 89:5839 (1967).
8. G. Cauquis and D. Serve, Anodic Voltammetry of some Aromatic Amines in Chloroform, Anal. Chem. 44:2222 (1972).
9. W. S. Huang, B. D. Humphrey and A. G. MacDiarmid, Polyaniline, a Novel Conducting Polymer, J. Chem. Soc. Faraday Trans. II 82:2385 (1986).
10. W. W. Focke, G. E. Wnek and Y. Wei, The Influence of Oxidation State, pH and Counterion on the Conductivity of Polyaniline, Submitted for publication.
11. A. G. MacDiarmid, J. C. Chiang, A. F. Richter and N. L. D. Somasiri, Polyaniline: Synthesis and Characterization of the Emeraldine Oxidation State by Elemental Analysis, in: "Conducting Polymers",L. Alcacer, ed., Reidel Publications, Dordecht (1986).
12. A. Kachalsky, J. Mazur and P. Spitnik, Polybase Properties of Polyvinylamine, J. Pol. Sci. 23:5   17 (1957).
13. A. H. Thompson, Instrumentation for Incremental Voltage Step Electrochemical Measurements, Rev. Sci. Instrum. 54:229 (1983).
14. F. M. Albert, M. Giurginca, A. Meghea and G. Ivan, Basicity Constants of some N,N'-disubstituted p-Phenylenediamines, Rev. Roum. Chim. 25:1543 (1980)

A PPP VARIATIONAL-PERTURBATION CALCULATION OF HYPERPOLARIZABILITIES OF CONJUGATED CHAINS

C. P. de Melo* and R. Silbey

Department of Chemistry and Center for
Materials Science and Engineering
Massachusetts Institute of Technology
Cambridge, MA 02139 USA

I. INTRODUCTION

It is well known that conjugated systems are promising materials for the development of nonlinear optical devices due to the anisotropy and high polarizability of their π-electron distribution [1-2]. In addition, in recent years the investigation of the effect of conformational changes on the optical properties of polymers has become a point of interest: for example, the dramatic color change observed in polydiacetylene (PDA) solutions as a result of variations of concentration or temperature has been attributed to a planar-nonplanar conformational transition involving individual polymer chains.[3]

We have recently implemented a method for the calculation of linear and nonlinear polarizabilities of conjugated molecules described by a Pariser-Parr-Pople (PPP) hamiltonian.[4-5] In this article we present a brief summary of the corresponding results for the hyperpolarizabilities of finite polyenic (of regular structure as well as those conformational defects of soliton and polaron nature) and polydiacetylene chains. We begin with a bird's eye view of the procedure utilized. The interested reader can find a detailed presentation of the method as well as a discussion of its relationship to standard variational-perturbation treatments in Ref. 5, where we have shown the results for the linear polarizabilities of finite polyenic chains.

I. THE PERTURBATION-EXPANSION FOR THE DENSITY MATRIX TREATMENT FOR THE POLARIZABILITIES OF CONJUGATED SYSTEMS

Our treatment is based on a perturbative expansion of the Hartree-Fock (HF) density matrix R of the system[6] and does <u>not</u>

---

*On leave of absence from the Departamento de Fisica, Universidade Federal de Pernambuco, 50000 Recife PE Brazil.

involve a summation over a small number of excited states; rather, it implicitly takes into account all excitations from the HF ground-state. When an external electric field of strength F is applied upon a molecular system, it produces a distortion of the electronic charge distribution. For large values of F, the induced dipole moment deviates from a simple linear dependence on the field and the interaction energy can be expanded as powers of F. On the other hand, the changes in the electronic distribution lead to successive orders of correction to the electronic energy. As a consequence, the linear polarizability α and the first and second hyperpolarizabilities β and γ can be determined in an unequivocal manner, as field-independent quantities. Instrumental for that purpose is the calculation of the changes in R required to guarantee in all orders its idempotency and commutation with the total hamiltonian.

An equivalent formalism has been used previously at the CNDO level for the calculation of α and γ for small polyenes and aromatic molecules.[7] In this series of work we were concerned with the polarization response of finite conjugated chains. Of special interest to us was the study of the evolu-evolution of the linear and nonlinear polarizabilities as the size of the molecules within a given family of conjugated polymers increases, and the investigation of the effect of different conformations on the polarization response of conjugated chains. Since we had set the initial goal of examining chains as large as 20 carbon atoms, we decided to adopt the PPP approximation, widely used on the treatment of the electronic structure of conjugated systems.[8] In this approximation, the one-electron perturbation term can be written as $f_{ij}=eFx\delta_{ij}$ and the subsequent changes in the two-electron part give origin to two independent self-consistent equations[5] for $R^{(1)}$ and $R^{(2)}$, respectively the first and second order corrections to the density matrix. It can be easily shown that the knowledge of $R^{(1)}$ suffices for the determination of α and β, while on has to calculate $R^{(2)}$ in order to obtain the second hyperpolarizabilities γ.

II. FIRST AND SECOND HYPERPOLARIZABILITIES OF POLYENIC CHAINS

We compute optical properties for conjugated chains with the geometries of regular polyenes, soliton, and polaron chains. The complete specification of the geometric parameters utilized can be found in Refs. 4-5.

As observed before for the linear polarizabilities, the higher order terms of the polarization response also depend on the charge state of the molecule considered. Due to symmetry requirements,[9] from all molecules examined only soliton chains could have nonzero first hyperpolarizabilities. We found that although β is exceedingly small for neutral soliton chains, positive (S^+) and negative (S^-) charged solitons have large xyy components for the first hyperpolarizabilities, with opposite signs; a linear least-square fit indicates that $\beta_{xyy}^{\pm}$ should scale as $\mp 16.4 \ N^{3.05}$, where N is the number of carbon atoms in the chain.[10] For the largest soliton chain considered (N=21), the corresponding absolute value is 1.79×10^6 a.u.

Our results have also confirmed the expectation that the

second hyperpolarizabilities γ of conjugated chains are extremely sensitive to conformational changes. Free-electron model[11] and simple Hückel[12] results have predicted that γ should evolve, respectively, as the 5.0 and the 5.26 powers of N. When our results for the longitudinal hyperpolarizability γ_{xxxx} of neutral and charged polyenic chains are plotted[4] one one can see that up to the N~20 range (with the possible exception of neutral polyenes) there is no sign of saturation of the polarization response. In fact, the calculated values can be reasonably well fitted through a linear regression formula. The appropriate fitting coefficients can be found in Table I. The calculated values for the largest chains considered in each case are presented in Table II. Singly charged polaron chains are predicted to have the largest absolute value of γ_{xxxx} of all cases examined. It should also be noted that once again the charge state of the system has an important effect on the sign of the polarization response, as exemplified by the fact that γ_{xxxx} is negative for singly charged solitons (in this case, either S^+ or S^-) and polarons, while doubly charged polarons (bipolarons) have positive second hyperpolarizabilities.

III. SECOND HYPERPOLARIZABILITIES OF POLYDIACETYLENES

The nonlinear optical properties of polydiacetylene have been the subject of interest since work with good optical quality samples of solid-state polymerized diacetylenes had shown that the optical nonlinear coefficients for conjugated polymers and crystals can be at least as large as those of traditional inorganic semiconductors.[13] More recently, the use of Langmuir-Blodgett (L-B) techniques has resulted in the preparation of highly oriented thin films of large surface area which are finding widespread application in different fields of optics.[14]

Different experimental techniques have been used for the determination of the nonlinear susceptibilities of polydiacetylene compounds yielding sometimes conflicting results: for example, third-harmonic generation measurements indicate that the second hyperpolarizability of polydiacetylene solutions should be positive,[15] while a negative sign is obtained from

TABLE I

		r.p.	n.s.	n.p.	c.s.	c.p.	bp
γ_{xxxx}	a	52.0	146	42.0	-27.0	-2.66	0.247
	b	4.25	4.05	4.60	4.79	6.57	6.04

a and b parameters derived from linear-least square fitting aN^b for calculated values (in a.u.) of the longitudinal component (γ_{xxxx}) of the second hyperpolarizability of finite chains of regular polyenes (r.p.), neutral solitons (n.s.), charged solitons (c.s.), singly charged polarons (c.p.) and bipolarons (bp).

Intensity Dependent Nonlinear Coupling experiments on L-B multilayers of PDA.[16] If the polarization response of PDA compounds present the same sensitivity to conformational changes along the conjugated chain as that found in polyenic molecules, the above results can be reconciled by assuming the presence of defects or other kind of disruption on the conjugation pattern. Here we will examine the second hyperpolarizabilities of regular PDA chain in the acetylenic and butatrienic structures. We have adopted the optimized ab initio geometries obtained by Karpfen.[17] As before, the molecules are considered to be oriented along the x axis. The calculated values indicate a completely different behavior for the longitudinal component of γ of the two conformations considered. Again, still up to the N~20 range no saturation effect can be seen; a linear least-square fitting reasonably fits the data to $(-2.0 \times 10^{-3} N^{8.78})$ and $(6.81 N^{5.23})$ for the polydiacetylenic and butatrienic structures, respectively. The values of γ_{xxxx} for the N=20 PDA chains examined can be found in Table II.

CONCLUSIONS

The above results confirm the expectation that the polarization response of conjugated compounds is extremely sensitive to conformational changes along the chain. In addition, we have found that extremely different behavior of the optical response (including opposite signs for the hyperpolarizabilities) can be associated to distinct charge states of conjugation defects in polyenic chains. As a rule we have found that the hyperpolarizabilities of the defects evolve more rapidly than those of regular chains; hence, the above results are suggestive that in polyacetylene chains with nonnegligible concentration of defects, solitons and polarons could dominate the optical response of the sample. On the other hand, more complete investigation of the optical properties of polydiacetylene is required before one could try to explain the recent experimental data.[15-16] It would be especially interesting to analyze the effect that conjugation defects[16] could have on the nonlinear optical properties of these systems.

We believe that the above results reinforce the idea that conjugated polymers form a promising class of materials for the construction of optical devices. If conjugation effects in fact dominate the optical response of these systems, one can foresee the possibility that, eventually, by suitable chemical methods one could control the delocalization length along the

TABLE II

r.p.	n.s.	n.p.	c.s.	c.p.	bp	pda	pbt
1.38	2.86	3.74	-5.79	-103	1.73	-12.1	4.23

Values (in 10^7 a.u.) of γ_{xxxx} for the largest chains considered (N=20(21) for the $C_{2h}(C_{2v})$ group) regular polyenes (r.p.), neutral solitons (n.s.), charged solitons (c.s.), singly charged polarons (c.p.), bipolarons (bp), and PDA in acetylenic (pda) and butatrienic structures (pbt).

conjugated chain in such degree as to permit the actual design of organic-based optical devices by the appropriate 'fine-tuning' of the polarization response of the material.

ACKNOWLEDGEMENTS

One of us (CPM) would like to acknowledge the support from the Brazilian Agency CAPES and from CIES for the Fulbright visiting fellowship under the tenure of which this work was performed. This work was supported in part by an NSF grant to RS (DMR 84 18718). In addition, we acknowledge the donors of the Petroleum Research Fund of the ACS for partial support of this work.

REFERENCES

1. "Nonlinear Optical Properties of Organic and Polymeric Materials, ed. by D.J. Williams, ACS Symposium Series 233, ACS, Washington, (1983).
2. "Polydiacetylenes," ed. by D. Bloor and R.R. Chance, NATO ASI Series E102, Nijhoff, Dordrecht, 1985.
3. See, for example, R.R. Chance, M.W. Washabaugh and D.J. Hupe, in Ref. 2, p. 239.
4. C.P. de Melo and R. Silbey, "Nonlinear Polarizabilities of Conjugated Chains: Regular Polyenes, Solitons and Polarons," submitted to Phys. Rev. Letters.
5. C.P. de Melo and R. Silbey, "Variational-Perturbational Treatment for the Polarizabilities of Conjugated Chains I. Theory and Linear Polarizabilities Results for Polyenes," to be submitted to J. Chem. Phys.
6. G. Diercksen and R. McWeeny, J. Chem. Phys. 44:3554 (1966) (1966); R. McWeeny and G. Diercksen, J. Chem. Phys. 49:4852 (1968).
7. C.A. Nicolaides, M. Papadopoulos and J. Waite, Theoret. Chim. Acta(Berl.) 61:427 (1982); M. Papadopoulos, J. Waite and C.A. Nicolaides, J. Chem. Phys. 77:2527 (1982); J. Waite, M. Papadopoulos and C.A. Nicolaides, J. Chem. Phys. 77:2536 (1982).
8. L. Salem, "The Molecular Orbital Theory of Conjugated Systems," Benjamin, New York (1966).
9. S.J. Cyvin, J.E. Rauch and J.C. Decius, J. Chem. Phys. 43:4083 (1965).
10. C.P. de Melo and R. Silbey, "Variational-Perturbational Treatment for the Polarizabilities of Conjugated Chains. II. Hyperpolarizabilities of Polyenic Chains," in preparation.
11. K.C. Rustagi and J. Ducuing, Opt. Commun. 10:258 (1974).
12. E.F. McIntyre and H.F. Hameka, J. Chem. Phys. 68:3481 (1978).
13. C. Sauteret, J.P. Herman, R. Frey, F. Pradere, J. Ducuing, R.H. Baughman and R.R. Chance, J. Chem. Phys. 69:4482 (1978).
14. A. Barraud and M. Vandervyver, in: "Nonlinear Optical Properties of Organic Molecules and Crystals," ed. by D.S. Chemla and J. Zyss, Academic, Orlando, (1987), vol. 1, p. 357.
15. F. Kajzar and J. Messier, in Ref. 2, p. 325.
16. D.J. Sandman, G.M. Carter, Y.J. Shen, B.S. Elman, M.K. Thakur and S.K. Tripathy, in Ref. 2, p. 299.
17. A. Karpfen, J. Phys. C13:5673 (1980).

OPTICAL SPECTRA OF FLEXIBLE CONJUGATED POLYMERS

Z. G. Soos and K. S. Schweizer*

Dept. of Chemistry, Princeton University, Princeton, NJ 08544
*Sandia National Laboratories, Albuquerque, NM 87185

INTRODUCTION

Organic dyes illustrate the close connection between π-electron delocalization and optical spectra. Both conjugation length and geometry are important[1]. Conjugated polymers are still longer and in highly ordered forms may be idealized as linear crystals. While dyes and polyenes are essentially planar at ambient temperatures, polymeric strands may only be planar locally. For instance, π-conjugated polydiacetylenes[2,3] and polythiophenes[4] or σ-conjugated[5] polysilanes[6,7] and polygermanes[8] can exist as random coils in either dilute solutions or solid films. Conformational degrees of freedom associated with bond rotations are coupled to the conjugation and result in significant thermochromism. Neither molecules nor crystals are apt models for a random coil. We develop in this paper the general features of optical transitions in flexible conjugated polymers.

Conformational changes that break the conjugation reduce a long polymer to many short ones. Kuzmany[9] and Mulazzi et al.[10] have interpreted Resonance Raman data on polyacetylene (PA) films in terms of a bimodal distribution of $(CH)_x$ conjugation lengths with $x_1 \sim 100$ and $x_2 \sim 10$. The spectral blue shifts observed by Patel et al.[2,3] at the rod-to-coil transition of several polydiacetylenes (PDAs) in dilute solution have been interpreted as shorter π-electron delocalization in the coil. Similar blue shifts at rod-to-coil transitions of polysilanes $(SiRR')_n$ in dilute solution were observed by Harrah and Zeigler and ascribed to shorter σ-conjugation[6]. On the theoretical side, Dobrosavljevic and Stratt[11] emphasize the low barrier (< 0.1 eV) for rotation about a single bond in PDA and the resulting localization of the one-dimensional states due to decreased and disordered transfer integrals. A $\pi/2$ rotation completely breaks π-conjugation. Schweizer has used decoupled segments to construct a unified theory of the optical properties[12] and rod-to-coil[13] transitions of conjugated polymers, based on the higher polarizability of extended segments and the configurational entropy of random coils. Complete conjugation breaks correspond to a "strong disorder" limit whose virtues are simplicity and tractability rather than any obvious realism. The true nature of rotational defects and their quantitative role in absorption spectra remain largely a matter for speculation.

To emphasize purely 3-dimensional conformational contributions instead of 1-dimensional localization, we adopt the opposite limit for flexible polymers. The transfer integrals t_p between nearest-neighbor orbitals are taken to be independent of the conformation. The physical picture is a worm-like chain with local planarity and a persistence length $\xi_c = (2\lambda)^{-1}$ $\gg 1$ for rotation by $\pi/2$. Such a model has been suggested[14] for interpreting light-scattering and spectroscopic measurements on soluble PDAs. On the theoretical side, Haddon has recently developed[15] a 3-dimensional Hückel theory that emphasizes σ-π mixing (rehybridization) for maintaining conjugation even for large deviations from planarity. The axial symmetry of σ-bonds in $(SiRR')_n$ suggest even smaller variations of the relevant transfer integrals with backbone conformation. In this "weak disorder" limit, the ground-state $|g\rangle$, excited states $|k\rangle$, and excitation energies $\epsilon_k > E_g$ are independent of conformation. But the transition dipole matrix elements $\langle k|\mu|g\rangle$ are sensitive to conformational changes, as previously noted within segments[16] or molecules.[1] The spectrum absorption $I(\omega)$ is strongly peaked at the optical gap E_g in extended (rod) conformations, while $\epsilon_k > E_g$ transitions become important in coils. The component of the field along the polymer's backbone determines which transitions are intense.

The experimental evidence for the importance of a one-dimensional density of states efects is mixed. Sharp excitonic absorptions have been resolved[3] in single crystals of several PDAs. In other cases, however, $I(\omega)$ consists of a broad (> 0.2 eV) asymmetric absorption for either rods or coils. Thus "rod like" structures conspicuously fail to show the sharp features associated with a crystalline density of states and inhomogeneous broadening is often invoked in segment models[6,9,10,12]. Site disorder, electron-phonon coupling, and environmental differences are all plausible contributors. They also tend to localize the electronic states, thereby undermining the premise of an extended wave function in rods. A physically attractive alternative is to invoke some conformtional broadening (i.e., $\lambda > 0$) in extended conformations before considering inhomogeneities. The alternation and persistence length then delineate the spectral features of flexible worm-like polymers.

SUM RULE FOR QUANTUM CELL MODELS

Both PA and PDA have half-filled π bands associated with a $2p_z$ orbital ϕ_p that is odd under reflection in the molecular (xy) plane. Two sp^3 hybrids per Si or Ge are invoked in σ-conjugated strands, with interatomic transfer integrals $t(1 + \delta)$ and intraatomic transfer $t(1-\delta)$; the result is again a half-filled valence shell. The restriction to a finite basis $\{\phi_p\}$ is typical of quantum cell models. Electron transfers between neighbors along the backbone define the Hückel Hamiltonian,

$$H_o = \sum_{p\sigma} t_p \ (a^+_{p\sigma} a_{p+1\sigma} + a^+_{p+1\sigma} a_{p\sigma}) \tag{1}$$

Here $a^+_{p\sigma}$, $a_{o\sigma}$ create, annihilate an electron with spin σ in ϕ_p. Each C in PA of PDA provides a π-electron, each Si in $(SiRR')_n$ provides two σ-electrons. On-site (Hubbard) or long-range Pariser-Parr-Pople (PPP) interactions are denoted as $V(\{n_p\})$; they depend only on the occupation numbers $n_p = 0,1,2$ of ϕ_p. Hubbard contributions are clearly independent of conformation, as are nearest-neighbor PPP contributions when only rotations about bonds are considered. Long-range ionic contributions have weak conformational dependence[17] in neutral molecules with standard PPP parameters.

To obtain the absorption spectrum, we need the dipole-moment operator

$$\vec{\mu} = e \sum_p \vec{r}_p \ (n_p - z_p) \tag{2}$$

where $z_p = 1$ is the charge associated with an empty ϕ_p and the origin of $\vec{r}_p$ is arbitrary for a neutral system. The oscillator strength f_k for dipole-allowed transitions at $h\omega = \varepsilon_k$ is

$$f_k = \varepsilon_k \ |\langle k| \ \mu_x \ |g\rangle|^2 = \varepsilon_k \ M^2_{kg} \tag{3}$$

for an applied field along x. The one-electron operator μ is site diagonal and commutes with arbitrary spin-independent $V(\{n_p\})$. The total absorption from the ground state $|g\rangle$ obeys a sum rule that, aside from constants, is[18]

$$\sum_k f_k = \langle g | [\vec{\mu}, H], \vec{\mu}] | g \rangle = - \sum_p |b_p|^2 t_p \, \partial E / \partial t_p \qquad (4)$$

Here $\vec{b}_p = \vec{r}_{p+1} - \vec{r}_p$ is the pth bond length, E is the exact ground state energy for $H_o + V(\{n_p\})$, t_p is negative, and $2p_n = \partial E / \partial t_n$ is the bond order. The total intensity reflects the kinetic energy and is rigorously independent of conformation in Hückel or Hubbard models. Since Hubbard or PPP bond orders in polyenes are within 10-20% of the Hückel values[17], the total intensity depends weakly on electron correlations. Individual f_k, by contrast, are sensitive to both conformation and correlation. The latter determine whether an excitonic level falls below E_g.

The sum rule (4) for quantum cell models shows that conformational changes merely redistribute intensity. This is consistent with recent experiments by Chance et al.[19] on soluble PDAs, both above and below the rod-to-coil transition, that show no change in the integrated oscillator strength. The following results are less general. We examine electron-hole excitations in the Hückel limit (1) for alternating transfer integrals $t_p/t = 1 - (-1)^p \delta$. Finite δ leads to an optical gap $E_g = 4|t|\delta$ and strongly alters the density of states near the Fermi energy ϵ_F. Now (4) shows that the total intensity hardly increases for small alternation, since bond-order changes are small. But correlations change the Hückel relation between E_g and δ and complicate direct analysis of conformational contributions.

FLEXIBLE REGULAR HUCKEL CHAINS

Identical $t_p = t$ in (1) and bond lengths $b = 1$ define a regular Hückel chain. Exact results are available for chains, rings, or continuum versions. For even $N_e = N$, the lowest N/2 MOs are filled in $|g\rangle$. An electron-hole pair with $r < N/2$ and $r' > N/2$ has excitation energy

$$\epsilon_{r'r} = 4|t| \sin \frac{\pi(r'+r)}{2(N+1)} \, \sin \frac{\pi(r'-r)}{2(N+1)} \qquad (5)$$

for a regular chain of N sites. The optical gap E_g for $r = N/2$, $r' = N/2 + 1$ vanishes as N^{-1}. Low-lying excitations with $Q = r' - r$ have a linear dispersion for large N. The expansion coefficients C_{pr} for the rth MO are[20]

$$C_{pr} = \left(\frac{2}{N+1} \right)^{1/2} \sin \frac{\pi pr}{N+1} \qquad (6)$$

and are closely related to states in a box with L = N+1. The transition moment is diagonal in the zero-differential-overlap[20] limit typical of quantum cell models,

$$\langle r'r | \mu | g \rangle = e \sqrt{2} \sum_p r_p C_{pr} C_{pr'} \tag{7}$$

where the $\sqrt{2}$ factor comes from the spin part. The continuum limit of (7) for a worm-like chain is

$$\vec{M}_{r'r} = \frac{2}{L} \int_o^L ds \ \vec{r}(s) \sin \frac{\pi rs}{L} \sin \frac{\pi r's}{L} \tag{8}$$

for e = 1 and contour length L = (N-1)b ~ N. The cos $\pi s(r' \pm r)/L$ factors in (8) are integrated by parts to obtain

$$\vec{M}_{r'r} = - \frac{1}{\pi} \int_o^L ds \ \frac{d\vec{r}}{ds} \left[\frac{1}{Q} \sin \frac{\pi sQ}{L} - \frac{1}{P} \sin \frac{\pi sP}{L} \right] \tag{9}$$

with Q = r'-r and P = r'+r ~ N. The unit tangent vector $\vec{u}(s) = d\vec{r}/ds$ defines the persistence length $\xi_c = (2\lambda)^{-1}$ for a worm-like chain through the configurational average[21]

$$\langle \vec{u}(s) \cdot \vec{u}(s') \rangle = \exp(-2\lambda |s-s'|) \tag{10}$$

The resulting integrals for $\langle M_{r'r}^2 \rangle$ can be done analytically, while parallel development[22] for discrete regular chains is also straightforward.

The exact analysis of $\langle M_{r'r}^2 \rangle$ leads to cumbersome expressions in λL. We retain below only the leading terms and show[22] separately that N > 10^2 chains entail corrections of a few percent. We note first that the Q=r'-r terms dominate in (9) for long chains. The rod ($\lambda \to 0$) limit immediately yields the even-odd selection rule by restricting Q to be an odd integer and leads to

$$\langle M_{r'r}^2 (0) \rangle = \frac{L^2}{2\pi^4} \sum_{Q \ odd} Q^{-4} \tag{11}$$

As expected on physical grounds the transition moment is of order N for an extended conformation. Since $\varepsilon_{r'r}$ is linear in Q and there are Q ways of obtaining r'-r, the absorption spectrum I(Q) for a rod goes as LQ^{-2} and is strongly peaked at Q=1. The rod absorption per site (per unit length) is sketched in Fig. 1 against x= $\pi Q/L = \varepsilon/2 |t|$ and normalized to unit total absorption.

The dominant $Q=r'-r > 0$ terms in (8) for $L\lambda \gg 1$ lead to

$$\langle M^2_{r'r} (\lambda)\rangle = \frac{1}{2\pi^2} \frac{L}{2\lambda} \sum_{Q=1} Q^{-2} \left[1 + \left(\frac{\pi Q}{2\lambda L}\right)^2\right]^{-1} \tag{12}$$

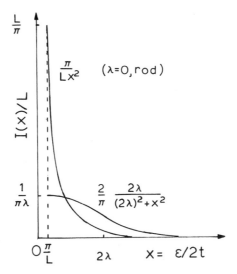

Fig. 1. Schematic absorption spectrum of a regular flexible Hückel chain of length L and persistence length $(2\lambda)^{-1}$.

We have left out correction terms[22] in $(\lambda L)^{-1}$ and note that the even-odd selection rule disappears for the flexible polymer. The transition moment cannot exceed R_g on physical grounds, where $R_g \sim (L/\lambda)^{1/2}$ is the radius of gyration. But Fourier components up to $Q \sim \lambda L$ become important in (12). The absorption spectrum again has a Q^2 factor from $\varepsilon_{r'r}$ and degeneracy. The resulting $I(x)$ curve is shown schematically in Fig. 1 and is half a Lorentzian with half-width $x=2\lambda$ and the same area, as required by the sum rule (4). The spectrum is broadened to the high-energy side.

The gapless spectrum in Fig. 1 is not appropriate for real polymers. For illustrative purposes, we shift the spectrum to finite E_g and introduce inhomogeneous broadening through a normalized Gaussian. The resulting convoluted spectrum is [22]

$$I(x, \lambda, \sigma) = (2\pi\sigma^2)^{-1/2} \int dx' I(x', \lambda) \quad \exp \left(- (x-x')^2/2\sigma^2\right) \tag{13}$$

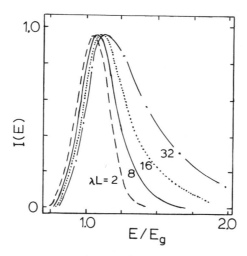

Fig. 2. Absorption spectrum of an N=400 site regular Hückel chain with inhomogeneous broadening $\sigma = 0.1E_g$ in eq. (13) and indicated λL.

where $I(x', \lambda)$ is essentially the $Q^2 <M^2_{r'r}(\lambda)>$ in (12) and $x = E/2|t|$. The spectra in Fig. 2 are based on the complete result [22] for N = 400, $E_g/4|t|=0.2$, $\sigma = 0.1E_g$, and various choices for λL.   Their peaks are normalized to the same intensity. Their detailed dependence on N, σ, and λL will be discussed elsewhere [22]. Both the magnitude and conformation dependence of the blue shifts increase with N and σ.

The salient features of the absorption curves in Fig. 2 are, first, that both the maximum and the width increases monotonically with λ; second, that as expected from Fig. 1 the high-energy width or asymmetry is sensitive to λ. The asymmetry consequently decreases with increasing ξ_c and practically disappears in long rods with inhomogeneous broadening. The principal results in Fig. 1 and 2 for flexible regular Hückel chains are, as already noted, only suggestive for alternating chains with different electron-hole excitations and density of states. The occurrence of Fourier components up to $Q \sim \lambda L$, on the other hand, is a natural consequence of flexibility that appears in modified form in alternating chains.

ALTERNATING FLEXIBLE POLYMERS

The prototypical conjugated polymer with even $N_e = N$ has alternating transfer integrals $t(1+\delta)$ and $t(1-\delta)$ in (1). The larger PDA repeat unit does not affect the region around E_g. As sketched for half-filled bands in Fig. 3, vertical electron-hole excitations occur at

$$\varepsilon(q)/4|t| = (\cos^2 q + \delta^2 \sin^2 q)^{1/2} \tag{14}$$

The q values for an even chain of N sites lie in $0 < q < \pi/2$ and are roots of a transcendental equation[23]. Their density matches the q values for a (2N+2)-site ring, where $q = 2\pi m/(N+1)$, with $m=0 \pm 1, \ldots, \pm N/2$, is the wavevector in the first Brillouin zone. The density of states in Fig. 3 has a gap $E_g = 4|t|\delta$ for $\delta > 0$. States around $\varepsilon_F \sim 0$ are strongly perturbed, while those around $q=0$ are hardly changed. The conventional explanation for the absorption peak at $E_g = 4|t|\delta$ is in terms of the density of states for vertical excitations.

The crystal selection rule of vertical excitations reflects the small momenta of optical photons, or equivalently their long wavelength compared to lattice spacings. Atoms in a crystal or an extended polymer are excited in phase. Such a picture fails for a flexible polymer with ξ_c shorter than the photon's wavelength. In addition to the density of states, we must include the q-dependence of the transition moment for vertical excitations and the breakdown of the crystal selection rule in random coils. We focus on the e-h excitations (14) near $q=\pi/2$.

Finite alternation produces a finite gap in the half-filled system and thus automatically introduces a correlation length ξ_a of order π/δ. For finite rings with q within π/N of the band edge, (14) effectively reduces to $4|t|\delta$. Alternating rods longer than $\xi_a = \pi/\delta$ consequently behave like infinite systems, in contrast to the N^{-1} behavior of $\varepsilon_{rr'}$ in (5) for regu-

lar chains. Conformational contributions in flexible alternating chains
are negligible for $\lambda \ll \delta$, when the persistence length exceeds the alternation correlation length. Such finite-size effects are also seen for
correlated spin or electronic systems[24].

Our qualitative approach relates physical ideas for alternating rings
to explicit results for regular chains. A general e-h pair near E_g has q_1
$= \pi/2 - q + k/2$ and $q_2 = \pi/2 - q - k/2$. The mean wavevector is $q \sim 0$, while
$k = q_1 - q_2 \sim 0$ is the deviation from a vertical transition. As has
already been shown in Fig. 1 for regular chains, nonvertical transitions
with $|k| < \lambda$ are important and can be modeled by a normalized Lorentzian
distribution

$$g(k, \lambda) = \frac{2}{\pi} \frac{2\lambda}{(2\lambda)^2 + k^2} \qquad (15)$$

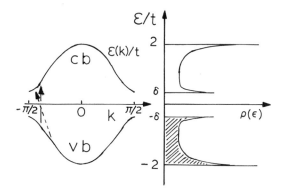

Fig. 3.   Schematic representation of the dispersion and density-of-states
of an alternating Hückel chain. Both vertical and nonvertical
excitations are indicated from the valence to the conduction band.

The limit $\lambda \to 0$ regains the crystal selection rule of vertical transitions.
Other choices for $g(k, \lambda)$ such as a Gaussian lead to similar results. The
principal effect of nonvertical e-h transitions near E_g is to increase the
phase space.

The oscillator strengths $f(q, \delta)$ for vertical transitions in alternating Hückel rings may be found[25] explicitly from the velocity-operator

representation of (2); $f(q)$ goes as ε^{-3} near E_g. Kuzmany et al.[16] found the same behavior in numerical studies of alternating Hückel chains representing PA. We retain the ε^{-3} behavior for nonvertical transitions near E_g and expand (14) to lowest order in q,k (from the band edge)

$$\varepsilon(q,k) = 4|t|(\delta^2 + q^2 + k^2/4)^{1/2} \tag{16}$$

Correction terms are of order $(qk)^2/(\delta^2 + q^2 + k^2/4)^2$. The oscillator strength is strongly peaked at E_g for small δ. The simple choice $f = (E_g/\varepsilon)^3$ regains below the previous result[26] for rods in the limit $\lambda \to 0$, when (15) gives a δ-function in k.

These considerations yield a new analytical expression for the absorption spectrum of alternating Hückel chains longer than $\xi_a = \pi/\delta$. For fixed δ we have

$$I(E,\lambda,\delta) \quad \alpha \quad \int dk dq \left[E_g/\varepsilon(q,k)\right]^3 g(k,\lambda)\delta\left[E - \varepsilon(q,k)\right] \tag{17}$$

for excitation energy (16), conformational contribution $g(k,\lambda)$ in (15), and $f \alpha \varepsilon^{-3}$. The proportionality constant in (17) depends on δ, but not on λ, and is fixed by the sum rule (4). Only the region around E_g is quantitative, but it dominates for small δ. Again neglecting a multiplicative constant independent of λ, we obtain from (17) the simple result

$$I(y,c) = y^{-2}\left[c^{-2} + (y^2-1)\right]^{-1/2} \tag{18}$$

with $y = E/E_g \geqslant 1$, $E_g = 4|t|\delta$, and $c = \delta/\lambda \sim \xi_c/\xi_a$. As expected in long alternating chains, the important parameter c is the ratio of the conformational (or conjugation) persistence length to the alternation correlation length. In the rod limit ($\lambda \to 0$, $c \to \infty$), (18) reduces to the standard result[26] with a square-root singularity typical of 1-dimensional system. We expect finite ξ_c ($\lambda > 0$) even for extended configurations in solution or in amorphous solids. Then (18) clearly displays the competition between flexibility and alternation.

The absorption maximum of $I(y,c)$ is at the band gap, with $I(1,c) = c = \delta/\lambda$. The $I(y,c)/c$ curves in Fig. 4 consequently have equal intensity at E_g. The absorption broadens on the high-energy side with decreasing $c = 2\delta\xi_c$. The conformational contributions in Fig. 4 for flexible alternating Hückel chains are analogous to the $\delta = 0$ results in Fig. 1.

To make contact with real systems, we convolute (18) with a Gaussian broadening function, as discussed in (13) for regular chains. The optical gap $E_g = 4|t|\delta$ now arises naturally. The convolution integral over $1 \leqslant y \leqslant \delta^{-1}$ acquires an explicit δ dependence. Representative results for $\delta = 0.07$, a typical PA value, are shown in Fig. 5 for the same broadening ($\sigma = 0.1 E_g$) used in Fig. 2. The spectral features of Fig. 5 are qualitively similar to Fig. 2, with $c = \delta/\lambda$ analogous to $(\lambda L)^{-1}$ for finite regular chains. The same asymmetric broadening above E_g is seen with increasing λ (decreasing c). A systematic analysis of parametric trends will be presented elsewhere[22].

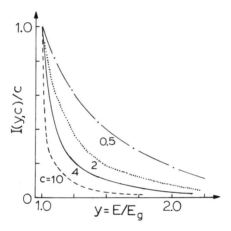

Fig. 4. Absorption spectrum of flexible alternating chains for different values of $c = \delta/\lambda$ and $I(y,c)$ from eq. (18). There is no absorption for $E < E_g$.

DISCUSSION

We have explicitly found 3-dimensional conformational contributions to the optical absorption of flexible conjugated polymers. The sum rule (4) is general for quantum cell models. The spectrum (18) is restricted to alternating Hückel models and a worm-like description of polymer structure. We are unaware of previous work in the "weak disorder" limit. As shown in Figs. 4 and 5, purely conformational changes can broaden and blue shift the absorption, with the relevant parameter $c = \delta/\lambda$ describing the ratio between conformational and alternation correlation lengths. We emphasize that such conformational contributions are intrinsic to any worm-like model.

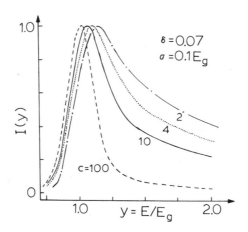

Fig. 5. Same as in Fig. 4, but including an inhomogeneous broadening $\sigma = 0.1 E_g$; alternation $\delta = 0.07$ is typical for PA.

Many other factors remain to be considered for applications to PA, PDA, or $(SiRR')_n$ systems. The present analysis is an alternative to complete conjugation breaks in "strong disorder" models. Both approaches require microscopic parameters that are not known accurately and both ignore electron correlation and vibronic coupling. Since Hückel models were originally parametrized to fit strong dipole-allowed excitations in molecules, we may hope correlation effects to be secondary. The overall intensity, for example, was shown to be insensitive to $V(\{n_p\})$. Correlations are certainly important for states below E_g, including solitons in PA, polarons[27] in polysilanes, and excitons[3] in PDAs. Correlations and vibronic contributions pose general problems for both conjugated molecules and polymers.

In the context of conjugated polymers, the central questions are the modification of t_p at conformational defects, the resulting localization of electronic states, the assessment of various inhomogeneous broadening mechanisms, and the choice of appropriate values for transfer integrals t, alternation δ, and persistence length $\xi_c = (2\lambda)^{-1}$. For random-coil PDAs or $(SiRR')_n$ in solution at relatively high T, the conjugation length is about $\xi_c \sim 6\text{-}10$ and ξ_c increases with decreasing T. The alternation is $\sim 0.25 - 0.40$ in polysilanes[28], and ~ 0.10 in[3] PDAs. The fundamental ratio $c = \delta/\lambda$ is consequently 3-8 in $(SiRR')_n$ and ~ 2 in soluble PDAs. Since purely conformational effects are irrelevant for large c, we conclude immediately that such contributions are less important in the strongly-alternating σ-conjugated polysilanes.

Any comparison with Fig. 5 also requires the inhomogeneous broadening σ. In alkyl-substituted polysilanes, theoretical[29] and experimental[6] work suggest adiabatic electron-phonon modulation of t_p to be important, thus leading to a temperature dependent σ. Its 300K value is ~ 0.06eV. The predicted blue shifts of ~ 0.05eV are almost an order of magnitude smaller than observed, but the calculated and observed linewidths are comparable. Furthermore, increasing c at fixed σ leads in Fig. 5 to a more symmetric spectrum, contrary to the observed behavior of polysilanes or PDAs on lowering T and thus increasing ξ_c. The data consequently cannot be fit by varying ξ_c only and holding all other parameters constant in the "weak disorder" limit. On the other hand, conformational contributions cannot be neglected for small δ or for linewidths. The absorption of PA films can readily be modeled: $\delta = 0.07$, $E_g = 1.8$eV, $\sigma = 0.1E_g$, $\xi_c = 14$ fits the peak position, breadth, and asymmetry of Fig. 4 in ref. 16. The parameter choice is not unique and a bimodal distribution[16] for crystalline and amorphous regions provides additional possibilities. Neither strong nor weak disorder can be conclusively established solely by optical absorption in solution or films.

Possible energy and temperature dependences of the inhomogeneous broadening further complicate detailed comparisons in either limit. Contributions due to solvent organization, side chain disorder, or atacticity are nearly independent of T. Electron-phonon contributions to σ increase with T. The backbone conformational disorder, the effective conjugation ξ_c considered above, also affects σ in a presently unknown way.

Rotations about PA or PDA single bonds necessarily leave an even number of π-electrons between defects at which $t(1-\delta)$ is reduced. Since each Si or Ge supplies two σ-electron in (1), gauche defects with reduced $t(1-\delta)$ again produce even segments with terminal $t(1+\delta)$ transfers.

Conformational changes consequently weaken the smaller transfer integrals without, in the first approximation, changing the strong ones. This has been the focus of previous theoretical work[9-13] with strong disorder corresponding to complete breaks (t=0 at defects).

Just from placing defects at weak bonds, however, we conclude that the average alternation increases with decreasing persistence length. The mean-field or virtual-crystal approximation then leads to an effective alternation

$$\delta_e(\lambda) \;=\; \delta + 2A\lambda \;=\; \delta + A/\xi_c \tag{19}$$

where the coefficient A is of order unity and requires a precise understanding of the reduced t at rotational defects. The effective alternation (19) may immediately and naturally be used in the spectrum (18), thereby introducing electronic disorder in our limiting picture of $\{t_p\}$ independent of conformation. The optical gap $E_g = 4|t|\delta_e$ then depends on conformation while $c = \delta_e/\lambda$ becomes less sensitive to λ: larger blue shifts and smaller broadening are obtained; (19) will be incorporated in future applications[22] of weak disorder. We have excluded conformational effects on $\{t_p\}$ for simplicity to emphasize purely conformational contributions.

In summary, we have introduced a weak-disorder model for the electronic absorption of flexible conjugated polymers. The coupling of rotational defects to the transfer integrals along the backbone leads to a conformation-dependent effective alternation. The crystal selection rule of vertical excitations is relaxed in random coils, thereby redistributing the oscillator strength to higher energy. Both of these physical effects, as well as correlations, vibronic contributions, and inhomogeneous broadening contributions, must eventually be incorporated in a general theory of the optical properties of conjugated polymers.

Acknowledgments: We gratefully acknowledge National Science Foundation support for work at Princeton under DMR-8403819 and U. S. Department of Energy support for the work at Sandia under DE-AC04-76DP00789.

REFERENCES

1. H. Suzuki, "Electronic Absorption Spectra and Geometry of Organic Molecules", (Academic Press, New York, 1967); A.P. Piechowski and G.R. Bird, Optics Comm., 50, 386 (1984).

2. G.N. Patel, R.R. Chance, and J.D. Witt, J. Chem. Phys., 70, 4387 (1979).

3. D. Bloor and R. Chance, eds., Polydiacetylenes, NATO ASI Series E-102 (Martinus Nijhoff, Dordrech, The Netherlands, 1985). Macromolecules, 20, 212 (1987).

4. S. Hotta, S. D. D. V. Rughooputh, A. J. Heeger, and F. Wudl, Macromolecules, 20, 212 (1987).

5. M.J.S. Dewar, J. Am. Chem. Soc., 106, 669 (1984).

6. L.A. Harrah and J.M. Zeigler, J. Poly. Sci. Poly. Lett. 23, 209 (1985); Polymer Preprints, 27, (2), 356 (1986); Macromolecules, 20, 601 (1987).

7. R.D. Miller, D. Hofer, J. Rabolt, and G.N. Ficknes, J. Am. Chem. Soc., 107, 2172 (1985); P. Trefonas, J.R. Damewood, R. West, and R. D. Miller, Organometallics, 4, 1318 (1985).

8. R.D. Miller and R. Sooriyakumaran, J. Polym. Sci. Poly. Chem. Ed., 25 111 (1987).

9. H. Kuzmany, Pure and App. Chem., 57, 235 (1985).

10. G. Brivio and E. Mulazzi, Phys. Rev. B., 30, 876 (1984).

11. V. Dobrosavljevic and R.M. Stratt, Phys. Rev. B., 30, 2781 (1987); R.M. Stratt and S.J. Smithline, J. Chem. Phys., 79, 3928 (1983).

12. K.S. Schweizer, J. Chem. Phys., 85, 4181 (1986).

13. K.S. Schweizer, Chem. Phys. Lett., 125, 118 (1986); J. Chem. Phys., 85, 1156 and 1176 (1986); Polymer Preprints, (2), 27, 354 (1986).

14. G. Wenz, M.A. Muller, M. Schmidt, and G. Wegner, Macromolecules, 17, 837 (1984); G. Allegra, S. Bruckner, M. Schmidt, and G. Wegner, Macromolecules, 19, 399 (1986).

15. R.C. Haddon, J. Am. Chem. Soc., 108, 2837 (1986).

16. H. Kuzmany, P.R. Surjan, and M. Kertesz, Sol. State Comm., 48, 243 (1983); P. Surjan and H. Kuzmany, Phys. Rev. B., 33, 2615 (1986).

17. Z.G. Soos and S. Ramasesha, Phys. Rev., B29, 5410 (1984); S. Ramasesha and Z.G. Soos, J. Chem. Phys., 80, 3278 (1984).

18. S. Mazumdar and Z.G. Soos, Phys. Rev., B23, 2810 (1981).

19. R.R. Chance, J.M. Sowa, H. Eckhardt, and M. Schott, J. Phys. Chem., 90, 3031 (1986).

20. L. Salem, "The Molecular Orbital Theory of Conjugated Systems" (Benjamin, New York, 1966).

21. H. Yamakawa, "Modern Theory of Polymer Solutions" (Harper and Row, New York, 1971).

22. K.S. Schweizer and Z.G. Soos, Phys. Rev. B., to be submitted.

23. H.C. Longuet-Higgins and L. Salem, Proc. R. Soc. London, Ser. A. 251, 172 (1959).

24. Z.G. Soos, S. Kuwajima, and J.E. Mihalick, Phys. Rev., B32, 3124 (1985).

25. Z.G. Soos, S. Kuwajima, and R.H. Harding, J. Chem. Phys., 85, 601 (1986).

26. K. Fesser, A.R. Bishop, and D.K. Campbell, Phys. Rev. B., 27, 4804 (1983); S. Kivelson, T.K. Lee, Y.R. Lin-Liu, I. Peschel, L. Yu, Phys. Rev. B., 25, 4173 (1982).

27. M.J. Rice and S.R. Phillpot, Phys. Rev. Lett., 58, 937 (1987).

28. W.G. Boberski and A.L. Allred, J. Organometal Chem., 88, 65 (1975).

29. K.S. Schweizer, Bull. Am. Phys. Soc., 32(3), 885 (1987).

EXCITATIONS IN CONJUGATED POLYMERS

E.J. Mele and G.W. Hayden

Department of Physics and
Laboratory for Research on the Structure of Matter
University of Pennsylvania
Philadelphia, PA 19104

Abstract

We apply a Hubbard Peierls Hamiltonian to study the spectrum of low lying excitations for conjugated polymers. This study makes use of a renormalization group approach for the many particle Hamiltonian to accurately describe electron correlation effects in the excited states, and also explicitly treats coupling of the electrons to the lattice degrees of freedom. We find that the noninteracting U=0 theories provide a reasonable starting point from which to interpret the excitations; however, the interaction effects are seen play a very important role, introducing an important class of low lying neutral excitations in the structure. The low lying electronic excitations in the model are self trapped triplet excitons, and the low lying singlet excitations are found to include a "bi-exciton" constructed from a bound pair of the low lying triplet excitations.

I. Introduction

In this paper we will review the results of some recent theoretical work which is directed towards understanding the physics of elementary excitations in conjugated polymers. The results relate to the excited states populated by optical excitation in a broad class of conjugated systems. However, we will focus on the case of polyactylene which has the simplest structure in this class of systems, yet the excited state dynamics are rich enough to provide for some very interesting and subtle phenomena in the excited states. Our motivation for this work is to assess the relative importance of two general kinds of interactions which

347

affect the excited state properties.   The first interaction is the repulsion between the π electrons. The importance of the resulting dynamical correlation ofthe electron motion on the backbone for ordering the excited states in finite molecular analogs of the extended systems has been thoroughly explored through a number of theoretical and experimental studies over the last decade [1-3].   The second general kind of interaction involves the coupling of the propagating π electrons to the structural degrees of freedom (i.e. bond lengths) on the polymer backbone.   In a mean field theory this provides an effective one body term in the π electron Hamiltonian, and is known to lead to the observed bond alternation in the ground state of both finite and extended versions of these systems.   This latter interaction has elicited considerable interest in this problem among the theoretical physics community [4-7], as it provides an example of a spontaneous symmetry breaking due to the internal dynamics of the system, and provides an experimental realization of a model devloped in the mid 1970's to describe the interesting physics of this situation.   This connection is important since it makes a nontrivial prediction about the *excitations* of such a system, namely, that the elementary excitations include a class of topological excitations, or solitons [4-7], similar to the appearance of Bloch walls in a one dimensional ferromagnet.   These two general effects lead to somewhat different models for the excitation spectrum of a one dimensional conjugated system; it is the goal of the present work to understand the connection between these ideas.

In this paper we will proceed in the following way.   First, in section II, we briefly review some of the general ideas which have been put forth concerning electronic excitations in these systems, and contrast the effects of the two interactions mentioned above.   In section III, we present a model we have recently developed which allows us to study simultaneously the effect of electron correlation, and lattice relaxation on the excited state properties of a model conjugated polymer.   This leads to an interesting and reasonably simple picture of the modes of excitation of the conjugated polymer, which is presented in section IV.

II. Models for Excitations

A sensible model for the excitations, requires some picture for the underlying ground state.   For the case of polyacetylene, the π electrons form a nearly one dimensional electron gas which can be described as a half filled one dimensional tight binding band;   and following a well known theorem due to Peierls, the one dimensional metal is unstable to the formation of a structure with lower translational symmetry, but with a gap separating the filled and empty one electron states. This instability manifests itself in the appearance of bonds with alternating lengths on the polymer backbone, so that the appearance of bond alternation may be understood as a result of the $2k_f$ instability of the π electron gas. Of course the translational symmetry can be broken in two ways, corresponding to the two distinct but energtically equivalent bond alternation patterns on the long chain trans- polyene.

The simplest model for an electronic excitation of this system is the promotion of an electron from the valence to conduction bands; however, a number of arguments can be put forth to demonstrate that this is not the lowest electronic excitation of the system. The relevant observation is that the appearance of the gap is a response of the system to the ground state electronic density; hence a perturbation to the density should trigger a concomitant adjustment of the structure. In fact the lowest excitation of the broken symmetry state described above is known to be a topological excitation which corresponds to an interface between the two degenerate ground states, where the interface describes a smooth interpolation between these structures [4,7]. The topological nature of these excitations, or solitons, requires that they be created in pairs however. A continuum theory of the situation described above has been developed [5] and demonstrates that the pair creation energy $E = 2E_g/\pi$ so that the creation of the soliton anti soliton pair is favored over the creation of free electron and hole, by an energy corresponding to 30% of the optical gap, which amounts to approximately 0.5 eV for the case of polyacetylene. Furthermore, the spontanenous unbinding of an oppositely charged kink antikink pair from an initial configuration corresponding to single particle excitation across the gap has been demonstrated in a number of molecular dynamics simulations [8,9], so that the relaxation is not restricted by kinetic considerations.

Alternatively, one may focus on the influence of direct electron electron interactions on the excited state properties. Minimally, starting from the independent particle picture this should provide some residual binding of the photoexcited pair [10], but the effects are much more subtle. In fact with relatively strong electron electron interactions, such as suggested in the Pariser-Parr-Pople model [2], the odd parity (optically exicted) single particle excitation is not the lowest singlet excitation, but lies above an excitation which best expressed as a two particle excited state [1-3,11]; i.e. it involves the coherent excitation of a pair of electrons from the filled Fermi sea. The appearance of this two particle excitation as a low lying excited state occurs in the theory once doubly excited configurations are included in a CI expansion for the excited states, and the effect is not small. The binding energy of the even parity two particle singlet below the "independent" particle singlet extrapolates to nearly 1 eV in the long chain limit. It is interesting to note that a similar extrapolation of the optical one particle threshhold falls near 2 eV [1], i.e. near the onset of the strong absorption edge of trans-(CH)x [12].

There are a number of important questions raised by this observation, which has been successfully applied to interpret a number of phenomena related to the photoexcitations of the *finite* polyenes. We are interested in the effect of lattice relaxations on these charge correlated states, how the finite chain results can sensibly be extrapolated to describe behavior on the extended polymer, and whether any connection between this behavior, and the interesting behavior predicted in the SSH (electron phonon) theory can be made. The extrapolation to larger systems becomes a crucial issue, since one only expects to see the nucleation of the topological defects of the SSH model once the systems size is expanded beyond the defect size, which is though to span some 10-15 bond lengths [4,6]. This is somewhat larger than the sizes which are conveniently studied by present valence bond or configuration interaction methods [1,11].

III. Correlation Effects and Excitations

Over the last several years there have been a number of studies directed towards gauging the strength of the effective electron electron interaction in simple conjugated polymers [13-24]. In this work we make use of the so called Hubbard-Peierls Hamiltonian, with an effective noninteracting electron phonon (SSH) model augmented with an on site repulsion term:

$$H = \Sigma_{i,\sigma} \, [t_0 + \alpha \, (x_{i+1} - x_i)] \, c^+_{i+1,\sigma} \, c_{i,\sigma} + h.c.$$

$$+ \, \Sigma_i \, K/2 \, (x_{i+1} - x_i)^2$$

$$+ \, \Sigma_i \, U/2 \, n_{i,\sigma} \, n_{i,-\sigma} \qquad [1]$$

This is the simplest model, containing an explicit two body repulsive potential which can capture the physics of the interacting π electron system. The model can be generalized to include off site interactions (the extended Hubbard Peierls Hamiltonian) [15,18], or even repulsive terms coupling off diagonal terms in the one body density matrix [24] (e.g. bond orders), but the essential effects are contained in (1).

The first question to be clarified is the relative strength of the effective repulsion U appearing in the model. Even for the ground state properties the effects of U are subtle. Intuitively, one expects the repulsion term to suppress density fluctuations of the π electrons, and therefore to compete with the Peierls instability. This is certainly true for U/W >> 1, where W is the bandwidth. Here the effect of the electron electron repulsion is to exclude configurations with double site occupancy from the ground state wavefunction. It is well known that in this limit, the model reduces to a Heisenberg spin 1/2 Hamiltonian with antiferromagnetic nearest neighbor coupling. We return to this limit below, but for the present note that we do not expect this interaction dominated regime to be directly relevant for the observed properties of the conjugated polymers. For weak repulsion strengths (i.e. U/W <<1) the situation is less clear. If we make a mean field approximation on the Hamiltonian in equation (1), it is found that even for weak coupling strengths, the effect of U is to favor a partial spin polarization of the system which competes with the Peierls instability, so that the dimerization amplitude is weakened monotonically as U increases [13]. In fact, in the mean field theory the dimerization if found to vanish above a critical repulsion strength, so that the observation of bond alternation in the ground state of the extended polymers suggests the presence of very weak two body repulsive interactions. The argument does not survive an analysis beyond the mean field theory. Using a number of theoretical approaches, (valence bond methods [16,17], lattice Monte Carlo method [14,15,18], and variational forms for two body correlations in the ground state wavefunction [19] have been applied to the problem), it has been demonstrated that weak U actually *enhances* the dimerization. The most intuitive arguments for this surprising result have been put forth by Mazumdar [17] who observes that for weak U, the removal of doubly occupied configurations from the ground state wavefunction removes a barrier to resonance, and thereby increases the bond order and promotes dimerization. The effect appears to saturate for U = W, after which the system rapidly approaches the asymptotic behavior associated with the strongly correlated spin Peierls limit. The result is that interactions U<W have only a weak (and may enhance) the dimerization amplitude, so that

the presence of the Peierls distorted ground state can not be used to further identify the repulsion strength.

More information about the effective repulsion strength can be extracted from electron nuclear double resonance (ENDOR) data on undoped (CH)x. [25,26] Polyacetylene is known to possess a residual density of neutral unpaired spins in the ground state, and these defects are thought to be associated with solitons quenched into the polymer as it is formed. An interesting prediction of the SSH theory [4] is that the unpaired spin, associated with the topological defect should be distributed through the domain wall, but resides on alternate lattice sites near the defect center, i.e. the wavefunction has an exact node on alternate atomic sites through the domain wall. This prediction is modified in the presence of the repulsion term, which produces an additional spin polarization within the defect. The effect is to produce an oscillating spin density within the defect, with both positive and negative components on alternate sites. This occurs even within a mean field approximation, and resembles the formation of a local spin density wave within the domain wall. The amplitude of these negative spin densities have been measured in ENDOR studies, and show a peak is the "negative" spin density of roughly 0.3 the amplitude of the peak in the "positive" spin density. Within the Hamiltonian of equation (1), such a strong spin polarization is obtained only for a rather strong two body repulsive potential. For example, using a highly accurate numerical procedure described below, we find that $\rho-/\rho+$ = .3 requires $U \geq 0.5$ W. In reference 25 a mean field analysis similarly yields $U \approx 1.4$ t. The results are reassuring; using the PPP Hamiltonian [23] the effective repulsion term (i.e. the difference between the on site repulsion strength and the intersite term) also amounts to of order half the bandwidth, which agrees with the estimate extracted from comparison with the ENDOR data. However, the result is also unsettling since it indicates that interation effects are occuring on the same energy scale as the one body "band" terms in the theory; any sort of perturbative treatment around either the weakly interacting (U=0) or strongly interacting (U/W >>1) regimes is not expected to accurately describe the physics of this situation.

The difficultly of correctly describing the ground state in this intermediate coupling regime has been appreciated , and over the last several years a number of theoretical numerical approaches have been directed at this problem, including the lattice Monte Carlo method, and variational representations for the ground state wavefunctions. The treatment of the excited states poses an additional complexity since one requires an approach with variational flexibility sufficient to describe both correlations in the ground state, *and* the low lying excitations out of the ground state. The most thorough studies using valence bond [11] or configuration interaction methods [reviewed in 1] to study this problem have consequently been restricted to relatively small system sizes (typically $N<12$). A natural approach to this problem, which is made difficult by the appearance of many interacting degrees of freedom for the system, is to make use of a renormalization of the many particle Hamiltonian to restrict our attention to the "important" degrees of freedom for the extended chain. The strategy is to identify and eliminate spurious, short wavelength degrees of freedom by studies on small clusters, so that the basis size used to study the extended system can be controlled. In the literature there have been reported several efforts applying this scheme, with varying success, to the pure one dimensional Hubbard model [27,28] . For example Hirsch [27] used a applied a scaling technique to the pure one dimensional Hubbard model which he used to study the critical properties of the theory as a function of the repulsion strength. Bray and Chui [28] reported a somewhat more controlled version of this theory, by retaining a sufficiently large basis set to give excellent convergence for the ground state energy as compared to the exact Lieb Wu results for this model [29]. This latter application did not attempt to iterate the renormalization to large system sizes, but simply used the scheme to provide a controlled basis set expansion for the many body eigenfunctions in a series of finite chains. Our approach to this problem follows this strategy, but introduces into the study the effect of coupling to the lattice degrees of freedom.

The computational strategy adopted in this approach is described schematically in Figure 1. We wish to study the interacting many electron eigenfunctions in a relatively large one dimensional chain, such as is sketched in the top panel. We proceed by deleting alternate bonds from the full structure which then yields an array of decoupled "dimers", the resulting micro-Hamiltonians within each dimer are mutually commuting, and the many electron states in each cluster are calculated independently.

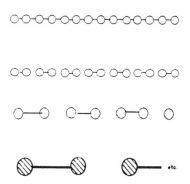

Figure 1: Schematic representation of the "renormalization" scheme. Many particle eigenfunctions from clusters of the full system are combined in properly symmetryized products to obtain the many particle states of an inflated cluster [for details of the implementation of this method see ref. 20-22,28].

These many particle eigenstates are indexed by quantum numbers describing the total number N, total spin, S, and z projection of the spin, S_z in each cluster. We then describe the effects of the deleted bonds, by combining these cluster eigenfunctions in properly symmetrized products, which are in turn mixed by the previously deleted one body hopping amplitudes on the chain. If all the cluster eigenfunctions are retained in this calculation the method is exact, though very cumbersome, since the interactions between the clusters grows increasingly complex as the system size expands. However the utility of this construction is that the low lying eigenfunctions of the composite clusters should be dominated by the low lying eigenfunctions from each cluster, and we therefore seek to truncate the expansion after including some maximum number of cell eigenfunctions. In Hirsch's original work with this method, very few functions were retained, in an effort to exactly map the composite Hamiltonian on to a renormalized version of the cell Hamiltonian. In our work with this method we found, as did Bray and Chui, that this method does not preserve quantitative accuracy, and considerable care must be taken to include a large number of expansion functions from each such cluster, allowing for fluctuations in particle number and spin.

By tests with this method, we found [21,22] an accurate expansion for chains of intermediate size required that typically 200 product states be directly retained in each N, S, S_z manifold, and of order 10^3 states included in second order perturbation theory [30]. This was found to give very accurate results on N=16 chains, and reasonably accurate data for several N=32 studies which we undertook. It is straightforward to demonstrate, however, that any such finite truncation of the exansion set in this procedure will not quantitatively describe the properties of the interacting Hamiltonian in the large N limit. However, for chains of intermediate size, the procedure proved to be both quite fast and accurate. We were able to calibrate the method by direct comparison with results in the noninteracting U=0 limit on N=16 chains. Here we found excellent convergence for both the ground state energy as a function of dimerization amplitude, and the "vertical" excitation energy to the rigid lattice singlet excited state. In the large U limit, we were able to reproduce the Bonner Fisher results [31] for the spin spin correlation function in the undimerized chains. One expects this theory which builds in the short range correlation of the electrons to be better suited to the large U limit of this problem than the small U "Bloch" limit; the tests on the noninteracting Hamiltonian demonstrate that sufficient variational flexibility to describe the band character of the system is retained.

We have applied this method to self consistently relax the bond lengths of a chain described by the Hamiltonian (1). The procedure is to apply the Hellman Feynman theorem to extract instantaneous forces on the ions, and use these to drive an overdamped molecular dynamics simulation to relax the structure to its equilibrium configuration. The relaxation may require of order 50-100 steps driven with the forces updated in each instantaneous configuration, so that the scheme for obtaining the interacting many particle states is employed repeatedly during the simulation. It is therefore essential to have a very efficient scheme for updating the forces. To interpret the results, it is useful to first examine the relaxed structures in the noninteracting U=0 limit. The data are presented in Figure 2. The results presented here are obtained for a finite N=16 chain, with fixed end boundary conditions. The parameters in the Hamiltonian (1) are: t_o = -3.0 eV, α = 8.0 eV- A^{-1}, K = 68.6 eV-A^{-2} [32].

The ordinate is the staggered displacement field, measuring the net compression or expansion of alternate bonds on the chain, (the ordinate can be interpreted as a measure of the local bond alternation amplitude) and the abcissa indexes the bonds in the structure. The four panels in the figure correspond to the four lowest singlet states in the theory; and the results clearly demonstrate that we recover the known behavior of the noninteracting Hamiltonian by this method. The ground state is uniformly dimerized, with slight ringing in the bond alternation amplitude due to the finite system size. The first singlet excited state is an odd parity state,

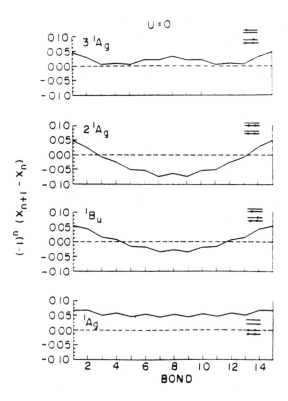

Figure 2: Relaxed lattice configuration for the four lowest singlet states of a N=16 chain (fixed end boundary conditions) for the case U = 0. The ground state is uniformly dimerized, and the lowest excitations unbind kinks in the bond alternation field.

and it relaxes by the unbinding of two oppositely charged kinks, shown as the zero crossings of the order parameter. Note that the kinks are not "sharp"; they are expected to be extended over roughly 10 lattice sites, and in fact even the N=16 chain is hardly large enough for us to observe the separation of these defects. For example, the dimerization amplitude at the center of the chain is approximately 25% smaller than the ground state equibilibrium value in the lowest panel. The results suggest that even N=16 is barely large enough to reveal relaxations of the photoexcitations which would be representative of the long chain behavior, and results for substantially smaller chains would be dominated by the finite size effects. The next two singlet excited states also demonstrate the well known behavior of the noninteracting Hamiltonian: the first even parity state is associated with kink unbinding on a slightly more repulsive potential surface, and the third even parity state corresponds to a "band" excitation of the polymer.

The analogous results for the case U = W/2 are given in Figure 3. Several interesting phenomena are introduced by the repulsive term in the Hamiltonian.

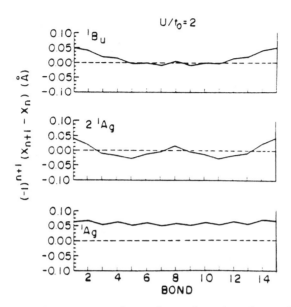

Figure 3: Relaxed lattice configurations for the three lowest singlet states of a N=16 chain (fixed end boundary conditions) for the case U = 2t. The ground state is uniformly dimerized, the lowest odd parity singlet relaxes to a solitonic "exciton', and the lowest even parity singlet is well described as a bound pair of triplet excitations.

For example we find a modest enhancement of the equilibrium dimerization amplitude in the ground state (the effect appears to be considerably weaker than predicted by the Gutzwiller wavefunction for the ground state [19]). Stronger effects are observed in the relaxed excited states however. For example, the lowest odd parity excited state, analogous to the state which released the oppositely charged kink-antikink pair in the noninteracting limit, is found to yield a "confined" kink-antikink pair. Notice that the order parameter is strongly suppressed near the center of the chain (barely crossing through zero). The resulting relaxed structure can be interpreted as a "solitonic exciton", and resembles a simple self trapped singlet exciton on the chain. The binding of the pair in the presence of an on site repulsion is at first surprising; in the noninteracting limit the kink and antikink are known to distribute charge on opposite sublattices, so that in lowest order the on site repulsion term would not yield a binding of the exciton , as noted by

Grabowski, Hone and Schrieffer, who found such a binding only once an off site repulsion terms is included in the theory [10]. Closer analysis shows that the binding is an indirect effect, resulting from the polarization of the scattering states near the charged defect. In fact a configuration interaction study shows that the binding occurs as soon as configurations including excitations into the low lying states in the continuum are included [22]. Though weak, the binding is sufficient to yield a relaxed structure which differs qualitatively from that obtained in Figure 2.

Perhaps the most intriguing result of the study is that we indeed obtain a singlet excitation below this optically allowed odd parity singlet. The lattice relaxation surrounding this excitation is given in the central panel; it provides some insight about the microscopic electronic structure of this excited state, and suggests how this result may be extrapolated to the extended chain. The relaxed structure in the central panel reveals the appearance of *two* "dips" in the bond alternation amplitude; i.e. the bond alternation amplitude is weakened on both the left and right hand segments of the polyene. Here we find that this weakening is quite strong, and yields a pair of zero crossings of the order parameter in the left and right hand regions. This behavior can be interpreted simply as follows. The lowest lying electronic excitation of this system in the presence of intermediate repulsion strength is not a singlet excitation, but a triplet excitation. The triplet excited state relaxes by the unbinding of a kink anti-kink pair, but here because the particle are excited in a triplet state, the residual "Coulomb" attraction is strongly reduced, and the resulting kinks are free to separate. The reduced charge transfer character of the excitation is responsible for the very low energy of this excitation relative to the optically allowed singlet excitation. This reduction of the repulsive energy is a very strong effect, in fact the lowest even parity singlet excitation can be reconstructed from a *pair* of triplet excitations out of the ground state, with the triplets coupled to an overall spin singlet, and a residual binding of the pair. There are a number of observations which support this interpretation. For example, a decomposition of the 2^1A_g wavefunction which we obtain in this study shows a very large admixture of the excited triplet configurations in the left and right segments, which are coupled to describe this singlet excited state in the renormalization scheme. The lattice relaxation in the central panel of Figure 3, in fact shows a superposition of the kink structures calculated for the relaxed triplets for the left and right hand 8 site segments separately (this feature, can be compared with equilbrium structures calculated for the eight atom segments, and is examined in more detail in references 21,22). Finally this observation correlates well with an empirical rule deduced by Schulten [2] years ago who observed that in his calculations on finite polyenes, the excitation energy to the even parity singlet correlated with twice the energy for excitation to the triplet. There is apparently precedent for this sort of composition rule for the low lying singlet excitations of conjugated systems; for example an analogous situation for anthracene is cited by C. Kittel is one of the older editions of his classic text, to demonstrate the ordering of optical excitations in an organic crystal [33].

The physical picture is that for intermediate repulsion strength, the triplet excitation is very strongly energetically favored over the odd parity singlet excited state. The stabilization energy appears to be sufficiently large, that the lowest lying singlet can be reconstructed from a coherent excitation of two such local triplet excitations. The resulting relaxed structure can be interpreted as a bound state of two "triplet" polarons, or a bound complex consisting of four neutral solitons (a soliton "quartet" !). The odd parity charge transfer excitation relaxes to a confined "solitonic" exciton, and is raised in energy relative to the even parity state owing to the larger charge transfer character of this excitation, which is apparently well described in a one body picture.

IV. Discussion

The results presented in section III demonstrate that electron electron interactions can play an important role in determining the character of the low lying excitations of the extended conjugated polymers. For the case where the on site repulsion strength is of order half the bandwidth, a class of neutral excited states is found to characterize the low lying excitation spectrum of the system. As in the noninteracting theory, these excitations are "self-localized" by the subsequent relaxation of the lattice, in fact the picture of low lying excited states in the form of neutral solitons and polarons is quite helpful for interpreting the nature of these self localized excitations. However, unlike the general picture anticipated in the noninteracting model, the picture of the ordering of these excited states must be modified. We find a very substantial reduction of the repulsion energy in the triplet excited state (relative to the optically excited singlet), so that the true low lying singlet excitations are even parity states which are better interpreted as resulting froma bi-exciton, i.e. the coherent excitation of triplet pairs.

Interestingly enough, while this sort of behavior is not readily anticipated from the small U end of this problem, it is precisely the behavior which is obtained in the highly correlated large U limit [34]. For example, in the large U limit, the Hamiltonian of equation (1) is known to map onto an effective Heisenberg spin 1/2 Hamiltonian with antiferromagnetic nearest neighbor exchange. With coupling of the exchange integrals to the lattice as in (1) this is the spin Peierls model, which exhibits a spontaneous bond alternation in the ground state. The electronic ground state in this configuration corresponds closely to a "Bethe singlet" with nearest neighbor electrons singlet paired across the strong (short) bonds on the structure, and with the structural relaxation restricting the resonance of the system to a configuration with singlet coupling across the weak (long) bonds. The lowest triplet excitation of this structure breaks such a singlet pair, weakening (or breaking) a bond on the structure. The resulting triplet excitation can then relax, with the mode of relaxation depending on the effective interaction between the "unpaired" spins. If the interaction in repulsive, as is the case in the simple spin Peierls model, the triplet dissociates leaving two unpaired spins in the structure. As the triplet dissociates, the order of the singlet pairing on the chain, and hence the order of the bond alternation of the structure is reversed along a length of the chain through which the kinks separate. The result is two spin 1/2 neutral excitations which can best be characterized as kinks or phase slippages of the spin Peierls ground state. In this limit of course, this triplet excitation is degenerate with the singlet combination of these two isolated kinks, i.e. the spin 1/2 defects should be understood as the true elementary excitations of the structure. These defects are loosely analogous to the neutral spin 1/2 kinks which occur in the SSH model for (CH)x.

The behavior we have found in the calculations described in section III, does *not* correspond to this simple situation, but instead correspond more closely to a situation in which the residual interaction between the unpaired spins is weakly attractive. The resulting triplet excitation does not dissociate, but remains confined over several bond lengths on the structure. The elementary singlet excitation is then characterized as an excitation of two such overlapping triplet pairs. It appears that the microscopic triplet character of these excitations remains relatively intact when the excitations are combined in the singlet state, so that the singlet excitation resembles a "molecule" condensed from the triplets. In this limit, it appears appropriate to regard the triplet exciton as the fundamental neutral excitation of the system. We believe that this description applies to the low lying neutral excitations in a large class of conjugated polymers, including nondegenerate ground state polymers such as cis CHx , polydiacetylene, and even for finite trans segments of PA in a cis structure. The situation for a truly degenerate structure such as a extended trans CHx chain remains unclear because of the residual finite size effects in the data which we have been able to extract form our simulations. On the basis of much more approximate configuration interation studies we speculate that the low lying neutral excitations of such a system would resemble more closely the isolated spin 1/2 kinks of the spin Peierls model. While this spin analog provides a useful model for the neutral excitations, it should also be noted that it clearly does not properly address the nature of the *charge transfer* excitation, i.e. the odd parity singlet excited state. Interestingly this optically allowed state, even in the presence of a moderately strong on site repulsion is relatively well described within the standard SSH model.

We close this section by bringing these observations together in a schematic level diagram for the low lying excitations of the conjugated polymers.

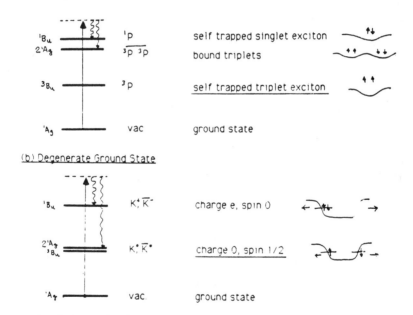

Figure 4: Schematic level diagrams for excitations in nondegenerate conjugated polymers (top panel) and degenerate ground state polymers (lower panel). In the nondegenerate case the triplet exciton unbinds a pair of neutral spin 1/2 excitations.

As we noted above, the situation is slightly different for the polymers with a nondegenerate ground state, and those with degenerate broken symmetry ground states. The former situation is somehat more general and is addressed more directly by our calculations (in which the system size breaks the degeneracy between the two structures) and is presented in the top panel of Figure 4. The ground state corresponds to an even parity singlet, and the interesting excitations are created following dipole allowed excitation to the rigid lattice odd parity singlet excited state indicated by the dashed line. There are several modes of relaxation for this excitation. The standard SSH theory suggests a self localization of the excitation , to form a self trapped exciton (a singlet polaron) as shown. However, the lowest electronic excitation corresponds to the low lying odd parity triplet exciton which self localizes in a similar manner. The triplet exciton lies quite low in energy so that the lowest singlet excitation can be reconstructed as an excitonic "molecule" formed by the condensation of two such triplet excitations. Interestingly, starting with the notion of relatively extended excitations of the structure, this results returns us quite closely to an interpretation which was suggested some time ago to describe the even parity excitations of finite molecular analogs of these systems.

The level ordering for the degenerate ground state polymers, which should be regarded as more of a special case, is given in the lower panel of this figure. The level ordering is similar in spirit to that of the nondegenerate system but different in detail. Again the elementary optical excitation corresponds to promotion from the ground state even parity singlet, to the rigid lattice odd parity singlet (dashed). For the case of polyacetylene, this corresponds to the strong absorption peaked near 1.9eV. The SSH relaxation pathway for this excitation yields a pair of oppositely charged kinks, with a relaxation energy $\approx$ 0.5 eV. As noted by several workers, this relaxation can be associated with the broad intrinsic linewidth of the optical excitation in trans (CH)x [35-37]. The role of the Coulomb term is to produce a weak residual binding of this pair to form a solitonic exciton. We can relax to a lower lying excited state by having the kinks dissociate in an even parity state, which yields two neutral spin 1/2 defects in the polymer. While direct relaxation from the odd parity excited state singlet to the even parity manifold is forbidden in the strict SSH model, Wu and Kivelson [38] have noted that weak departure from the simple model due to three dimensional interactions, or even non nearest neighbor in chain interactions, open this relaxation channel for the odd parity excited state singlet, so that one should expect rapid relaxation back into the even parity manifold. Once dissociated the neutral singlet and triplet excitations are degenerate, so that the neutral spin 1/2 kink is the elementary (and perhaps long lived) excitation of this structure.

There have been a number of experiments directed at probing the transient changes in optical, infrared and magnetic properties following photoexcitation, in an effort to directly study the properties of these photoexcited states [39,40]. A detailed understanding of that data will require a more thorough understanding of the kinetics of the relaxation of excitations through this spectrum of excited states, and may well also require a thorough understanding of three dimensional effects, trapping of photoexcitations at defects in the structure, etc. Modulo these difficulties we expect that the excitation spectrum outlined above should provide useful input for interpreting observations regarding excitations of this class of polymers.

Acknowledgements

This work was supported by NSF under grant DMR 84-05524 and through the MRL at the University of Pennsylvania under grant DMR 84-14640. I would like to thank the organizers for their additional support, and for their efforts organizing this Symposium. We thank J. Orenstein, S. Mazumdar, D. Campbell, G. Baker, S. Etemad, and S.T. Chui for essential input, advice and encouragement at various stages of this study.

References

1. B. Hudson and B. Kohler, *Synthetic Metals* **9**, 241 (1984)

2. K. Schulten, I. Ohmine and M. Karplus, *J. Chem. Phys.* **64**, 4422 (1976)

3. P. Tavan and K. Schulten, *J. Chem. Phys.* **70**, 5407 (1979)

4. W. P. Su, J. R. Schrieffer and A. J. Heeger, *Phys. Rev. B* **22**, 2099 (1980)

5. H. Takayama, Y. R. Lin Liu and K. Maki, *Phys. Rev. B* **21**, 2388 (1980)

6. W. P. Su and J. R. Schrieffer, *Phys. Rev. Lett.* **46**, 738 (1981)

7. M. J. Rice, *Phys. Lett.* **71A**, 152 (1979)

8. W. P. Su and J. R. Schrieffer, *Proc. Nat. Acad. Sci (USA)* **77**, 5626 (1980)

9. E. J. Mele, *Phys. Rev. B* **26**, 6901 (1982)

10. M. Grabowksi, D. Hone and J. R. Schrieffer, *Phys. Rev. B* **31**, 7850 (1985)

11. Z. G. Zoos and S. Ramasesha, *Phys. Rev. B* **29**, 5410 (1984)

12. N. Suzuki, M. Ozaki, A. J. Heeger and A. G. MacDiarmid, *Phys. Rev. Lett.* **45**, 1209 (1980)

13. S. Kivelson and D. Heim, *Phys. Rev. B* **26**, 4278 (1982)

14. J. Hirsch, *Phys. Rev. Lett.* **51**, 296 (1983)

15. J. Hirsh and M. Grabowski, *Phys. Rev. Lett.* **52**, 1713 (1984)

16. S. N. Dixit and S. Mazumdar, *Phys. Rev. B* **29**, 1824 (1984)

17. S. Mazumdar and S. N. Dixit, *Synthetic Metals* **9**, 275 (1984)

18. D. K. Campbell, T. A. deGrand and S. Mazumdar, *Phys. Rev. Lett.* **52**, 1717 (1984)

19. D. Baeriswyl and K. Maki, *Phys. Rev. B* **31**, 6633 (1985)

20. G. W. Hayden and E. J. Mele, *Phys. Rev. B* **34**, 5484 (1986)

21. G. W. Hayden (Ph. D. thesis, University of Pennsylvania, unpublished)

22. G. W. Hayden and E. J. Mele, *Phys. Rev. B* **32**, 6527 (1985)

23. H. Fukutome and M. Sasai, *Prog. Th. Phys.* **69**, 1 (1983); **69**, 373 (1983)

24. S. Kivelson, W. P. Su, A. J. Heeger and J. R. Schrieffer (preprint, 1987)

25. H. Thomann, L.R. Dalton, Y. Tomkiewicz and T. C. Clarke, *Phys. Rev. Lett.* **50**, 533 (1983), H. Thomann, L. R. Dalton, M. Grabowski, and T. C. Clarke, *Phys. Rev. B* **31**, 3141 (1985).

26. H. Thomann and L. R. Dalton in <u>Handbook on Conducting Polymers</u> (T. Skotheim, ed., Marcel Dekker, 1986) p. 1157

27. J. Hirsch, *Phys. Rev. B* **22**, 5259 (1980)

28. J. W. Bray and S. T. Chui, *Phys. Rev. B* **19**, 4876 (1979)

29. E.H. Lieb and F. Y. Wu, *Phys. Rev. Lett.* **20**, 1445 (1968)

30. P. Lowdin, *J. Chem. Phys.* **19**, 1396 (1951)

31. J. C. Bonner and M. E. Fisher, *Phys. Rev A* **135,** 640 (1964)

32. D. Vanderbilt and E. J. Mele, *Phys. Rev. B* **22**, 3939 (1980)

33. C. Kittel in <u>Introduction to Solid State Physics</u> (Wiley, 4th edition, 1969)

34. F. D. M. Haldane, *Phys. Rev. B* **25**, 4925 (1984)

35. J. Sethna and S. Kivelson, *Phys. Rev. B* **26**, 3513 (1982)

36. E. J. Mele, *Synthetic Metals* **9**, 207 (1984)

37. L. Yu (preprint, 1986)

38. S. Kivelson and W.-K. Wu, *Phys. Rev. B* **34**, 5423 (1986)

39. a good review may be found in J. Orenstein, <u>Handbook of Conducting Polymers</u> (T. Skotheim, ed; Marcel Dekker, 1986) p. 1297

40. J Orenstein, Z. Vardeny, S. Etemad, G. Baker and G. Eagle, *Phys. Rev. B* **30**, 786 (1984)

TOWARD NEW ELECTRONIC STRUCTURES IN
CRYSTALLOGRAPHICALLY ORDERED CONJUGATED POLYMERS

Daniel J. Sandman and James W. Shepherd III

GTE Laboratories Incorporated
40 Sylvan Road
Waltham, Massachusetts 02254

M. Thomas Jones

Department of Chemistry
University of Missouri - St. Louis
St. Louis, Missouri 63121

INTRODUCTION

The polydiacetylenes[1-3] (PDA, $\underline{1}$) are a class of conjugated polymers obtainable in crystallographically ordered form, a consequence of their synthesis via topochemical solid state polymerization. PDA have at least two physical properties, electronic carrier mobility and third-order nonlinear optical processes, whose coefficients have magnitudes reminiscent of prototypical atomic inorganic semiconductors.[4] Current quantum chemical treatments[5,6] are compatible with a description of PDA as broad-band one-dimensional semiconductors.

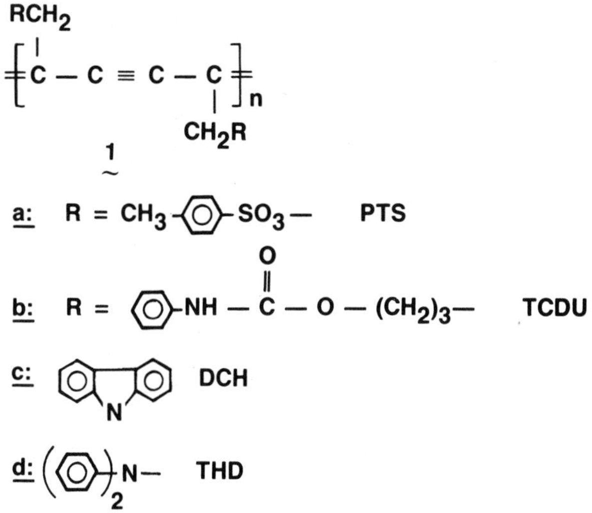

The perspective of the preceding paragraph argues not only for detailed physical and theoretical study of known PDA but also for a broadly based materials research initiative into possible modification of existing PDA, synthesis and characterization of new PDA, and new approaches to other ordered conjugated polymers with broad-band electronic structures, such as polyacetylenes.[5,7,8] Accordingly, herein are summarized experimental studies of the degenerate third-order nonlinear optical susceptibility and its temporal response in well-defined PDA materials and of the chemical modification of a PDA in which the conversion of a PDA structure to that of a mixed polyacetylene is deduced. Since PDA crystallography[9] indicates strong covalent forces along the conjugated chain and van der Waals interactions in other directions, analogy to graphite and other layered solids invites experimental approaches to the diffusion of reagents into these polymers.[4] Our experiments in the chemical modification of PDA are the first experimental examples of this situation.

NONLINEAR OPTICS IN POLYDIACETYLENE WAVEGUIDES

The current interest in third-order nonlinear optical phenomena in PDA had its genesis in the report of Sauteret et al.,[10] which disclosed large values, comparable to those of Ge and GaAs, of the third-order coefficient ($\chi^{(3)}$) measured along the chain axis in poly-PTS (1a) and -TCDU (1b) free-standing crystals from frequency tripling experiments. These investigators[10] inferred that if the nonlinearity originated in the intrinsic PDA electronic structure, it might have a response time in the femtosecond regime. Since several ultrafast nonlinear optical signal processing schemes rely on a material's intensity-dependent index of refraction (proportional to the degenerate third-order susceptibility, $\chi^{(3)}(\omega)$, as the basic nonlinear mechanism),[11] research at GTE Laboratories has focused on the evaluation of the magnitude and sign of $\chi^{(3)}(\omega)$, as well as its temporal response, in both transparent and absorbing regions using well-defined planar waveguide PDA samples. The latter are either Langmuir-Blodgett (L-B) multilayers or single-crystal thin films with optical-quality surfaces. Recent summaries of this work have appeared.[12-14]

Degenerate four-wave mixing experiments in poly-PTS waveguides established[13,14] that the magnitude of $\chi^{(3)}(\omega)$ in the transparent region is comparable to that of $\chi^{(3)}$ (3ω) previously reported.[10] Using tunable dye laser pulses of 300 femtosecond duration, measurement of the excited-state lifetime at 6520 Å revealed a value of 1.8 picoseconds, while off-resonance (7230 Å), the nonlinear response was found to be limited by the optical pulse width.[15] The latter observation is not compatible with Flytzanis' conjecture that strong electron-lattice coupling would limit the response time to 1–4 picoseconds due to formation of conjugation defects.[16]

Other researchers have reported four-wave mixing studies of PDA in solution[17-19] and third harmonic generation in L-B multilayer PDA films.[19]

CHEMICAL MODIFICATION OF POLY-1,6- Di-N-CARBAZOLYL-2,4-HEXADIYNE

The interaction of charge transfer reagents with the partially crystalline polyacetylene resulting in conductive materials raises the question of chemical reactivity of PDA. Since diffusion of gases and liquids in partially crystalline polymers is facile only in the amorphous regions of the polymer, crystals of PDA which have tight-packed structures would be expected to be relatively unreactive. Such was the state of affairs[12] prior to our observations[4,20-23] that assorted electrophilic reagents react with poly-1,6-di-N-carbazolyl-2,4-hexadiyne (DCH, 1c) to give new materials which are homogeneous upon examination by electron microscopy.[24]

Of particular interest are the reactions of poly-DCH with bromine which produce materials which have gained 3–8 Br atoms per polymer repeat unit and have retained crystallographic order.[22,23] These reactions are anisotropic and in their initial stage proceed more rapidly as temperature is lowered.[20] The latter observation is noteworthy because diffusion in solids is typically an activated process.

X-ray powder and photographic studies[21,22] revealed that crystallographic order is retained, especially in the chain direction. The introduction of 6 Br per repeat unit expands the unit cell volume by ca. 27 percent. The poor correlation of the crystallites perpendicular to the chain[21,22] precludes a structure proof by single-crystal x-ray diffraction techniques.

The general insolubility of the derivatives of poly-DCH, especially the brominated materials, suggested the application of solid state ^{13}C NMR using cross polarization and magic angle spinning techniques (CP-MAS) for the structure proof. The ^{13}C CP-MAS NMR studies relied on known assignments of shifts for carbazole derivatives and model molecular compounds.[22,25,26] The ^{13}C CP-MAS spectra indicate that bromine reacts initially with the carbazole groups to bring about 3,6-dibromo substitution. Subsequently, the backbone becomes involved, and at the level of introduction of 6 Br per repeat unit, the conversion of the PDA structure to a mixed polyacetylene, illustrated in Figure 1, is deduced. The deduction of such a mixed polyacetylene structure is noteworthy because of the diversity of mechanistic situations noted in the reported solid state reactivity of crystalline monoacetylenes,[7,8] the obvious approach to an ordered polyacetylene. The spectra of materials which have gained 7–8 Br atoms per repeat unit indicate that carbazole substitution occurred at the 3 and 6 positions. When the spectra of materials which have gained 7–8 Br atoms per repeat are recorded using the dipolar delay technique,[25] additional weak peaks at 165.5 and 154.5 ppm are detected. These chemical shifts are typical values for the β-carbons of butatrienes in the liquid[27] and solid states,[4,12] and this assignment is offered.

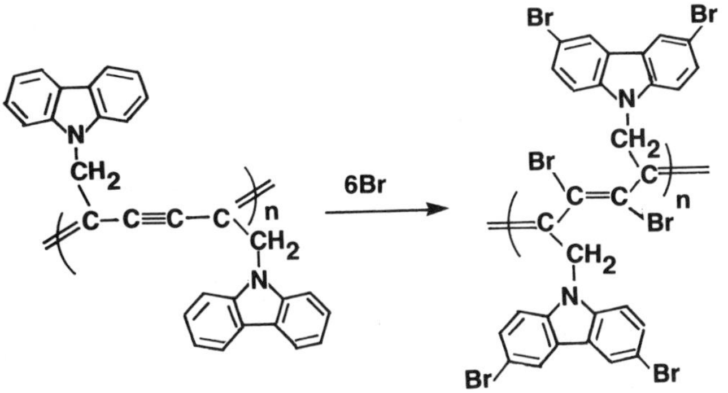

Figure 1. Conversion of poly-DCH to the mixed polyacetylene structure deduced for material which gains 6 Br atoms per repeat unit, from reference 23 with permission.

The ^{13}C CP-MAS NMR studies deduced covalent bond formation involving the conjugated backbone. This conclusion raises the issue of the mechanistic processes involved in reactions with the backbone. Three possibilities have been considered: (a) reaction involving electron transfer, (b) "classical" addition processes, and (c) reaction via conjugation defects.[23] Possibility (a) has been experimentally excluded,[21,23] and possibility (b) cannot be rigorously excluded based on our present knowledge of these materials. The association of the peaks at 165.5 and 154.5 ppm in the ^{13}C CP-MAS spectra with butatriene segments prompts us to give serious consideration to conjugation defects as reactive intermediates. A reaction pathway involving a six-carbon defect leading to a butatriene segment is given in Figure 2. Butatriene segments readily account for the disruption of conjugation observed[23] in materials with 7–8 Br atoms per repeat unit although alternative possibilities cannot be ruled out at present.

The introduction of bromine into poly-DCH does not lead to a detectable enhancement of dc conductivity. The brominated polymers are largely diamagnetic, and the low concentrations of paramagnetic species have been probed by both static susceptibility and electron spin resonance (esr) spectroscopy.[21,23,28,29] In the material which has gained 6 Br atoms per repeat, it is conceivable that the bulk of the paramagnetism is associated with conjugation defects ("solitons") of a mixed polyacetylene.[23]

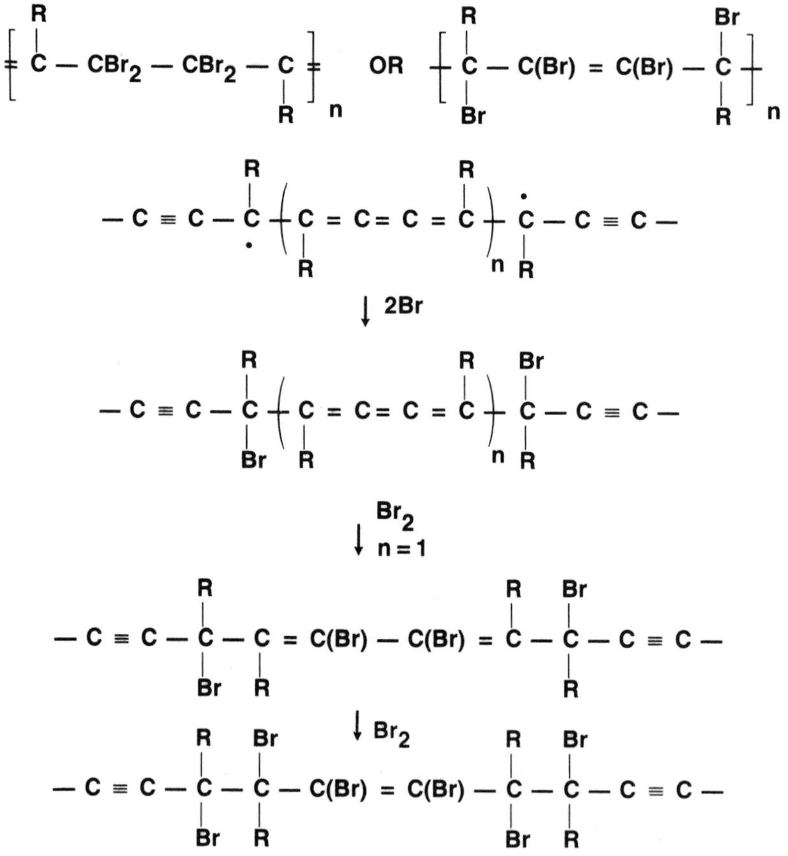

Figure 2. Possible scheme for reaction with PDA backbone involving conjugation defects leading to disruption of the conjugated system.

REACTIVITY OF POLY-1,1,6,6-TETRAPHENYLHEXADIYNEDIAMINE (THD)

In seeking further examples of chemical reactivity in PDA other than poly-DCH, several considerations seem relevant. For reaction with external reagents, the PDA should be one in which the monomer is completely converted to polymer so that the reagents do not have large vacancies through which to enter the polymer. Additionally, it is desirable for the side chain energy levels to have a relationship to those of the carbazoles in poly-DCH. Both of these considerations are well met by poly-1,1,6,6-tetraphenylhexa-diynediamine (THD, 1d).[30]

Prior to discussing the issues relevant to reactivity of poly-THD with external reagents and noting the presence of the diphenylamine side chain in poly-THD, we recall the solution photochemical conversion of diphenylamines to carbazoles (Figure 3).[31] It is clearly of interest to attempt an ultraviolet-light-initiated conversion of dephenylamine groups in poly-THD to carbazole groups with the ultimate hope of modifying the spectrum associated with the backbone.

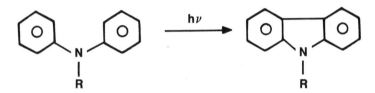

Figure 3. Ultraviolet light initiated conversion of a diphenylamine to a carbazole.

Irradiation of poly-THD crystals with assorted lamps with output ranging from 253 to 400 nm for 16–24 hours did not lead to detectable reaction as judged by FTIR spectra. Moreover, irradiation of THD monomer crystals under conditions analogous to those used for the polymer, or with ^{60}Co gamma radiation, resulted in initiation of THD polymerization as the only detectable reaction. Finally, irradiation of crystalline diphenylamine revealed it to be unreactive to ultraviolet light. The difficulties associated with diphenylamine photochemistry in rigid environments have been noted by others.[32] The possibility of doing side-chain-initiated photochemistry in other PDA remains intriguing.

Our approach to comparing side-chain-controlled thermal reactivity in poly-THD to that of poly-DCH has been to compare probes of the highest occupied molecular orbital (HOMO) for isolated model molecular compounds. This is summarized in Table 1, where the gas-phase vertical ionization potentials (I_G)[33,34] and first electrochemical oxidation potentials (E_1)[35] for N-ethylcarbazole and N-methyldiphenylamine, respectively, are listed. The question as to whether I_G or E_1 is a relevant figure of merit, other things being equal, for such reactivity leads to a straightforward testable prediction. The I_G values

Table 1. Side-chain energy levels from model compounds

PDA	SIDE-CHAIN MODEL	I_G (eV)	E_1 (Volts vs SCE)
DCH	N-Ethylcarbazole	7.29	+ 1.12
THD	N-Methyldiphenylamine	7.33	+ 0.84

for the model compounds are the same, within experimental error, leading to the expectation of similar reactivity, while the diphenylamine is more easily oxidized at an electrode, leading to the expectation that poly-THD should be more reactive. Experimentally, poly-DCH is inert to bromine vapor[20] and solutions of bromine in CCl_4 less than 75% bromine by weight. In contrast, poly-THD readily reacts at 20 °C with either a dense bromine vapor or a CCl_4 solution 10% by weight in bromine to give a black, largely amorphous solid which has gained 8–9 Br atoms per repeat unit as judged by either weight gain or complete elemental analysis. An electron-transfer process has clearly occurred, and the solid state spectrum of the brominated poly-THD is shown in Figure 4. Analogous to the bromination of poly-DCH,[23] the FTIR spectrum of brominated poly-THD reveals aromatic ring substitution. In particular, the strong absorption at 750, 705, and 690 cm^{-1} in the pristine polymer is markedly weaker in the brominated material, and new strong absorption at 815 cm^{-1} is observed. The ESR spectra of samples of brominated poly-THD reveal a single line of width 11.5–17.3 gauss with a g value (at crossover) of 2.0045–2.0050. The spin count ($1-8 \times 10^{18}$ spins/gm) is approximately an order of magnitude greater than that of poly-DCH which has gained 6 Br per repeat unit, yet the dc resistivity of samples of brominated poly-THD measured to data is not less than 10^8 Ω-cm.

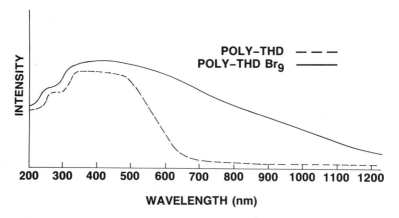

Figure 4. Solid state spectra, recorded by diffuse reflectance, of pristine and brominated poly-THD.

CONCLUSIONS

PDA, as a class of electroactive polymers, have much promise for their electronic and nonlinear optical properties. Progress in this general area requires judicious synthetic work as well as careful materials processing and characterization. While conjugation defects do not appear to limit the PDA temporal response, they merit serious consideration as reactive intermediates involving covalent bond formation on the PDA backbone. The chemical reactions of PDA described herein allow the synthesis of materials which have not been, and likely cannot be, synthesized by other methods. For PDA, the energy levels associated with the conjugated backbone are the highest occupied and lowest unoccupied; yet, for morphological reasons, initial reactivity is controlled by side-chain energy levels. The redox potentials of relevant model compounds for PDA side chains appear to be a useful figure of merit for the types of PDA reactivity described herein.

ACKNOWLEDGMENTS

James W. Shepherd III was an Industrial Undergraduate Research Participant from Brandeis University at GTE Laboratories in the summer of 1984. The authors thank L.A. Samuelson, C.S. Velazquez, and M. Downey for technical assistance. The photochemical experiments involving diphenylamine and THD monomer and polymer were performed by Dr. R. Haaksma.

REFERENCES

1. H.-J. Cantow, ed., "Polydiacetylenes," *Advances in Polymer Science, Vol. 63*, Springer Verlag, Berlin (1984).
2. D. Bloor and R.R. Chance, ed., "Polydiacetylenes," Martinus Nijhoff, Dordrecht, Boston (1985).
3. D.J. Sandman, ed., "Crystallographically Ordered Polymers," *Am. Chem. Soc. Symp. Series No. 337*, Washington, DC (1987).
4. D.J. Sandman, S.K. Tripathy, B.S. Elman, and L.A. Samuelson, *Synthet. Met.* 15:229 (1986).
5. R. Silbey, in reference 2, pp. 93–104.
6. A. Karpfen, in reference 2, pp. 115–124.
7. D.J. Sandman, in reference 3, pp. 1–10.
8. D.J. Sandman, C.S. Velazquez, B.M. Foxman, J.M. Preses, and R.E. Weston, Jr., National Synchroton Light Source Annual Report 1986, Brookhaven National Laboratory, pp. 157–158.
9. V. Enkelmann, in reference 1, pp. 91–136.
10. C. Sauteret, J.P. Hermann, R. Frey, F. Pradiere, J. Ducuing, R.H. Baughman, and R.R. Chance, *Phys. Rev. Lett.* 36:959 (1976).
11. P.W. Smith, *Bell. Syst. Tech. J.* 61:1975 (1982).
12. D.J. Sandman, G.M. Carter, Y.J. Chen, B.S. Elman, M.K. Thakur, and S.K. Tripathy, in reference 2, pp. 299–316.
13. G.M. Carter, M.K. Thakur, J.V. Hryniewicz, Y.J. Chen, and S.E. Meyler, in reference 3, pp. 168–176.
14. G.M. Carter, Y.J. Chen, M.F. Rubner, D.J. Sandman, M.K. Thakur, and S.K. Tripathy, in "Nonlinear Optical Properties of Organic Molecules and Crystals," D.S. Chemla and J. Zyss, ed., Academic Press, Orlando, FL, Vol. 2, pp. 85–120 (1987).
15. G.M. Carter, J.V. Hryniewicz, M.K. Thakur, Y.J Chen, and S.E. Meyler, *Appl. Phys. Lett.* 49:988 (1986).
16. C. Flytzanis, in "Nonlinear Optical Properties of Organic and Polymeric Materials," *Am. Chem. Soc. Symp. Series No. 233*, Washington, DC, ed., D.J. Williams, pp. 167–185 (1983).
17. W.M. Dennis, W. Blau, and D.J. Bradley, *Appl. Phys. Lett.* 47:200 (1985).
18. P. Cong, Y. Pang, and P.N. Prasad, *J. Chem. Phys.* 85:1077 (1986).
19. P.-A. Chollet, F. Kajzar, and J. Messier, *Synthet. Met.* 18:459 (1987).
20. D.J. Sandman, B.S. Elman, G.P. Hamill, C.S. Velazquez, and L.A. Samuelson, *Mol. Cryst. Liq. Cryst.* 134:89 (1986).
21. D.J. Sandman, B.S. Elman, G.P. Hamill, J. Hefter, C.S. Velazquez, and M.T. Jones, *Mol. Cryst. Liq. Cryst.* 134:109 (1986).
22. D.J. Sandman, B.S. Elman, G.P. Hamill, C.S. Velazquez, J.P. Yesinowski, and H. Eckert, *Mol. Cryst. Liq. Cryst. Lett.* 4:77 (1987).

23. D.J. Sandman, B.S. Elman, G.P. Hamill, J. Hefter, and C.S. Velazquez, in reference 3, pp. 118–127.

24. J. Hefter, *Scanning Microsc.* 1:13 (1987).

25. H. Eckert, J.P. Yesinowski, D.J. Sandman, and C.S. Velazquez, *J. Am. Chem. Soc.* 109:761 (1987).

26. J.P. Yesinowski, H. Eckert, D.J. Sandman, and C.S. Velazquez, in reference 3, pp. 230–252.

27. J.P.C.M. Van Dongen, M.J.A. DeBie, and R. Steur, *Tetrahedron Lett.* 16:371 (1973).

28. M.T. Jones, J. Roble, and D.J. Sandman, in reference 3, pp. 253–264.

29. M.T. Jones, S. Jansen, J. Roble, D.J. Sandman, and C.S. Velazquez, *Synthet. Met.* 18:427 (1987).

30. V. Enkelmann and G. Schleier, *Acta Crystallogr.* B36:1954 (1980).

31. E.W. Forster, K.H. Grellmann, and H. Linschitz, *J. Am. Chem. Soc.* 95:3108 (1973).

32. E.J. Bowen and J.H.D. Eland, *Proc. Chem. Soc.* 202 (1963); K.S. Sidhu, W.R. Bansal, and S.K. Jaswal, *Indian J. Chem.* 25B:910 (1986).

33. R.W. Bigelow and G.P. Ceasar, *J. Phys. Chem.* 83:1790 (1979).

34. H.J. Haink, J.E. Adams, and J.R. Huber, *Ber. Bunsenges. Physikal. Chem.* 78:436 (1974).

35. D.J. Sandman and G.P. Caesar, *Isr. J. Chem.,* in press.

SIDE CHAIN LIQUID CRYSTALLINE COPOLYMERS FOR NLO RESPONSE

Anselm C. Griffin, Amjad M. Bhatti and
Robert S. L. Hung

Departments of Chemistry and Polymer Science
University of Southern Mississippi
Hattiesburg, MS 39406

INTRODUCTION

Interest is currently high in the utility of organic compounds and, in particular, polymeric organics as nonlinear optical (nlo) materials. Several recent publications[1-3] have described the advantages and uniqueness of these carbon based species. Polymeric organic materials offer additional key features such as fabricability, ease of processing and the ability to form thin films of high optical quality. Very recently reports have appeared on side chain liquid crystalline polymers as candidate nlo materials.[4-7] Rationale for use of side chain mesogenic polymers as nlo materials includes their ease of dipolar orientation in an external electric field; formation of a poled, transparent, glassy solid state having liquid crystalline orientation of the pendant groups; and the possibility of designing and incorporating π-electron conjugation in the pendant moiety.

We wish to report here the design, synthesis and characterization of a series of chiral nitroaromatic copolymers as potential nlo materials. The conceptual approach we have taken has its origins in the chemistry of low molar mass liquid crystals. It has been established that binary mixtures of mesogenic (or potentially mesogenic) compounds having one component with a π-electron deficient aromatic ring (such as p-nitrophenyl) and the other component having only π-electron rich aromatic rings (such as p-alkoxyphenyl) usually exhibit highly nonlinear thermal behavior.[8] That is, an orthogonal smectic phase, usually smectic A, is either injected into the binary phase diagram or the thermal stability of an existing smectic A phase is dramatically enhanced over the linear rule-of-mixtures values. There has been considerable work done toward understanding the microscopic chemical origins of this phenomenon. For some such mixed systems it has been possible to detect the appearance of charge transfer bands in the electronic spectrum. In other systems the appearance of an enhanced (or induced) smectic phase is accompanied by formation of a molecular complex strong enough to produce a double eutectic in the melting point curve. Often the maximum in smectic phase stability occurs at a 1:1 molar ratio of the pure components. It is commonly accepted that formation of a π-molecular complex between the two components is responsible for this nonlinear thermal behavior.

As centrosymmetry precludes second harmonic generation, we felt it advantageous to incorporate chirality into our polymers. Chirality provides a formally noncentrosymmetric material. Removal of centrosymmetry by dipolar alignment is also possible by electric field poling of polymers.[5,6] In our previous work[7] we located the chiral center in the main chain of a side chain aliphatic polyester mesogen. In an attempt to produce a more effective interaction between the nlo π-electron system and the chiral center, the formation of a polymeric analogue to the π-molecular complex described above for low molar mass liquid crystals was envisioned. The polymeric π-molecular complex should transfer the chirality of a second chemical species to the nlo π-system as a result of the electronic complexation of the two species. It was therefore decided to design a copolymer system in which the two comonomers had opposite π-electronic character; one, the nlo species, having a nitroaromatic moiety; and the other, the chiral comonomer, having only π-electron rich aromatic rings. Intimate association of the two comonomer aromatic cores in a side chain liquid crystalline copolymer should result, by analogy with small molecular species, in effective chirality transfer to the π-system of the nlo comonomer.

In order to insure that this approach was reasonable for the chemical species thought desirable for incorporation into a polymer, it was decided to synthesize one prototypical set of non-polymerizable monomeric species which had exactly the same chemical constitution as the desired pendant moieties. A binary phase diagram was constructed composed of these two low molar mass compounds; one, a nitroaromatic compound; the other, a chiral π-electron rich compound. Upon finding a large enhancement of the smectic A phase in this mixed system, five copolymer systems were designed and synthesized. Each of these copolymers has a π-donor/π-acceptor conjugated system with a nitroaromatic moiety and also a chiral alkoxy moiety attached to π-electron rich aromatic rings. These five target copolymers (a-e) are shown on the following page. Four are of the acrylate type and one is an aliphatic polyester.

EXPERIMENTAL

Melting points are uncorrected. Elemental analyses were performed by Galbraith Laboratories, Inc., Knoxville, Tennessee. 'H-N.M.R. spectra were recorded on a Varian EM360A 60 MHz spectrometer, U.V.-Vis spectra on a Perkin-Elmer 552 spectrophotometer, and I.R. spectra on a Perkin-Elmer 567 instrument. Differential scanning calorimetry (DSC) analysis was carried out on a DuPont 910 DSC with a model 990 Thermal Analyzer and/or DuPont 9900 Thermal Analysis system using model 910 DSC cell. Mesomorphic transition temperatures and phase identifications were observed microscopically using a Reichert Thermovar microscope with a Mettler FP5/52 microfurnace. Identifiable optical textures from polarized light microscopy were obtained by cooling the polymer sample slowly from its isotropic phase into the mesophase. The sample was maintained at a temperature a few degrees below the isotropic temperature for periods of up to several hours. Slowly the mesophase texture developed into a recognizable one. For the polymers the mesophase-isotropic transition was somewhat broad and the transition temperatures given in the results and discussion section reflect the temperature at which 90% of the sample has undergone a mesophase-isotropic transition. Polymer viscosities were obtained using a constant temperature (25°C) water bath and an Ubbelohde viscometer.

376

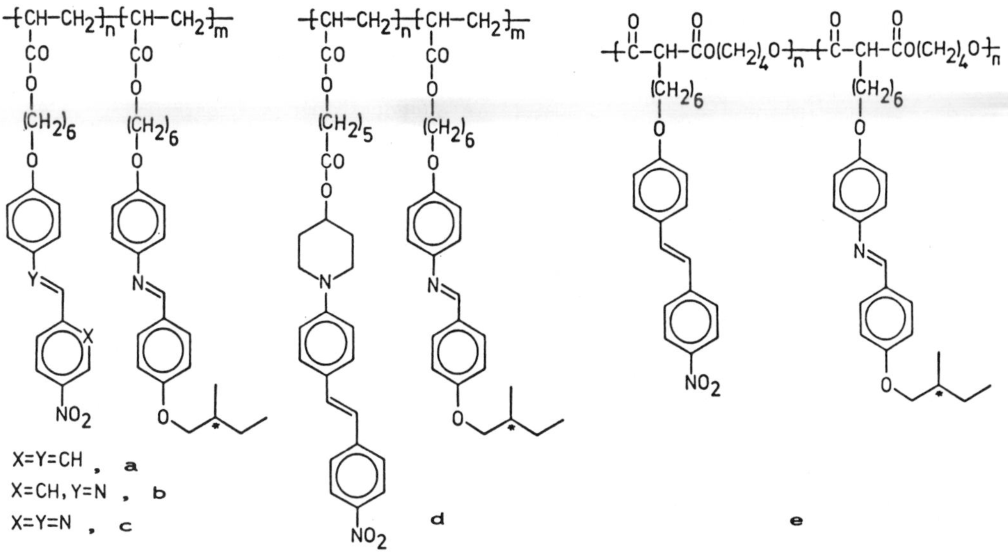

Figure 1. Target Copolymers

For the phase diagram, mixtures of nine different compositions (by molar ratio) of compounds A and B were prepared by weighing on a Perkin-Elmer AD-2 Autobalance. Complete mixing was achieved by repeatedly melting the mixtures and vigorously stirring. After keeping in a refrigerator for two days, the mesophase properties were determined microscopically as for other samples.

Nitro monomer incorporation in the copolymers was calculated from NMR spectra using the alkyl-to-alkoxy proton ratios. For example, in the copolymer $(NSV)_n(C*SBV)_m$; 0% nitro monomer incorporation corresponds to the homopolymer $(C*SBV)_m$ which has alkyl-to-alkoxy proton ratio of 3.333 (the three vinyl protons move over to the alkyl region after polymerization); 100% nitro monomer incorporation corresponds to the homopolymer $(NSV)_n$ which has alkyl-to-alkoxy proton ratio of 2.750. For the copolymer itself, the observed ratio (3.146) lies between these two ratio limits. Assuming a linear relationship, the percentage of nitro monomer incorporation is found to be 32%.

4-(6-Hydroxyhexyl)-4'-nitrostilbene (9)

First, 4-hydroxy-4'-nitrostilbene (8) was prepared by a reported method.[9] Then a mixture of 8 (10 g, 41.49 mmol), 1-chloro-6-hydroxyhexane (5.95 g, 43.56 mmol), powdered anhydrous potassium carbonate (7.16 g, 51.88 mmol), and dry dimethylformamide (40 ml) was stirred and refluxed for 24 hours. After allowing to cool to room temperature, the reaction mixture was poured into water (4 L) and then refrigerated overnight. Thereafter the product was filtered, washed with water, and air dried. Recrystallization from acetone gave 13.14 g (93%) of pure product. It was found to be liquid crystalline; Polarizing microscope: K158.5N(137.4)I. This compound has been reported in the literature[4] but procedure for its preparation is not given. The finding[10] of liquid crystallinity in such ω-hydroxyalkyl aromatic compounds is of interest in its own right. It is felt that the presence of a primary hydroxyl group at the terminus of an alkyl chain off a rigid core can facilitate the formation of nematic phases through a hydrogen bonding interaction with a dipolar group (such as nitro) at the rigid core end of a second molecule. A ribbon-like nematic phase should result.

Monomer 1

The compound 9 (6.82 g, 20 mmol) was dissolved by stirring in dry tetrahydrofuran (130 ml) at 60°C under nitrogen atmosphere. Then dry triethylamine (2.22 g, 22 mmol) was added. This was followed by dropwise addition of acryloyl chloride (1.99 g, 22 mmol). The stirring was continued for 18 hours. TLC examination (alumina/ethyl acetate) of the reaction mixture showed complete conversion of the starting material. The solvent was evaporated and the residue extracted with chloroform and water. The chloroform layer was dried over anhydrous magnesium sulfate. Filtration, evaporation of the solvent, and recrystallization of the residue from ethanol/chloroform gave 5.41 g (69%) of yellow product crystals, mp 116.5-117.5°C. This compound has been reported in the literature[4] but procedure for its preparation and its melting point are not given.

Imine 11

First, 4-(6-hydroxyhexyloxy)aniline (10) was prepared by a known method.[10] Then a mixture of 4-nitrobenzaldehyde (6.04 g, 40 mmol) and 10 (8.36 g, 40 mmol) was refluxed in absolute ethanol (150 ml) for 2 hours. The product, which crystallized on cooling, was separated by filtration and recrystallized from chloroform and petroleum ether to give 10.8 g (79%) of pure product, mp 125-126°C.

Monomer 2

The imine 11 (6.0 g, 17.5 mmol) was dissolved by stirring in dry tetrahydrofuran (100 ml) at 60°C under nitrogen atmosphere. Then dry triethylamine (1.95 g, 19.3 mmol) was added, followed by dropwise addition of acryloyl chloride (1.75 g, 19.3 mmol). The stirring was continued for 20 hours and the reaction mixture was allowed to cool slowly to room temperature. TLC examination (alumina/50% ethyl acetate + 50% hexane) showed complete conversion of the starting material. The solvent was removed by evaporation and the residue extracted with chloroform and water. The chloroform layer was dried over anhydrous magnesium sulfate. After filtration and evaporation of the solvent, the crude product was recrystallized twice from ethyl acetate and hexane to yield 4.35 g (63%) of pure product, mp 126-126.5°C.
$C_{22}H_{24}N_2O_5$ Calc. C 66.65, H 6.10, N 7.07%
        Found C 66.45, H 6.23, N 6.83%
'H-N.M.R. ($CDCl_3$/TMS): δ 1.20-2.07 (m, 8H), 3.80-4.33 (m, 4H), 5.60-6.37 (m, 3H), 6.73-7.37 (m, 4H), 7.73-8.37 (m, 4H), 8.48 (s, 1H).

Imine 12

For this synthesis, it was necessary to prepare 5-nitropyridine-2-carboxaldehyde. This was accomplished, starting from 2-chloro-5-nitropyridine, by a reported method.[11] Then a mixture of 5-nitropyridine-2-carboxaldehyde (5.22 g, 34.3 mmol) and 10 (7.19 g, 34.4 mmol) was warmed and stirred in absolute ethanol (175 ml) for 24 hours. TLC examination (alumina/80% ethyl acetate + 20% hexane) showed complete disappearance of the starting material. The solvent was distilled off under reduced pressure and the crude product was recrystallized from ethanol and chloroform to give 10.35 g (88%) of pure product. It was found to be liquid crystalline; Polarizing microscope: K113.2N125.7I. Liquid crystallinity in this compound, as in the analogous stilbene compound 9, is in part related to the ω-hydroxyalkyl group. This pyridine imine has a stable, enantiotropic nematic phase. Of particular interest is that the mesophase is nematic and not smectic as is more common for pyridines having the nitrogen laterally situated in the molecule.   Hydrogen bonding between terminal units giving a ribbon like nematic apparently predominates over the smectic tendencies of the pyridine moiety.

Monomer 3

The imine 12 (5.0 g, 14.6 mmol) was dissolved in dry tetrahydrofuran (50 ml) at 60°C. Thereafter dry triethylamine (1.78 g, 17.6 mmol) was added, followed by dropwise addition of a solution of acryloyl chloride (1.59 g, 17.6 mmol) in tetrahydrofuran (5 ml). The reaction mixture was stirred for 22 hours while allowing it to cool slowly. TLC examination (alumina/50% ethyl acetate + 50% hexane) showed almost complete conversion. The solvent was evaporated and the residue extracted with chloroform and water. The chloroform layer was dried over anhydrous magnesium sulfate. Filtration of magnesium sulfate, followed by evaporation of the solvent gave the crude product, which was recrystallized twice from ethanol and chloroform to give 2.62 g (45%) of pure product, mp 121-122°C. $C_{21}H_{23}N_3O_5$ Calc. C 63.47, H 5.83, N 10.57%
Found C 63.22, H 6.14, N 10.40%
'H-N.M.R. (CDCl$_3$/TMS): δ 1.20-2.13 (m, 8H), 3.77-4.37 (m, 4H), 5.60-6.37 (m, 3H), 6.73-7.47 (m, 4H), 8.17-8.70 (m, 3H), 9.40 (s, 1H).

Sodium acrylate

While stirring a solution of acrylic acid (20 ml) in ether (100 ml), sodium carbonate (10.6 g) was added in small installments. This mixture was allowed to stand overnight. Ether was then removed by rotary evaporation and unreacted acrylic acid by applying high vacuum. The residual white powder was used as such.

Monomer 4

Compound 13 was first synthesized by following the literature procedures.[12,4] Then a mixture of 13 (7.5 g, 15.0 mmol) and sodium acrylate (1.48 g, 15.7 mmol) was stirred in dry hexamethylphosphoric triamide (30 ml) at 50-60°C for 15 hours. TLC examination (silica gel/50% ethyl acetate + 50% hexane) showed completion of reaction. The reaction mixture was cooled to room temperature and poured into water (600 ml). Filtration of the precipitated product was followed by washing with water and drying. The crude product so obtained was purified by dissolving in chloroform and passing through a column of basic alumina (75 g). The chloroform solution from the column was concentrated by evaporation and ethanol was added to reach the crystallization point when hot. The crystals were separated and subsequently dried at 60-62°C in a vacuum oven. The yield was 6.71 g (91%). This monomer was found to be liquid crystalline. Le Barny and coworkers[4] have synthesized this compound but the procedure is not reported.

Aldehyde 14

A mixture of S(+)-1-bromo-2-methylbutane (20.81 g, 137.81 mmol), 4-hydroxybenzaldehyde (17.82 g, 146.07 mmol), potassium carbonate (23.77 g, 172.25 mmol) was stirred and refluxed in dry acetone (250 ml) under nitrogen for 120 hours. Thereafter the reaction mixture was cooled to room temperature, filtered, and the residue washed with acetone. Rotary evaporation of acetone gave 31.18 g of crude product, which on vacuum distillation at 104-110°C/2 torr yielded 20.82 g (79%) of pure product (clear colorless liquid). Purity check was made by TLC on alumina/50% ethyl acetate + 50% hexane. This product should completely retain its optical activity because the chiral carbon in the starting S(+)-1-bromo-2-methylbutane is located adjacent to the reaction center and does not participate in any bond-breaking or -making step.
'H-N.M.R. (CDCl$_3$/TMS): δ 0.73-2.23 (m, 9H), 3.73-4.07 (d, 2H), 6.80-8.07 (m, 4H), 9.90 (s, 1H).

Imine 15

A mixture of 14 (8.0 g, 41.67 mmol), and 10 (8.70 g, 41.63 mmol) was refluxed in absolute ethanol (50 ml) for 2 hours. The product crystallized on cooling. It was recrystallized twice from ethanol. Yield of pure product was 12.35 g (77%), mp 107-108°C.

Monomer 5

The imine 15 (12.0 g, 31.33 mmol) was dissolved in dry tetrahydrofuran (75 ml) at 60°C under nitrogen atmosphere. Then dry triethylamine (3.96 g, 39.21 mmol) was added, followed by dropwise addition of a solution of acryloyl chloride (3.55 g, 39.23 mmol) in tetrahydrofuran (5 ml). The reaction mixture was allowed to cool to room temperature and stirring was continued for 22 hours. The solvent was removed by rotary evaporation and the residue extracted with chloroform and water. The chloroform layer was dried over anhydrous magnesium sulfate. After filtering off magnesium sulfate, the solvent was evaporated and the residue chromatographed over neutral alumina (300 g) using 25% ethyl acetate and 75% hexane as eluant. Yield of the product, after vacuum drying, was 9.14 g (67%), mp 62-63°C. It was found to be of single spot purity by TLC on alumina/50% ethyl acetate + 50% hexane.
'H-N.M.R. ($CDCl_3$/TMS): δ 0.73-2.10 (m, 17H), 3.65-4.33 (m, 6H), 5.60-6.38 (m, 3H), 6.72-7.95 (m, 8H), 8.35 (s, 1H).

Diethyl 6-bromohexylmalonate (16)

This compound was prepared differently than a reported procedure.[13] Sodium sand was made by melting sodium (9.0 g, 0.39 mol) in dry xylene (200 ml) followed by vigorous stirring. The resulting mixture was allowed to cool to room temperature while continuing stirring. Then diethyl malonate (62 ml, 0.41 mol) was added dropwise and the reaction flask cooled as necessary to keep the temperature below 30°C. After completion of the addition, the mixture was allowed to stand overnight at room temperature. Next morning, 1,6-dibromohexane (142.7 g, 0.58 mol) was added in one installment and the mixture gently heated with stirring. After the initial vigorous reaction subsided, the mixture was refluxed for a further 4 hours and then cooled to room temperature. Sodium bromide, which had settled down, was filtered off and xylene was subsequently removed by simple distillation. The residual product mixture was distilled under vacuum. The fraction coming at 157-160°C/3 torr corresponded to the expected product. Yield 57.3 g (46%). Lit.[13] bp 185-187°C/10 torr.

Monomer 6

A mixture of 8 (5.0 g, 20.75 mmol), 16 (7.11 g, 22.01 mmol), powdered anhydrous potassium carbonate (3.73 g, 27.03 mmol) and dry acetone (75 ml) was stirred and refluxed for 120 hours. After allowing to cool to room temperature, the reaction mixture was filtered. TLC of the filtrate on alumina/50% ethyl acetate + 50% hexane showed complete disappearance of 8 and appearance of a new spot (product). [Due to the facile decarboxylation of β-carboxy esters and β-diacids, it was anticipated that difficulties might arise with this Williamson ether synthesis reaction. The aliphatic diester moieties could be potentially subject to saponification under the basic conditions employed possibly leading to decarboxylation products. Fortunately, no evidence of saponification was detected either in this preparation or in the preparation of monomer 7.] Rotary evaporation of the filtrate to dryness and recrystallization of the yellow residue from ethyl acetate and hexane gave 9.2 g (92%) of pure product, mp 118-119°C.

$C_{27}H_{33}NO_7$ Calc.   C 67.06, H 6.88, N 2.90%
             Found   C 67.08, H 6.76, N 2.77%
'H-N.M.R. (CDCl₃/TMS):   δ 1.13-2.13 (m, 16H), 3.33 (t, 1H), 3.80-4.43 (m, 6H), 6.70-7.17 (m, 4H), 7.30-7.77 (m, 4H), 8.13 (d, 2H, J = 17Hz).

Diethyl 6-(4-acetamidophenoxy)hexylmalonate (17)

A mixture of 4-acetamidophenol (20.0 g, 0.13 mol), 16 (45.0 g, 0.14 mol), powdered anhydrous potassium carbonate (22.4 g, 0.16 mol), and dry acetone (200 ml) was stirred and refluxed for 120 hours. After cooling to room temperature, the reaction mixture was filtered. TLC of the filtrate on alumina/ethyl acetate showed complete disappearance of 4-acetamidophenol and appearance of a new spot (product). The filtrate was evaporated to dryness and the residue recrystallized from ethyl acetate/hexane to give 30.3 g (59%) of white product crystals, mp 59-60°C.

$C_{21}H_{31}NO_6$ Calc.   C 64.10, H 7.94, N 3.56%
             Found   C 64.15, H 7.82, N 3.36%
'H-N.M.R. (CDCl₃/TMS):   δ 1.13-2.03 (m, 16H), 2.12 (s, 1H), 3.34 (t, 1H), 3.78-4.43 (m, 6H), 6.70-7.53 (m, 4H), 8.04 (s, 1H).

Diethyl 6-(4-aminophenoxy)hexylmalonate (18)

The amide 17 (30.0 g, 76.3 mmol) was hydrolyzed by refluxing in 50:50 (v/v) concentrated hydrochloric acid/95% ethanol (100 ml) for 3 hours. The resulting solution was neutralized with slow addition of saturated sodium bicarbonate solution while stirring. It was then allowed to stand an additional 1 hour. The product was extracted with ether (3 x 50 ml). TLC of the extract (alumina/ethyl acetate) showed complete disappearance of the starting amide and appearance of a new spot (product). Removal of ether by rotary evaporation gave 12.1 g (49%) of the desired product. Rapid decomposition (oxidation) prevented standard characterization of this material. It was used immediately in the next reaction step.

Monomer 7

A mixture of 14 (1.92 g, 10 mmol) and 18 (3.51 g, 10 mmol) was refluxed in absolute ethanol (20 ml) for 2 hours. After this, the solvent was evaporated and the residue dissolved in a small amount of ethyl acetate and put on a column of neutral alumina (200 g). Elution with 25% ethyl acetate and 75% hexane gave 3.99 g of crude product, which on two recrystallizations from hexane gave 1.93 g (37%) of pure single-spot product, mp 46°C.

'H-N.M.R. (CDCl₃/TMS):   δ 0.73-2.20 (m, 25H), 3.28 (t, 1H), 3.63-4.40 (m, 8H), 6.68-7.90 (m, 8H), 8.30 (s, 1H).

4-n-Octyloxyaniline

A mixture of 4-acetamidophenol (10.57 g, 70 mmol), 1-bromooctane (13.51 g, 70 mmol), powdered anhydrous potassium carbonate (13.54 g, 98 mmol) and 50 ml of dry dimethylformamide was refluxed with stirring for 4 hours. Afterwards the reaction mixture was poured into ice cold water and allowed to stand overnight in a refrigerator. Then the mixture was filtered and washed thoroughly with water. It is imperative that the mixture be refrigerated overnight. The DMF/water solution clears and produces solid product only after extended periods at low temperatures. Apparently the desolvation of product from DMF is a slow process. Water washing of the crude product must be repeated numerous times to completely eliminate traces of DMF from the product. The

solid thus obtained was air dried for 4-6 hours. The amide, without
further purification, was hydrolyzed by refluxing in 50:50 (v/v)
mixture of concentrated hydrochloric acid and 95% ethanol (150 ml)
for 3 hours. After cooling to room temperature, the acid was
neutralized by slow addition of saturated sodium bicarbonate solution
while stirring, and then allowed to stand an additional one hour.
The product was filtered, washed with water, vacuum dried, and
recrystallized from heptane to yield 6.50 g (42%) of the desired
compound, mp 79-81°C. Lit.[14] bp 151-152°C/1 torr.
'H-N.M.R. (CDCl$_3$/TMS): δ 0.76-2.12 (m, 15H), 3.80 (t, 2H), 5.26 (s,
2H), 6.53-6.86 (m, 4H).

N-(4-Nitrobenzylidene)-4'-n-octyloxyaniline (Compound A)

4-Nitrobenzaldehyde (0.70 g, 4.66 mmol) and 4-octyloxyaniline
(1.03 g, 4.66 mmol) were refluxed in absolute ethanol (25 ml) for 2
hours. The product crystals were formed on slow cooling to room
temperature. Filtration and recrystallization from absolute ethanol
gave 0.98 g (59%) of the Schiff base.
C$_{21}$H$_{26}$N$_2$O$_3$ Calc. C 71.16, H 7.39, N 7.90%
         Found C 71.27, H 7.34, N 7.74%
Polarizing microscope: K62.8S$_A$72.8N84.9I.
'H-N.M.R. (CDCl$_3$/TMS): δ 0.87-2.13 (m, 15H), 4.10 (t, 2H), 6.93-7.53
(m, 4H), 8.00-8.53 (m, 4H), 8.67 (s, 1H).

The chiral carbon-containing Schiff base (Compound B) was
prepared from 14 and 4-octyloxyaniline in an analogous manner in 58%
yield.
Polarizing microscope: K89.0I
'H-N.M.R. (CDCl$_3$/TMS): δ 0.70-2.10 (m, 24H), 3.67-4.10 (m, 4H),
6.70-7.90 (m, 8H), 8.32 (s, 1H).

Polymerizations

A mixture of a nitro monomer and monomer 5 (50 mol % each) was
dissolved in a small volume of dry toluene by stirring at 95-100°C.
Then 2 mol % of AIBN was added at 2-minute intervals until ten
installments had been added. After this, the reaction was allowed to
proceed for a further 10 minutes. The reaction mixture was then
poured into a wide and shallow dish, and toluene was allowed to
evaporate completely. The residue was dissolved in chloroform and
the polymer precipitated by adding ether. One or two more
dissolutions in chloroform and precipitations with ether gave pure
polymer. Absence of monomers was tested by TLC on alumina/50% ethyl
acetate + 50% hexane. Polymers were dried by applying high vacuum.
Yields ranged from 15-65%.

For condensation polymerization, monomers 6 and 7 (50 mol %
each) and 1,4-butanediol (molar equivalent plus 5% excess) were mixed
in a small round-bottom flask. It was then swept with prepurified
nitrogen. The temperature was raised to 95°C and nitrogen kept
bubbling slowly through the melt. One drop of titanium (IV)
isopropoxide was added and the reaction run at atmospheric pressure
for 23 hours. During this period, initial transesterification takes
place converting the (relatively) low boiling 1,4-butanediol into an
ester and liberating ethanol in the process. This early reaction in-
corporates the 1,4-butanediol into a larger molecule rendering it less
volatile and therefore less easily lost when high vacuum is applied to
complete the polycondensation reaction. Polymerization stoichiometry
is preserved at 1:1 molar equivalents of alcohol and ester by this low
temperature, low vacuum prepolymer formation with subsequent vacuum
removal of excess diol. Then vacuum was applied gradually and the
temperature was raised 5°C at a time, keeping the reaction at various

temperature settings for a few hours. Maximum temperature setting was 135°C at 2 torr for 3 hours. After cooling, the product mixture was worked up as above. Schemes for these synthetic transformations and polymer structures are given on the next few pages.

RESULTS AND DISCUSSION

The isobaric binary phase diagram for model compounds (A and B) structurally analogous to one of our desired nlo monomer/chiral monomer copolymers is shown in Figure 2. It is readily seen that a significant enhancement of the smectic A phase occurs in this mixture. As mentioned previously, this enhancement can be taken as indicative of a possible π-electronic interaction in a copolymer with analogous structures to transfer chirality to the π-electron system of the nlo nitroaromatic moiety.

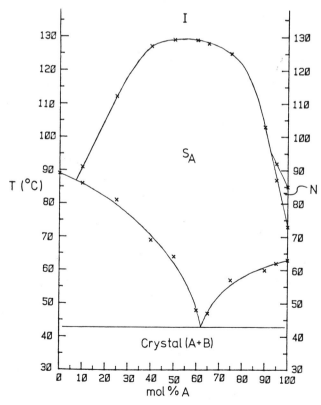

Figure 2. Binary Phase Diagram Showing Enhanced S_A Phase

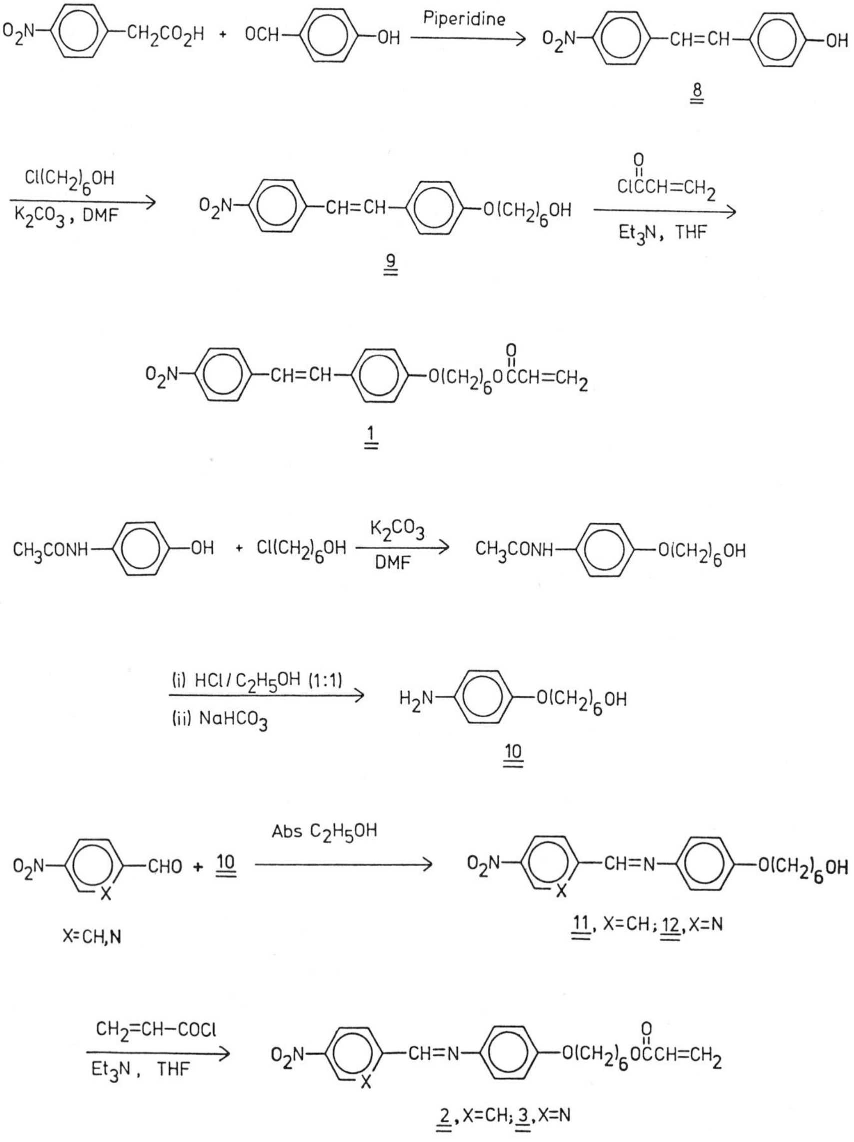

$O_2N-\langle\rangle-CH_2CO_2H$ + $OCH-\langle\rangle-OH$ $\xrightarrow{\text{Piperidine}}$ $O_2N-\langle\rangle-CH=CH-\langle\rangle-OH$

$\underline{\underline{8}}$

$\xrightarrow[\text{K}_2\text{CO}_3,\ \text{DMF}]{\text{Cl(CH}_2)_6\text{OH}}$ $O_2N-\langle\rangle-CH=CH-\langle\rangle-O(CH_2)_6OH$ $\xrightarrow[\text{Et}_3\text{N, THF}]{\overset{O}{\underset{}{Cl\overset{\|}{C}CH=CH_2}}}$

$\underline{\underline{9}}$

$O_2N-\langle\rangle-CH=CH-\langle\rangle-O(CH_2)_6O\overset{O}{\overset{\|}{C}}CH=CH_2$

$\underline{\underline{1}}$

$CH_3CONH-\langle\rangle-OH$ + $Cl(CH_2)_6OH$ $\xrightarrow[\text{DMF}]{\text{K}_2\text{CO}_3}$ $CH_3CONH-\langle\rangle-O(CH_2)_6OH$

$\xrightarrow[\text{(ii) NaHCO}_3]{\text{(i) HCl/C}_2\text{H}_5\text{OH (1:1)}}$ $H_2N-\langle\rangle-O(CH_2)_6OH$

$\underline{\underline{10}}$

$O_2N-\langle\overset{}{\underset{X}{}}\rangle-CHO$ + $\underline{\underline{10}}$ $\xrightarrow{\text{Abs C}_2\text{H}_5\text{OH}}$ $O_2N-\langle\overset{}{\underset{X}{}}\rangle-CH=N-\langle\rangle-O(CH_2)_6OH$

X=CH,N

$\underline{\underline{11}}$, X=CH; $\underline{\underline{12}}$, X=N

$\xrightarrow[\text{Et}_3\text{N, THF}]{\text{CH}_2\text{=CH-COCl}}$ $O_2N-\langle\overset{}{\underset{X}{}}\rangle-CH=N-\langle\rangle-O(CH_2)_6O\overset{O}{\overset{\|}{C}}CH=CH_2$

$\underline{\underline{2}}$, X=CH; $\underline{\underline{3}}$, X=N

SYNTHETIC SCHEME

SYNTHETIC SCHEME

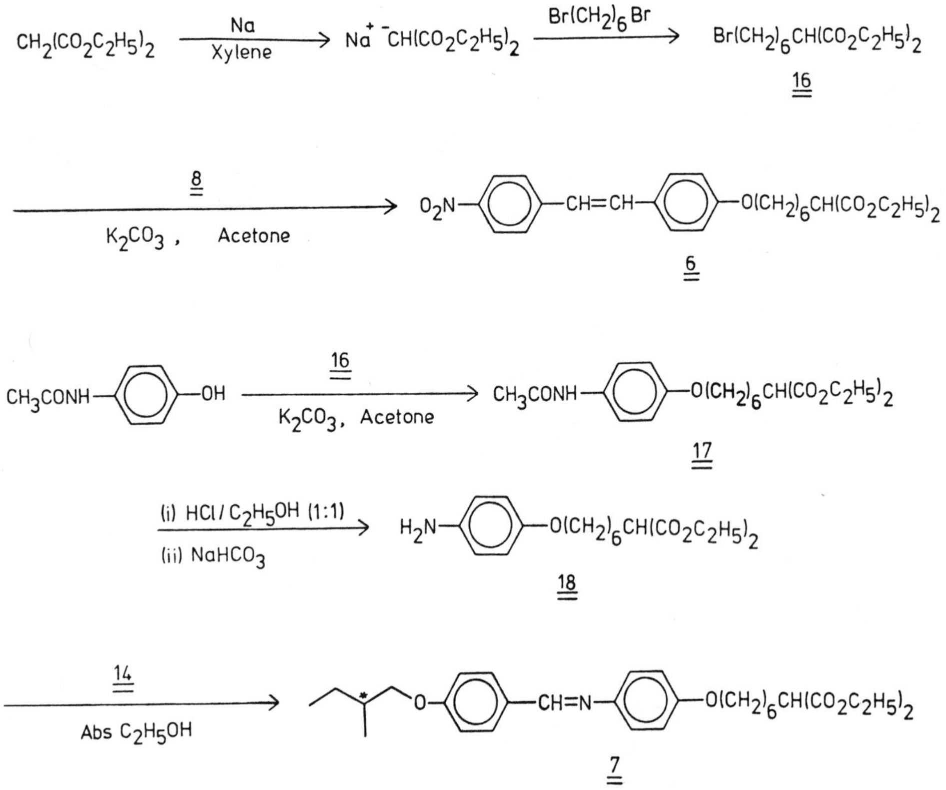

$$CH_2(CO_2C_2H_5)_2 \xrightarrow[\text{Xylene}]{\text{Na}} Na^+ {}^-CH(CO_2C_2H_5)_2 \xrightarrow{Br(CH_2)_6Br} Br(CH_2)_6CH(CO_2C_2H_5)_2$$

16

$$\xrightarrow[\text{K}_2\text{CO}_3, \quad \text{Acetone}]{\text{8}} O_2N\text{—}\langle\text{—}\rangle\text{—}CH{=}CH\text{—}\langle\text{—}\rangle\text{—}O(CH_2)_6CH(CO_2C_2H_5)_2$$

6

$$CH_3CONH\text{—}\langle\text{—}\rangle\text{—}OH \xrightarrow[\text{K}_2\text{CO}_3, \text{ Acetone}]{\text{16}} CH_3CONH\text{—}\langle\text{—}\rangle\text{—}O(CH_2)_6CH(CO_2C_2H_5)_2$$

17

$$\xrightarrow[\text{(ii) NaHCO}_3]{\text{(i) HCl}/C_2H_5OH\ (1:1)} H_2N\text{—}\langle\text{—}\rangle\text{—}O(CH_2)_6CH(CO_2C_2H_5)_2$$

18

$$\xrightarrow[\text{Abs } C_2H_5OH]{\text{14}} \ \text{—}O\text{—}\langle\text{—}\rangle\text{—}CH{=}N\text{—}\langle\text{—}\rangle\text{—}O(CH_2)_6CH(CO_2C_2H_5)_2$$

7

TARGET COPOLYMERS

X=Y=CH , a
X=CH, Y=N , b
X=Y=N , c

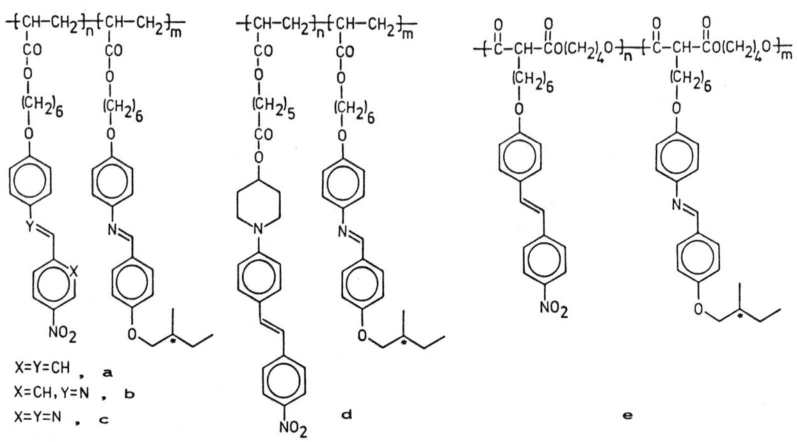

SYNTHETIC SCHEME AND POLYMER STRUCTURES

386

Table 1. Thermal and UV-Vis Data for Monomers

Monomer	Melting Point (°C)	U.V.-Vis λ_{max} (nm)	U.V.-Vis Cutoff (nm)
NSV, 1	K 117.1 I	256, 379	478
NBSBV, 2	K 125.9 I	263, 383	479
NPSBV, 3	K 124.4 I	255, 394	498
NSPV, 4	K 139.1 (S_A 137.3) N 153.4 I	295, 414	550
C*SBV, 5	K 63.2 I	239, 285, 335	428
NSME, 6	K 119.0 I	253, 379	474
C*SBME, 7	K 45.8 I	239, 285, 334	428

Table 2. Viscosity and UV-Vis Data for Polymers

Polymer	Inherent Viscosity dℓ/g in $CHCl_3$	uv-vis λ_{max} (nm)	uv-vis Cutoff (nm)
$(NSV)_n(C*SBV)_m$, a	0.149	240, 280, 352	472
$(NBSBV)_n(C*SBV)_m$, b	0.118	239, 282, 344	472
$(NPSBV)_n(C*SBV)_m$, c	0.170	245, 283, 349	498
$(NSPV)_n(C*SBV)_m$, d	0.264	239, 284, 352, 412	552
$(C*SBV)_n$	0.317	240, 284, 335	409
NSME-C*SBME-1,4-Bu, e	0.147	240, 281, 365	485

Thermal and uv-visible data for the seven monomers in this study
are given in Table 1. With the exception of the NSPV monomer (4),
all monomers are not enantiotropic liquid crystals. This piperidine
based monomer has both nematic and smectic A phases. Solution
viscosity values and uv-visible data are collected for the five
copolymers (and the chiral homopolymer) in Table 2. A presentation
of uv-visible absorption spectra for a nitroaromatic monomer, the
chiral monomer and the corresponding copolymer is given in Figure 3.
The presence of both monomer chromophores is seen in the composite
spectrum of the copolymer.

Although the molar feed ratio of monomers was 1:1 in our
copolymerizations, the polymers obtained did not always reflect this
ratio of comonomers in their compositions. A rationale for this
difference is not immediately obvious. It is assumed that incorporation
of the different comonomers is random along the polymer chain.
Assessment of percent incorporation of the two monomers in polymers can
be made by various methods. For example, Le Barny[4] has used the UV-
visible spectrum for nitroaromatic side chain copolymers. For our
polymers, however, we have determined relative incorporation of
comonomers by N.M.R. spectroscopic integrations as is described

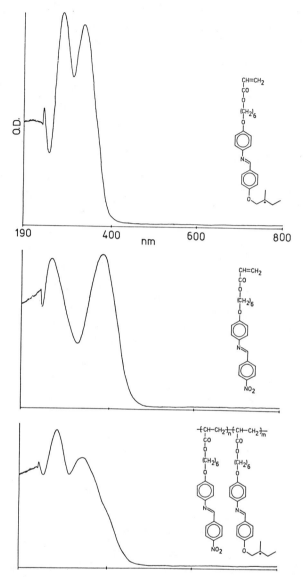

Figure 3. UV-Vis Spectra of Monomers <u>2</u> and <u>5</u> and Polymer b

in the experimental section. Table 3 gives the percentage
incorporation of the nitroaromatic monomer. The first two entries,
polymers a and b, have lower incorporation of the nitroaromatic

monomer than the feed ratio, whereas polymers c, d and e have nearly the same ratio as the feed. Also given in Table 3 are thermal data obtained by DSC and optical microscopy. A clearly defined glass transition was not evident for all these polymers. Instead it was decided to search by microscopy for the first evidence of fluidity in the polymer sample. In cases where both a Tg and a fluidity temperature were obtainable, the Tg was approximately 30°C below the fluidity temperature. The mesophase in polymers b, c and e is

Table 3. Thermal and Compositional Data for Polymers

Polymer	Fluidity Temp[1] (°C)	Mesophase-Isotropic Temp[2] (°C)	% Nitro Monomer Incorporated[3]
(NSV)$_n$(C*SBV)$_m$, a	101	184	32
(NBSBV)$_n$(C*SBV)$_m$, b	73	160	23
(NPSBV)$_n$(C*SBV)$_m$, c	64	162	47
(NSPV)$_n$(C*SBV)$_m$, d	105	147	45
(C*SBV)$_n$	[K-I @ 63.7]	–	–
NSME-C*SBME-1,4-Bu, e	95	138	50

1) glass transition approximately 30°C below this temperature

2) these transitions have wide biphasic region; temperatures shown correspond to approximately 90% isotropic in equlibrium with mesophase

3) by proton nmr

smectic A above the fluidity (glass) transition. Formation of batonettes which coalesced to a focal conic fan texture was seen by polarizing light microscopy for these polymers. For polymers a and d, both stilbene materials, there was evidence of gel formation which was particularly evident upon repeated thermal cycling of the sample. It is our opinion that these stilbene polymers are undergoing a crosslinking reaction through the Ar-CH=CH-Ar double bond. The optical appearance for these stilbene polymers a and d resembles a nematic Schlieren texture although the materials remained quite viscous above their nominal fluidity temperature. The crosslinking process, albeit perhaps involving only a small number of stilbene units, is felt to inhibit effective π-interactions between different monomer units in the copolymer and no smectic A phase is seen for copolymers a and d. For polymers b, c and e the smectic A phase thermal stability is quite pronounced suggesting strong π-interaction between comonomer units in these polymers. Based on current ideas it is most likely that the liquid crystalline interactions between pendant groups are interchain rather than intrachain. This implies that the chirality transferring π-interactions are also inter-chain in nature. It is of interest to note that a (chiral) smectic A phase can.be obtained for both vinyl copolymers and polyester copolymers. This finding further argues for an interchain interaction among pendant moieties since in the polyesters nine atoms intervene between pendant branch positions.

Mechanical shearing of these polymers in the mesophase between glass plates can produce optically transparent samples if the shearing force is maintained as the material cools below its glass transition. These materials are currently being evaluated for nonlinear optical response.

CONCLUSIONS

The formation of chiral smectic A phases with high thermal stability was found for several side chain liquid crystalline copolymers in which one comonomer was a nitroaromatic species and the other comonomer was a chiral alkoxyaromatic material. The presence of a π-electronic interaction between electronically dissimilar comonomer units was inferred. For comonomers having a stilbene chemical structure, crosslinking seemed to occur under the polymerization conditions leading to a gel with nematic-like optical textures. Optically transparent thin films can be formed by mechanical stress of samples between glass plates.

ACKNOWLEDGEMENT

This research was sponsored by the Air Force Office of Scientific Research, Air Force Systems Command, USAF, under Grant Number AFOSR 84-0249. The U. S. Government is authorized to reproduce and distribute reprints for Governmental purposes notwithstanding any copyright notation thereon.

REFERENCES

1. G. Khanarian, ed, "Molecular and Polymeric Optoelectronic Materials: Fundamentals and Applications," Proc. SPIE 682 (1987).
2. D. J. Williams; ed., "Nonlinear Optical Properties of Organic and Polymeric Materials," ACS Symp. Ser. No. 233, Washington, DC (1983).
3. D. J. Williams, Angew. Chem. Int. Ed. Engl., 23:690 (1984).
4. P. Le Barny, G. Ravaux, J. C. Dubois, J. P. Parneix, R. Njeumo, C. Legrand, and A. M. Levelut, in: "Molecular and Polymeric Optoelectronic Materials: Fundamentals and Applications," G. Khanarian, ed., Proc. SPIE 682, 56 (1987).
5. R. N. DeMartino, E. W. Choe, G. Khanarian, D. Haas, T. Leslie, G. Nelson, J. Stamatoff, D. Stuetz, C. C. Teng, and H. Yoon, this volume.
6. J. B. Stamatoff, A. Buckley, G. Calundann, E. W. Choe, R. DeMartino, G. Khanarian, T. Leslie, G. Nelson, D. Stuetz, C. C. Teng, and H. N. Yoon, in: "Molecular and Polymeric Optoelectronic Materials: Fundamentals and Applications," G. Khanarian, ed., Proc. SPIE 682, 85 (1987).
7. A. C. Griffin, A. M. Bhatti, and R. S. L. Hung, in: "Molecular and Polymeric Optoelectronic Materials: Fundamentals and Applications," G. Khanarian, ed., Proc. SPIE 682, 65 (1987).
8. A. C. Griffin, T. R. Britt, N. W. Buckley, R. F. Fisher, S. J. Havens, and D. W. Goodman, in: "Liquid Crystals and Ordered Fluids, Vol. 3," J. F. Johnson and R. S. Porter, eds., Plenum Press, NY, NY, 61 (1978).
9. M. M. Cullinae, J. Chem. Soc., 123:2056 (1923).
10. S. R. Vaidya, Ph.D. Dissertation, University of Southern Mississippi (1986).

11. L. Achremowicz and L. Syper, _Rocz. Chem._, 46:409 (1972).
12. E. C. Taylor and J. S. Skotnicki, _Synthesis_, 606 (1981).
13. E. Buchta and M. Fischer, _Chem. Ber._, 99:1509 (1966).
14. T. R. Criswell, B. H. Klanderman, and D. C. Batesky, _Mol. Cryst. Liq. Cryst._ 22:211 (1973).

SYNTHESIS OF CERTAIN SPECIFIC ELECTROACTIVE POLYMERS

Fred Wudl, Y. Ikenoue and A. O. Patil

Institute for Polymers and Organic Solids
Department of Physics
University of California
Santa Barbara, California 91036

INTRODUCTION

Among the many synthetic challenges in electroactive polymers we consider here those which we believe are the most important: (a) to prepare polymers which have a very small semiconductor bandgap and (b) to prepare polymers which are soluble and hence fabricable. In this presentation we describe our efforts as well as those of others to achieve these goals.

The main consequence of the former goal is the development of polymers which will not need to be doped to exhibit high conductivities since at room temperature thermal energy will be enough to excite electrons from the valence to the conduction band ($\pi \rightarrow \pi^*$). The second goal has already been achieved in several laboratories[1,2] by successful preparation of long alkyl-chain-substituted polythiophenes which are soluble in organic solvents. In this article we concentrate on our most recent results obtained with our water-soluble polythiophenes, a subset of the organic solvent-soluble conducting polymers. Specifically we report on the first verification of the "self doping mechanism".

BANDGAP REDUCTION

In very simple terms, the relatively large bandgaps of currently known conducting polymers can be ascribed to bond-alternation in the backbone; e.g., alternating "single" and "double" bonds in polyacetylene. In recent papers, Bredas[3] suggested an approach toward bandgap reduction in polyheterocycles by the stratagem of making the "quinoid" and "benzenoid" forms in the backbone essentially isoenergetic:

Benzenoid   Quinoid

There have been two approaches to date to implement this model (a) stabilize the quinoid form of polythiophene by incorporating the quinoid thiophene 3,4-"double bond" into a benzene ring i.e. as in poly(isothianaphthene) (PITN)[4]:

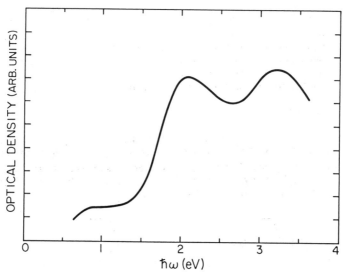

$$\xrightarrow{\text{FeCl}_3} \qquad (3)$$

occurred, raising the bandgap energy. To remove this possibility, an electrochemical polymerization-dehydrogenation (eq. 2) was carried out and again the same electronic spectrum was obtained (Fig. 1). Finally a chemical polymerization (eq. 3) also gave the same spectrum. As can be seen from Fig. 1, the electronic spectrum of the polymer showed **a clean gap**, indicating that there were no states in the gap which could have arisen from different forms of catenation of the polymer. The discrepancy between our experimental results and theoretical calculations is surprizing in view of the fact that the theory has a good "track record" in predicting the Eg of conjugated polymers.

In the second approach, In a series of papers dealing with a different way to implement current theoretical models for the design of small semiconductor bandgap ($0 <$ Eg > 1eV) conducting polymers, it was claimed that polymers **1-4** could be dehydrogenated to low gap polymeric semiconductors with E_g as low as 0.75 eV[8-10] by treatment with bromine vapor.

Fig. 1. Electronic spectrum of a thin film of a polymer derived from **I** on indium-tin oxide coated glass. The same spectrum could be obtained by three independent methods of polymer preparation.

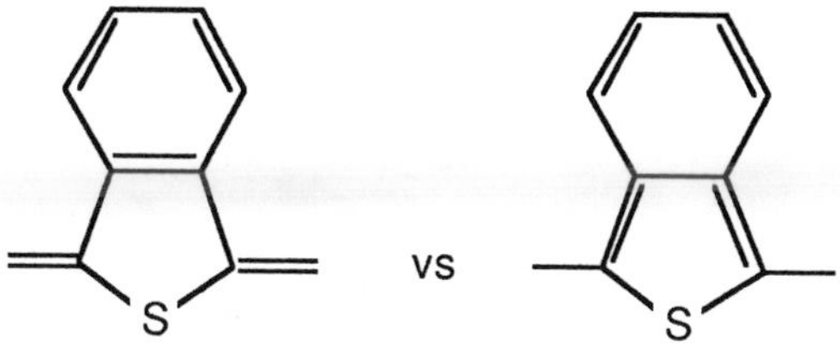

vs

and (b) to incorporate known amounts of quinoid and benzenoid forms into the backbone by insertion of a methine at specific locations[5] :

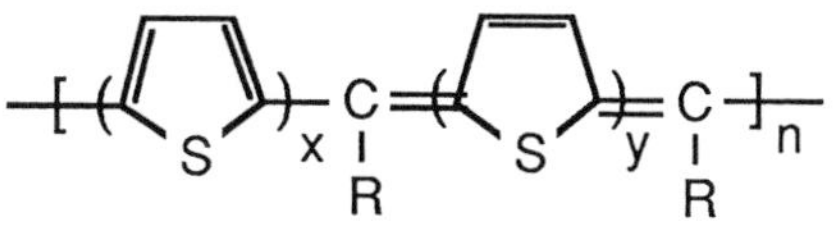

The first approach was relatively successful[4]; the bandgap of PITN (Eg = 1 eV) was found to be one half that of polythiophene (Eg = 2 eV). More recent calculations by Brédas led him to conclude that a second benzene annulation ; i.e., poly(isonaphthothiophene) (a polymer derived from structure I, below), should lead to further bandgap reduction to 0.01eV (practically zero). However, results from our laboratory employing three different syntheses: i) direct electrochemical polymerization of isonaphthothiophene (INT, I, equation 1), ii) electrochemical polymerization of 8,9-dihydroisonaphtho-thiophene (equation 2) and iii) a chemical dehydrogenative polymerization[6] of 1,3-dihydroisonaphtho-thiophene[7] (equation 3); revealed that the polymer had a bandgap of 1.4 eV. One could argue that polymerization of **I** may not have proceeded regiospecifically and that some random polymerization through the undesired positions (c and d , structure I)

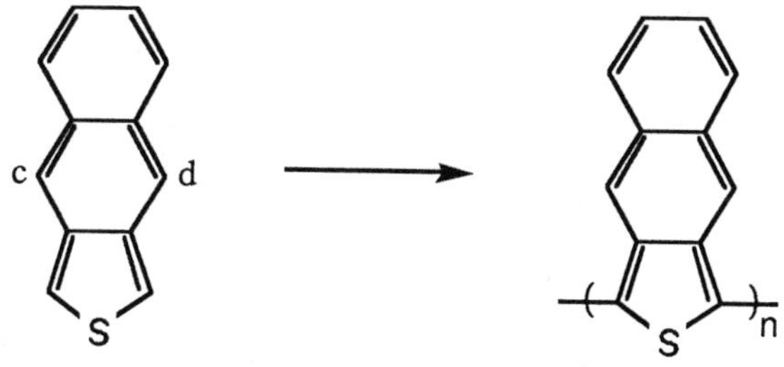

(1)

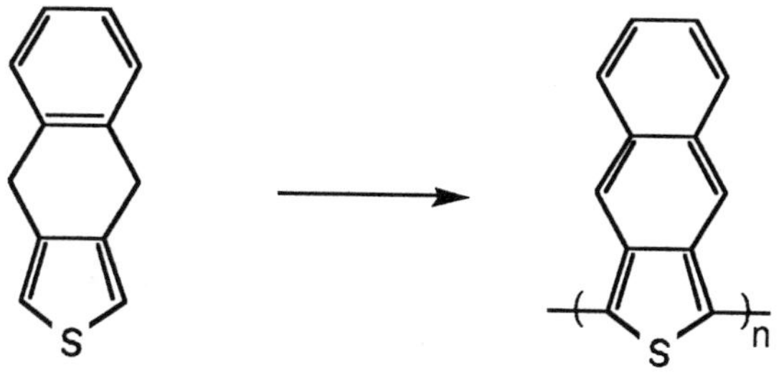

(2)

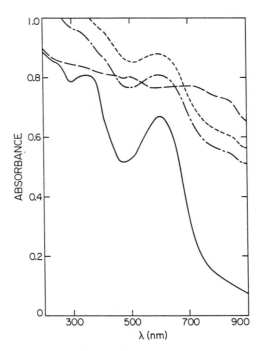

Fig. 2. Electronic spectrum of a thin film of impure polymer **2** on a quartz surface (___),
the same film exposed to bromine vapor for a ~ 1 minute (-----), the same
brominated film exposed to hydrazine vapor (_ . _ . _ .) and the same film
exposed to a second bromination-reduction cycle (........).

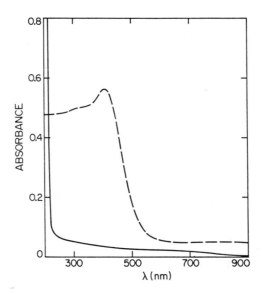

Fig. 3. Elctronic spectrum of a thin film of tetrabutylammonium bromide on quartz (___),
the same film exposed to bromine vapor (------).

The bromination of thin films was followed by electronic and FTIR spectroscopy but no nmr experiments nor elemental analyses were described[9,10]. It was also concluded that since there were no peaks due to C-Br stretching in the 500-650 cm^{-1} region of the infrared spectrum, no electrophilic substitution had taken place and that the action of bromine vapor was strictly a dehydrogenative process. The products were postulated to have the general structures **1A-4A**[6] and more specifically structure **3A**[9,10] shown below.

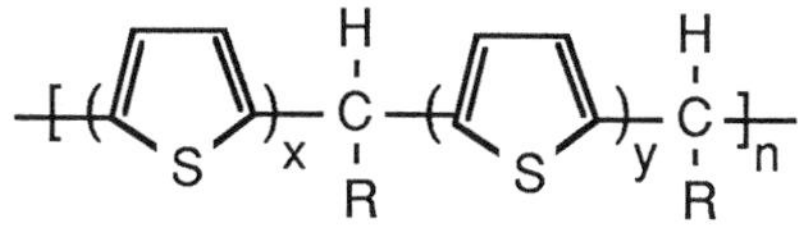

1	x = y =1, **R** = Ph
2	x = y = 2, **R** = Ph
3	x = y = 2, **R** = p-CH$_3$C(O)OC$_6$H$_4$
4	x = y = 3, **R** = Ph

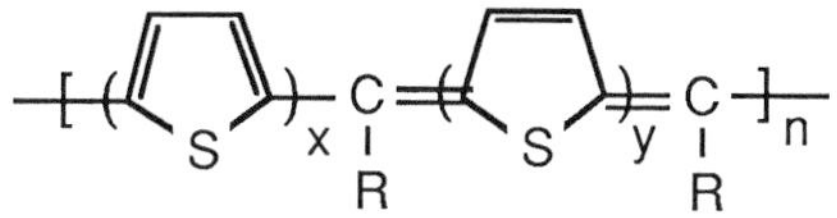

1A	x = y = 1, **R** = Ph
2A	x = y = 2, **R** = Ph
3A	x = y = 2, **R** = p-CH$_3$C(O)OC$_6$H$_4$
4A	x = y = 3, **R** = Ph

We have learned[11] that the reaction of both PBTB (**2**) and PBTAB (**3**) with bromine vapor depends on the phase (solid-gas vs. solution-gas) and the product is not the simple dehydrogenated product reported before.

When the UV-vis spectrum of an impure sample[12] of the title compounds in CDCl$_3$ solution was recorded during *in situ* bromination, the impurity 680nm and 810nm bands increased in intensity as shown in Figure 2. This result, in the absence of other data, could easily lead one to believe that a dehydrogenation had taken place and that a semiconductor with a decreased bandgap had been produced. However, when the ^{1}H nmr spectrum of the title compounds was recorded during *in situ* treatment with bromine in homogeneous solution in CDCl$_3$ at ambient temperature, quite unexpectedly, the methine hydrogen resonance **remained unchanged** as a function of bromine treatment even after a large excess of the reagent had been added and the reaction was allowed to proceed for hours at room temperature. The only noticeable changes were in the aromatic region, particularly a decrease in the ratio of the bithiophene resonances relative to the phenyl resonances.

When a thin film (~5-10μm thickness) of polymers **2** and **3** was exposed to bromine vapor, the observed changes in the electronic spectrum were qualitatively the same as described above for homogeneous solution (*cf.* Fig. 2); however, dissolution of the brominated film in deuterio chloroform, followed by ^{1}H nmr spectroscopy revealed that this time the methine hydrogen resonances had disappeared but the pattern of resonances in the aromatic region had also changed dramatically, similarly to the experiments described above for solution bromination.

Elemental analyses of products of both experiments (solids evacuated for 24 hr at 0.1

incorporated, indicating that possibly a **bromine-doped** product, with some covalently bonded bromine was produced.

Finally, when a tetrabutylammonium bromide film was treated with bromine under the above described conditions, an electronic spectrum remarkably similar to that of a "low bandgap semiconductor" ("Eg" ~ 2 eV, Fig. 3) could be recorded. Of course this is just the spectrum of Bu_4NBr_3 and points up the danger of attributing an extended solid state delocalized electronic structure ("bands" and "bandgaps") to an obvious **molecular** solid on the basis of only one probe; Uv-vis spectroscopy.

We conclude from the above that polymers reported to have structures **2A** and **3A** are in fact overbrominated, not easily characterizable solids of **unknown bandgap**.

SELF-DOPING MECHANISM

We recently reported on the development of conjugated polymers which are water soluble. These polymers can be processed in the form of films cast from aqueous solution. The polymers were developed to test a new concept, that of self-doped polymers[12] (SDP). These are polythiophenes with an alkanesulfonic acid (or salt derived therefrom) substituent which is expected to lose a proton (or cation if it is a salt) concomitant with p-doping:

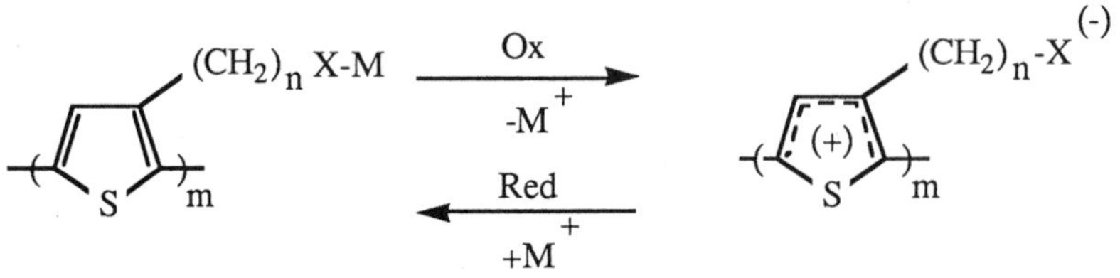

M = H, Li, Na, etc

$X = SO_3$

n = 2, 4

Whereas we reported on some of the properties of these polymers, we did not report on any evidence substantiating SDP. An immediate implication of this concept is a **potential dependent ion exchange**. For example if, in the above scheme, M = H; as the polymer is doped, the pH of the medium is expected to decrease and as the polymer is dedoped, the pH is expected to increase. This is precisely what is observed as shown in Fig. 4. The experiment on which the figure is based was the simultaneous recording of pH variation with cyclic voltammetry in acetonitrile at room temperature. Before the potential was cycled, the polymer film was allowed to come to "ionic equilibrium" with the medium (CH_3CN, Bu_4NClO_4); i.e. the "pH" (quotations are employed to denote the fact that pH always refers to water, whereas our experiments were carried out in a non-aqueous solvent) was allowed to settle to a reading of ca "1" (point α). As the polymer was oxidized, the "pH" was observed to decrease ($\alpha \rightarrow A$) and as the polymer was reduced, the "pH" increased ($A \rightarrow B$) and decreased ($B \rightarrow C$). A plausible explanation for this behavior instead of the expected continuous increase in "pH" with increasing reduction of the polymer backbone, is that initially, the fastest ions to move into the polymer are indeed protons, but before the polymer reduction is complete, ion exchange with the electrolyte (H^+ for Bu_4N^+) becomes a faster process than reduction and the "pH" decreases.

398

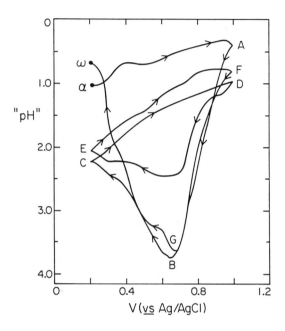

Fig. 4. Change in hydrogen ion concentration as detected by a hydrogen ion
electrode *in situ* during a cyclic voltammetry experiment in acetonitrile (0.1M-
Bu$_4$N$^+$BF$_4$$^-$ electrolyte) using Ag/Ag$^+$ as reference electrode

CONCLUSION

We have shown that, apparently, further benzannelation of isothianaphthene does not result in a further decrease in the bandgap of the polymer derived from this monomer and a previous claim of semiconductor bandgap reduction in poly(methinethiophenes) was in fact not the case but was the result of assignment of a UV-vis spectrum to a not-fully-characterized material. We have also shown that the newly discovered concept of self-doped polymers led to the discovery of a material which exhibits a potential dependent ion exchange property; i.e. change of pH with doping llevel of a polymer.

REFERENCES

1. K.-Y. Jen, R. Oboodi, and R. Elsenbaumer, Synth. Met. 15:169 (1986); M. A. Sato, S. Tanaka, and K. Kaeriyama, J. Chem. Soc. Chem. Commun. 873 (1986).
2. S. Hotta, S. D. D. V. Rughooputh, A. J. Heeger, and F. Wudl, Macromol. 20:212 (1987).
3. J. L. Brédas, Synth. Met. 17:115 (1974).
4. F. Wudl, M. Kobayashi, N. Colaneri, M. Boysel, and A. J. Heeger, Mol. Cryst. Liq. Cryst. 118:195 (1985).
5. S. A. Jenekhe, Nature 322:345 (1986). Note that in this case fusing both "quinoid" and "benzenoid" forms "statically" into the same backbone gives rise to a macromolecule with a degenerate ground state (analogous to polyacetylene) but cannot give a zero bandgap because the same bond alternation that exists in polyacetylene also exists in this case.
6. K.-Y. Jen and R. Elsenbaumer Synth. Met. 16:379 (1986).
7. M. P. Cava, N. M. Pollak, O. A. Mamer, and M. J. Mitchell, J. Org. Chem. 36:3932 (1971); J. Bornstein and R. P. Hardy J. Chem. Soc. Chem. Commun. 612 (1972).
8. S. A. Jenekhe, Nature 1986, 322, 345.
9. S. A. Jenekhe, Macromolecules 1986, 19, 2663.
10. S. A. Jenekhe, Polym. Preprints 1986, __, 74.
11. A. O Patil and F. Wudl, Polym. Preprints 1987, _, in press
12. In references 7 and 8 it is claimed that polymer 3 is blue with a λmax of 692nm and a bandgap of 810nm (1.53 eV). The pure polymer is actually off white with λmax of 332nm and a hint of absorption in the 700-800nm region (cf Fig. 2), indicating a small degree of unsaturation. However, both pure and impure polymers show qualitatively the same electronic spectroscopy behavior when brominated (the former less dramatically so than the latter).

RIGID AROMATIC HETEROCYCLIC POLYMERS: SYNTHESIS OF POLYMERS AND

OLIGOMERS CONTAINING BENZAZOLE UNITS FOR ELECTROOPTIC APPLICATIONS

James F. Wolfe and Steven P. Bitler

SRI International
333 Ravenswood Avenue
Menlo Park, California 94025

INTRODUCTION

Conjugated organic structures exhibit large, ultrafast nonlinear optical (NLO) responses that arise from excitation of highly charged correlated π-electron states. Highly aligned, high strength films of high molecular weight, rigid rod poly(benzobisazole) (PBZ) polymers have been shown recently to possess oustanding third-order NLO properties.[1,2] PBZ polymers, being comprised exclusively of aromatic rings, are conjugated structures with numerous possibilites for structural modification. In addition to ultrafast NLO response, PBZ polymers offer advantages of high damage thresholds, environmental stability, a wide variety of processing options, and excellent mechanical properties. Molecular structures with or without centers of symmetry can be prepared and the molecular weight distribution, electron density, and the molecular morphology can be varied over broad ranges. This tailorability in both molecular structure and supramolecular order provides promise for new materials with optimized NLO properties. This paper describes our recent work on the synthesis of PBZ polymers having controlled molecular weights. By controlling the molecular weight at relatively high levels, we aim to improve the processability and reproducibility of solutions for better optical quality high strength films. By preparing novel lower molecular weight materials, we aim to explore new structures that are specifically designed to maximize their NLO response and may require new methods for obtaining the high degree of molecular order that has become associated with PBZ materials.

Our polybenzazole synthesis research, which has been funded jointly by the Air Force Office of Scientific Research and the Air Force Materials Laboratory as part of the Air Force's Ordered Polymers Research Program, was originally aimed at materials for high performance structural applications. Our research focused initially on developing synthesis techniques to prepare particularly rigid polymer structures that would form lyotropic, liquid crystalline phases, when both the molecular weight and the solution concentration were above certain values. The basic synthesis methodology for preparing high molecular weight PBZ polymers has been described previously.[3,4] PBZ materials are prepared by condensation in poly(phosphoric acids) (PPAs) having high phosphorous pentoxide (P_2O_5) contents by an efficient process called the P_2O_5 Adjustment Method. The amount of P_2O_5 used depends on the concentration of the condensing species. This process

provides the basic condensation for the preparation of polymeric, oligomeric, substituted and unsusbstituted PBZ materials.

Our efforts were expanded to include NLO applications when it became
apparent that many of the same molecular design elements that resulted in
thermal stability, high mechanical properties and orderability also led to
desirable NLO properties.[5] The series of molecular design choices and
morphological options shown in Scheme 1 illustrates the similarities between structural and NLO applications. To obtain maximum specific tensile
properties and stability in a lightweight material for structural applications, the clear choices are those listed first on each line. Given the
same choices for ultrafast NLO applications, the choices are clearly the
same--except for those related to molecular weight and backbone substitution. The following two sections discuss the molecular design considerations relating to these two areas.

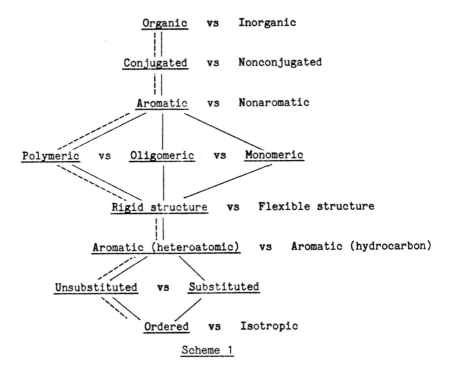

Scheme 1

Molecular Structure and Morphological Choices for Two Applications[*]

[*]High Strength Structural Applications   ------
[*]Nonlinear Optical Applicatons   _____

Molecular Weight Considerations

The measurement of electron delocalization lengths in PBZ polymers by
Dalton et al.[6] shows that delocalization does not extend beyond a few
repeat units, even though the molecular structure is formally conjugated
over the full length of the backbone. Because the NLO response is highly
dependent on the electron delocalization length, we can view a high molecular weight PBZ polymer chain, in particular a rigid rod polymer chain, as a
collinear string of separate NLO active sites. The question arises as to
the role of the covalent links between the sites. Are they really necessary? They are clearly responsible for achieving nematic ordering, that

is, the additional dimension of order beyond the collinear one-dimensional array afforded by a single chain. Spontaneous formation of an anisotropic phase results if the molecular weight is above a critical value and the solution concentration is sufficient. The molecular weight required for obtaining good mechanical properties in films of PBZ polymers is much higher than the critical molecular weight required to obtain anisotropic (nematic) ordering. When rigid rod PBZ polymers are formed from monomer at high concentrations, such as 15 wt%, the degree of polymerization at which the nematic phase obtains is about 15 repeat units, corresponding to an intrinsic viscosity of 1.5 dL/g.

This second dimension of order, however, can also be obtained by poling small highly polarizable molecules, which do not obtain the nematic phase spontaneously at any achievable concentration. With the possibility of obtaining supramolecular order with rigid lower molecular weight molecules, we began to investigate methods of controllably reducing the molecular weight of PBZ materials having a "handle" for attachment to a flexible polymer backbone. In these efforts, good mechanical properties are desirable but are no longer the driving force behind the molecular design.

Substituent Effects

This portion of the current effort attempts to establish the synthesis goals for low molecular weight PBZ materials having electron-withdrawing and/or electron-donating substituents. The ultimate goal in this program is to establish the structure-property relationships that allow the design of optimized NLO materials. We have begun this effort by studying the effect of various substituted monofunctional reagents on the normal condensation in PPA.

RESULTS AND DISCUSSION

The five systems that address the above goals and will be described in this paper are listed in Figure 1.

PBT Polymers with Controlled Molecular Weights

Poly(p-phenylene-benzobisthiazole) (PBT) can be prepared with intrinsic viscosities as high as 48 dL/g by the condensation of 2,5-diamino-1,4-benzenedithiol dihydrochloride (DABDT) with terephthalic acid in PPA. Chenevey has stated that the preferred range of PBT instrinsic viscosities for preparing films directly from PPA solutions, or dopes, is 20-30 dL/g.[7] A major goal of our synthesis research has been to define the variables of the PPA polymerization process sufficiently to allow not only the reproducible attainment of a specific intrinsic viscosity, but also control of all the key dope characteristics. The key dope characteristics, in addition to the molecular weight of the polymer, are the shear viscosity of the dope, and the morphology. The important synthesis variables to control are the DABDT purity, the concentration of the condensing species, the P_2O_5 contents during polymerization, and the amount of endcapping agent used.

We conducted a series of PBT polymerizations in which small percentages of benzoic acid were used as an endcapping agent. The percentage of benzoic acid, based on the molar quantity of DABDT employed, was varied from 0 to 1.0%. The equivalency of functional groups was maintained by reducing the molar amount of terephthalic acid by one-half of the molar amount of benzoic acid used. The final P_2O_5 contents of the dopes were maintained within the range determined previously to be optimal,[3] and the polymer concentrations employed were either 12.5 or 15 wt%. The intrinsic viscosities of the PBT isolated from the dope were measured and the shear

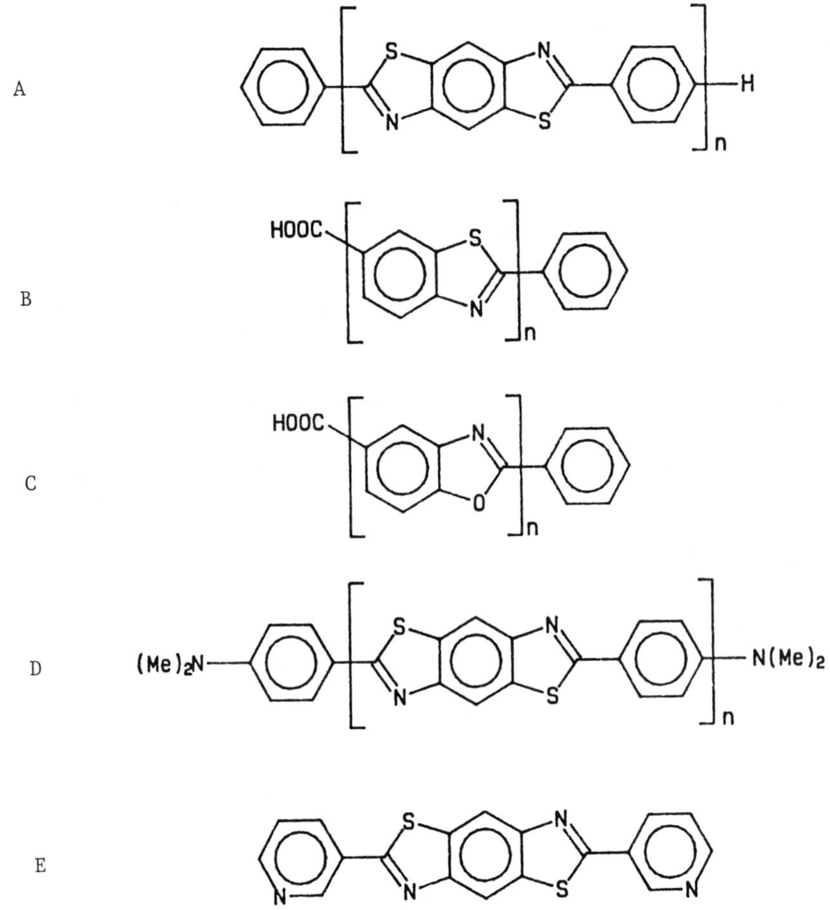

Figure 1. Chemical structures of (A) PBT, (B) ABPBT,
(C) ABPBO, (D) p-dimethylaminobenzoic acid
endcapped PBT, and (E) 2,6-di(3-pyridyl)-
benzo[1,2-d:4,5-d']bisthiazole

viscosities of the dopes were determined. The experimental results are
presented in Table 1.

The weight average molecular weights, Mw, were calculated from the
intrinsic viscosities using the following relationship for PBT determined
by Berry et al.:[8]

$$[\eta] = 1.65 \times 10^{-7} M_w^{1.8}$$

Flory's theory[9] for the depression of molecular weight by the nonequi-
valence of functional groups and the presence of monofunctional reagents
was used to analyze the results. Flory defined the number average degree
of polycondensation, X_n, as

$$X_n = \frac{1 + r}{2r(1 - p) + 1 - r}$$

where p is the extent of reaction of either of the functional groups, and r
is defined as

404

TABLE 1
PBT ENDCAPPING STUDY

| | Polymerization Conditions | | | | | Dope Properties | | Theory | |
Run No.	Endcapper %	r	Final P2O5 Content, %	PBT Conc., wt %	$[\eta]$, dL/g	PBT Mw	Shear Viscosity poise/1000	p = 0.993 PBT Mw	p = 0.9945 PBT Mw
1	0	1	82.7	15	44	48,000	–	38,000	48,400
2	0.1	0.9995	82.9	15	30	38,800	76	36,700	46,300
3	0.2	0.9990	82.9	15	29	38,100	–	35,500	44,400
4	0.25	0.9988	82.9	15	27	36,600	–	35,000	43,500
5	0.5	0.9975	82.9	15	25	35,100	63	32,300	39,500
6			82.7	15	29	38,100	78		
7			82.7	15	26	35,800	48		
8	0.6	0.9970	82.9	15	20.5	31,400	34	31,400	38,140
9			82.9	15	21.9	32,600	30		
10			82.9	15	22.8	33,300	38		
11			82.9	15	23.1	33,500	38		
12	0.75	0.9963	82.7	12.5	15	26,400	–	30,100	36,200
13			82.3	12.5	17	28,300	–		
14			82.6	15	22	32,600	–		
15			82.4	15	24	34,300	–		
16	1.0	0.995	82.7	12.5	18	29,200	–	28,100	33,400
17			82.7	12.5	21	31,800	–		

$r = 200/\{2\ [100-(\%\ endcap)/2]+2(\%\ endcap)\}$

$[\eta] = 1.65 \times 10^{-7} [Mw]^{1.8}$

Shear viscisities were measured at 0.1 sec^{-1} at 175°C with a Brookfield RVT viscometer, Spindle #7

p = theoretical extent of reaction of either functional group

Theoretical Mw = 2 x 266.33(xn/2), assuming Mn = Mw/2, where

xn = number average degree of polymerization = $[1+r]/[2r(1-p)+1-r]$

$$r = \frac{\text{moles of A groups}}{\text{moles of B groups} + (2 \times \text{moles of C groups})}$$

For these PBT polymerizations, the A groups are the ortho-mercaptoamino functions of DABDT, the B groups are the carboxyl functions of terephthalic acid, and the C groups are the carboxyl functions of benzoic acid.

Theoretical weight average molecular weights for two hypothetical cases can be calculated using Flory's equations, the appropriate values for r and assuming two extents of reaction, p, of 99.45% and 99.3%. These data and the experimental results are presented graphically in Figure 2. Nearly all the experimental results fall between the bounds defined by these two hypothetical cases, which indicates that polymerizations are fairly consistent in their effective extent of reaction. These data also show that the molecular weight is predicted to decrease by about 10,000, at r values near unity, for a decrease in the extent of reaction of only 0.15%.

The effective extent of reaction can be affected by many factors. First, impurities in either monomer could be the cause of a stoichiometric imbalance that would result not only in a lower extent of reaction but also in a lowered r value. Second, the extent of reaction could be lowered by the occurrence of side reactions, which could be caused by an improper heating schedule or an insufficient P_2O_5 content. We find that we achieve better reproducibility of results if the final P_2O_5 content of the polymerization is greater than 82.7% and preferrably 82.9% or higher. The shear viscosity is highly dependent on the P_2O_5 content. To maintain the shear viscosity at values less than 80,000 poise (at 0.1 sec^{-1}, 175°C) for

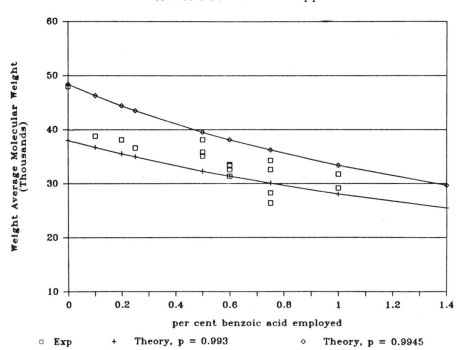

Mw of PBT vs % endcapper

Figure 2. Weight average molecular weight of PBT versus percent benzoic acid endcapper used in the polymerization

intrinsic viscosities near 30 dL/g, the final P_2O_5 content of the dope must be kept below about 83%.

Early in this study, we noticed occasionally that the shear viscosities of the PBT dopes were abnormally high compared with other dopes having similar intrinsic viscosities, PBT concentrations, and final P_2O_5 contents. For example, the shear viscosities of runs 5 and 7 (see Table 1) should be nearly the same, based on these three variables. However, run 5 is 30% higher. We have attibuted the nonreproducibility of the shear viscosities to small amounts of disulfide impurities in the DABDT monomer that would become incorporated into the backbone of the otherwise rigid rod polymer. The presence of small amounts of such "articulated PBT polymers" in the anisotropic dope is expected to raise the shear viscosity dramatically. We spent considerable effort trying to detect this disulfide impurity in DABDT samples that gave high shear viscosities to confirm this assumption. Elemental analysis was not expected to be sensitive enough to detect low levels of this impurity. We used Raman and FTIR spectroscopy, but could not detect assignable differences between appropriate DABDT samples. We have indirect evidence that the cause of abnormally high shear viscosities is such a species, because normal shear viscosities were obtained from DABDT samples that were recrystallized under reducing conditions. This reductive recrystallization, which is described below, has become standard procedure in preparing DABDT of monomer grade purity.

Reproducibility of the shear viscosity and the intrinsic viscosity has been achieved by using DABDT monomer that has been recrystallized from stannous chloride/hydrochloric acid and by polymerizing at a concentration of 15 wt% and a final P_2O_5 content of 82.9%. Using these guidelines, PBT with an intrinsic viscosity of 23 dL/g can be obtained by replacing 0.3 mol% of the terephthalic acid with 0.6 mol% of benzoic acid.

Poly(benzazole) Polymers with High Levels of Endcapping

We also studied the effect of endcapping on the molecular weight of the two AB polymers, poly(2,6-benzothiazole) (ABPBT) and poly(2,5-benzoxazole) (ABPBO). The percentage of endcapping agent was varied from 0 to 10%, which is a wider range than investigated in the PBT study. The experimental results are presented in Table 2. A theoretical extent of reaction was chosen for each experiment that gave a theoretical Mw, again using Flory's equations, that matched the observed molecular weight. One of the striking interpretations of this study is the apparent higher extent of reaction in ABPBO polycondensations relative to ABPBT. The extents of reaction of ABPBT are also higher than those believed to be operative in PBT polymerizations. We believe these interpretations are consistent with the electronic nature, which relates to relative oxidative stability, of the three monomers involved.

The weight average molecular weights of ABPBT and ABPBO were calculated from the experimentally determined intrinsic viscosities by using relationships derived by Chow et al.[10]

For 2,6-ABPBT, with Mw between 5,000 and 200,000

$[\eta] = 1.256 \times 10^{-4} \, Mw^{1.00}$.

For 2,5-ABPBO, within a similar molecular weight range,

$[\eta] = 1.089 \times 10^{-4} \, Mw^{1.02}$.

TABLE 2
TABLE 2
ABPBT AND ABPBO ENDCAPPING STUDY

	Experimental				Theory	
	% endcap	r	[η]	M_w	p	M_w
ABPBT Runs						
1	0	1	11.5	88,500	0.977	88,800
2	0.5	0.99	4.2	32,300	0.996	29,700
3	1.0	0.98	2.1	16,300	0.996	19,200
4	5.0	0.91	0.76	5,900	0.996	5,200
5	10.0	0.83	0.27	2,100	0.994	2,800
ABPBO Runs						
1	0	1	15.0	106,800	0.998	117,000
2	0.5	0.99	4.95	36,100	0.999	39,200
3	1.0	0.98	3.39	24,900	1.000	23,700
4	5.0	0.91	0.62	4,700	1.000	4,900

r = mol AB monomer/(mol AB monomer + 2 x mol of endcapper)
p = theor. extent of reaction = mol of A condensed/total mol of A

The plots of these equations and the relationships for the two rigid rod PBZ polymers, PBT and PBO, are shown in Figure 3.

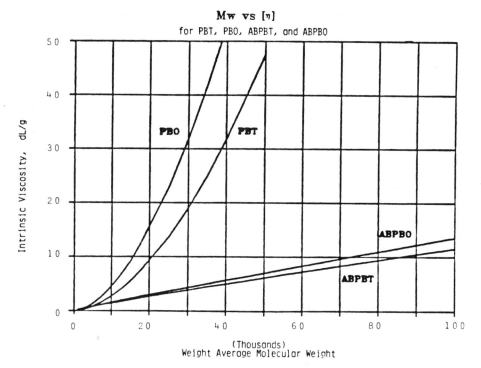

Mw vs [η]
for PBT, PBO, ABPBT, and ABPBO

Figure 3. Weight average molecular weight versus intrinsic viscosity for PBT, PBO, ABPBT, and ABPBO

Preparation of Substituted PBZ Oligomers and Model Compounds

In addition to the above studies in which relatively minor percentages of benzoic acid were added to PBT, ABPBT, and ABPBO polymerizations, we have prepared PBZ materials using various functionalized endcapping agents. The goal is to determine the reactivity, stability, and compatibility of these compounds in the polymerizing medium to assess their usefulness in preparing PBZ materials with tailored electron densities and polarizabilities. By attaching a functional group to the end of a PBZ chain, the compound can either be synthetically modified after the PBZ portion of the molecule has been formed, or attached to the backbone of a common polymer. The material could also possibly be used directly in device fabrication.

We are investigating the following six functionalized endcapping agents:[11]

    nicotinic acid (3-pyridinecarboxylic acid)
    p-(N,N-dimethylamino)benzoic acid
    4-nitrobenzoic acid
    2-amino-5-nitrophenol
    2-amino-4-nitrophenol
    1,2-diamino-4,5-dinitrobenzene

Only the first two compounds allow the PPA polymerization of the oxidatively sensitive PBZ monomers to proceed without significant side reactions. The nitro-group containing compounds are believed to cause electron transfer from the electron-rich monomers under the PPA polymerization conditions leading to significant decomposition. Alternative synthesis methods may be required to introduce the nitro-group, such as a two-step procedure of oxidizing a nitro-group precursor.

Nicotinic acid has been used to prepare the dipyridyl benzobisthiazole model compound. Materials of this type have the potential for functionalization through the pyridyl nitrogen. Capping AB polymers with dimethylaminobenzoic acid gives a functionalized material that can be attached to polymer backbones directly through the terminal carboxyl group. The dimethylaminobenzoic acid endcapping agent gave results comparable with the results given for benzoic acid endcapping of DABDT. A PBT polymerization was conducted with 0.5 mol% of terephthalic acid replaced by 1 mol% of dimethylaminobenzoic acid. An intrinsic viscosity of 16.6 dL/g was obtained for the product.

Monomer Syntheses

Several advances have been realized during the current study in the synthesis of one of the important monomers used to prepare benzobisazole structures, namely 2,5-diamino-1,4-benzenedithiol dihydrochloride (DABDT). The synthesis scheme for preparing DABDT, which is shown in Figure 4, is the same as that reported previously.[12] Several modifications have been found, however, to increase the yields and improve the reproducibility of the process. In the first step, p-phenylenediamine (PPD) dihydrochloride is dissolved in deaerated water and heated at $70^{\circ}C$ with excess ammonium thiocyanate (NH_4SCN). We found that p-phenylenebis-(thiourea) (BTU), thus formed, is essentially void of the p-thiourea-aniline intermediate if the reaction temperature is decreased from $100^{\circ}C$ to $70^{\circ}C$, the reaction time is increased to 48 h, and a sufficient excess of NH_4SCN is employed. The optimal ratio of NH_4SCN/PPD is 4.9/1 (mol/mol), which represents a 22.5% excess over the four moles required. The BTU formed under these conditions can be used in the next step without purification.

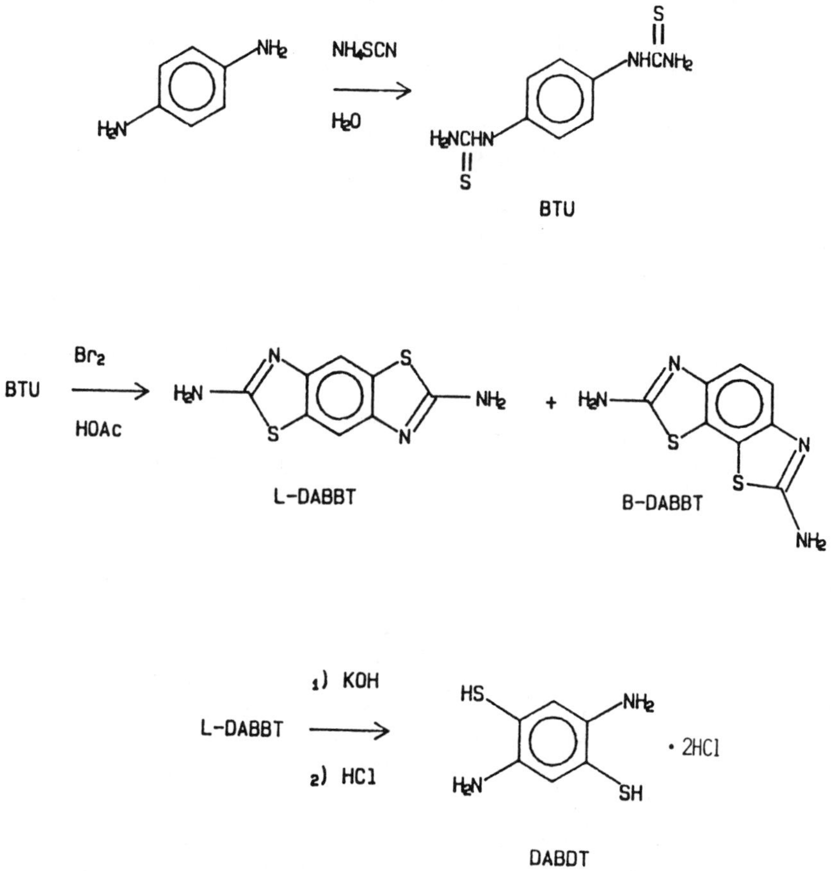

Figure 4. Synthesis of DABDT

The other modification of the published procedure is the addition of a hot stannous chloride recrystallization of the final product. This new procedure is described in the Experimental Section.

EXPERIMENTAL

Procedures

All intrinsic viscosities are reported in dL/g and are determined in distilled methanesulfonic acid (MSA) at 30.0°C by the extrapolation of reduced and inherent viscosities to zero concentration. Concentrations of the polymers are employed to give relative viscosities between 1.1 and 1.5.

Shear viscosities of dope samples were measured with a #7 spindle of a heavy duty Brookfield RVT Viscometer placed in a sample that had been immersed in an oil bath at 175°C.

Materials

Polyphosphoric acid (PPA). PPAs of various P_2O_5 contents are prepared by mixing commercially available 115% phosphoric acid (FMC) and 85% phosphoric acid (Baker Chemical Co.) and heating with stirring for 2 h at 115-120°C under reduced pressure.

MONOMERS

Terephthalic acid (TA). High purity TA was obtained from Amoco Chemical Co. and its particle size was reduced by an air impact method such that more than 95% of the particles were less than 10 um.

2,5-Diamino-1,4-benzenedithiol dihydrochloride (DABDT). Deaerated water (115 ml) was added to 27.5 g (0.124 mol) of 2,6-diaminobenzo[1,2-d:4,5-d']bisthiazole (DABBT) and 109 g (1.94 mol) of KOH in a 250 ml flask equipped with a reflux condenser and an overhead stirrer under inert atmosphere. The mixture was stirred and heated to 115-120°C for 7 h. The resulting clear amber solution was slowly cooled to 20-25°C overnight causing the dipotassium salt of DABDT to crystallize from solution. This white crystalline solid was collected by filtration under a carefully controlled nitrogen atmosphere, redissolved in 300 ml of deaerated water, and filtered under inert atmosphere into 300 ml of aqueous 37% HCl. The product was allowed to crystallize at least 6 h at 20-25°C then collected by filtration to yield 39 g of wet DABDT. The wet DABDT was recyrstallized in a reducing medium under inert atmosphere by adding the 39 g of wet DABDT to a hot (70-75°C) solution of 350 ml of water, 3.9 g of stannous chloride dihydrate, and 120 ml of aqueous 37% HCl. After stirring the solution for 5 min, 350 ml of deaerated water, preheated to 70-75°C, was added and the mixture was stirred at 70-80°C for 20 min to effect complete dissolution. The hot solution was filtered into 250 ml of aqueous 37% HCl causing immediate crystallization of the very light yellow DABDT. The product was allowed to crystallize at 10-20°C for at least 6 h and then collected by filtration. The DABDT was washed with methanol and then diethyl ether and then dried at 40-50°C under reduced pressure to yield 24.6 g (81%).

ENDCAPPING AGENTS

Benzoic acid, ACS reagent grade, was obtained from MCB Company and used as received.

p-(N,N-Dimethylamino)benzoic acid was obtained from Aldrich Chemical Company and recrystallized from toluene before use.

Nicotinic acid (3-pyridinecarboxylic acid) was obtained from Aldrich and used as received.

MODEL COMPOUNDS

2,6-Di(3-pyridyl)benzo[1,2-d:4,5-d']bisthiazole. A PPA solution was prepared by heating a mixture of 107.85 g of 115% phosphoric acid and 38.55 g of 85.3% phosphoric acid under reduced pressure for 2 h. DABDT (10.428 g, 42.52 mmol) was placed in a 100 ml resin kettle fitted with a mechanical stirrer and 36.54 g of the PPA solution was added. The mixture was heated at 50-60°C under reduced pressure to effect removal of the hydrogen chloride. Nicotinic acid (10.793 g, 87.67 mmol) was then added. P_2O_5 (24.49 g) was then added to give a calculated P_2O_5 content of the PPA of 82.7% after 100% of the theoretical condensation. The mixture was slowly heated to 170-175°C and maintained at that temperature for 4 h. The product was precipitated in water, washed with methanol and dried. The product was recrystallized from toluene to give a 49% isolated yield. Mass spectrum calc, 346; found, 346. Anal. ($C_{18}H_{10}N_4S_2$): C, 62.41; H, 2.91; N, 16.17; S, 18.51. Found: C, 62.20; H, 3.18; N, 16.02; S, 18.96.

POLYMERS

Dimethylamino capped PBT. DABDT (11.67481 g, 47.62 mmol) was placed in a 100 ml resin kettle equipped with a mechanical stirrer. A PPA solu-

411

tion (40.91 g), prepared from 228.91 g of 115% phosphoric acid and 81.79 g of 85.3% phosphoric aicd, was then added under a stream of dry nitrogen. The mixture was heated at 50-70°C under reduced pressure until a clear yellow solution was obtained (72 h). To the clear solution was added 7.87148 g (47.38 mmol) of terephthalic acid, 0.07720 g (0.632 mmol) of N,N-(dimethyl)aminobenzoic acid, and 27.57 g of P_2O_5. The mixture was calculated to have a P_2O_5 content of 82.7% after polymerization. The reaction mixture was heated to 95°C for 16 h followed by heating to 185°C over 2 h then heating at 180-185°C for 16 h. The capped PBT/PPA solution was stirred during the heated phase until the viscosity became so high that the mixture rode on the stirrer. The mixture became yellow-green opalescent. Fiber samples were drawn from the solution, precipitated in water, washed with water in a Soxhlet extractor, and dried at 160-165°C under reduced pressure. The intrinsic viscosity was measured to be 16.6 dL/g in methanesulfonic acid at 30°C. The bulk of the product was preserved as the liquid crystalline solution for subsequent processing.

Poly(2,6-benzothiazole) (ABPBT) with 1% Benzoic Acid Endcapper. A PPA solution (76.7% P_2O_5 content) was prepared by heating 20.04 g of 115% phosphoric acid and 9.07 g of 85% phosphoric acid at 100°C under reduced pressure for 2 h. A portion of this PPA (24.27 g) was added to a 100 ml resin kettle equipped with mechanical stirrer that contained 15.4 g (74.9 mmol) of 3-mercapto-4-aminobenzoic acid hydrochloride. This mixture was then heated at 65-70°C for 20 h to effect removal of the hydrogen chloride. To this solution was added 92 mg (0.75 mmol) of benzoic acid and 22.0 g of P_2O_5. The mixture was then stirred at 70-75°C for 2 h. The temperature was increased to 185°C over 6 h and maintained at that temperature overnight yielding 58.2 g of an opalescent golden-green ABPBT/PPA dope. The final dope was calculated to have an ABPBT concentration of 17 wt% and a P_2O_5 content of 83.0%. The intrinsic viscosity of fibers that were extracted with water and dried under reduced pressure at 160-165°C was 4.2 dL/g.

PBT Terminated with Benzoic Acid. A PPA solution, having a P_2O_5 content of 78.0%, was prepared from 247.17 g of 115% H_3PO_4 and 88.27 g of 85.3% H_3PO_4. DABDT (13.90951 g, 56.70 mmol) was added to a 100 ml resin kettle equipped with an overhead stirrer and then 47.97 g of the PPA was added. The mixture was heated at 50-70°C for 48 h under reduced pressure to effect elimination of the hydrogen chloride. To the clear yellow solution was added 9.40043 g (56.585 mmol) of terephthalic acid, 0.03471 g (0.2842 mmol) of benzoic acid, and 33.51 g of P_2O_5. The mixture was stirred at 90-95°C under inert atmosphere for 16 h. The yellow mixture was then heated to 180°C over a 1.5 h period and then heated at 180-185°C for 24 h. The resulting yellow-green opalescent PBT/PPA solution was calculated to be 15.0 wt% PBT and have a P_2O_5 content of 82.9%. Fibers were drawn from the solution, coagulated in water, washed with water in a Soxhlet extractor for 24 h, and dried at 160-165°C under reduced pressure for 8 h. The intrinsic viscosity in MSA was measured to be 25 dL/g. The shear viscosity of the PBT/PPA solution was 63,000 poise (0.1 sec^{-1}, 175°C).

REFERENCES

1. A. F. Garito, Nonlinear Optical Materials, Proc. Mat. Res. Soc., ed. D. A. B. Miller (Boston, MA, Nov. 1985); C. C. Teng and A. F. Garito, Nonlinear Optics, SPIE Proc., ed., P. Yeh (SPIE publ., San Diego, CA, Jan. 1986).
2. D. N. Rao, J. Swiatkiewicz, P. Chopra, S. K. Ghoshal, and P. N. Prasad, Appl. Phys. Lett., 48 (18), 1187 (1986).

3. J. F. Wolfe, P. D. Sybert, and J. R. Sybert, US Patent 4,533,693 (August 6, 1985) (to SRI International).

4. J. F. Wolfe, SPIE Proceedings, **682**, 70-76 (1986).

5. D. R. Ulrich, Polymer, **28**, 533-542 (April 1987).

6. L. R. Dalton, J. Thomson, and H. S. Nalwa, Polymer, **28**, 543-552 (1987).

7. E. C. Chenevey and T. E. Helminiak, US Patent 4,606,875 (1986) (to Celanese).

8. G. C. Berry, P. C. Metzger, S. Venkatraman, and D. B. Cotts, Am. Chem. Soc., Civ. Polym. Chem., **20**(1) 42 (1979).

9. P. J. Flory, "Principles of Polymer Chemistry," Cornell University Press, (1953) p. 92-93.

10. A. W. Chow, P. E. Penwell, S. P. Bitler, and J. F. Wolfe, Am. Chem. Soc., Div. Polym. Chem., **28**(1), 50-51 (1987).

11. Our work involving these encapping agents was not presented at the conference and is presented here to provide specific methodologies used to prepare functionalized poly(benzobisazole) materials.

12. J. F. Wolfe, B. H. Loo, and F. E. Arnold, Macromolecules, **4**, 915 (1981).

NONLINEAR AND ELECTRO-OPTIC ORGANIC DEVICES

R. Lytel, G. F. Lipscomb, J. Thackara, J. Altman,
P. Elizondo, M. Stiller, and B. Sullivan

Lockheed Research and Development Division
Lockheed Missiles and Space Company, Inc.
D-9720, B-202, 3251 Hanover St., Palo Alto, CA 94304

I. INTRODUCTION

Organic and polymeric materials have emerged in recent years as promising candidates for advanced device and system applications. This interest has arisen from the promise of extraordinary optical, structural, and mechanical properties of certain organic materials, and from the fundamental success of molecular design performed to create new kinds of materials[1]. From an optical standpoint, organics offer temporal responses ranging over fifteen orders of magnitude, including large nonresonant electronic nonlinearities (fsec-psec), thermal and motional nonlinearities (nsec-msec), configurational and orientational nonlinearities (µsec-sec), and photochemical nonlinearities (psec-sec). Additionally, organic and polymeric materials can exhibit high optical damage thresholds, broad transparency ranges, and can be polished or formed to high-optical quality surfaces. Structurally, materials can be made as thin or thick films, bulk crystals, or liquid and solid solutions, and can be formed into layered film structures, with molecular engineering providing different optical properties from layer to layer. Mechanically, the materials can be strong and resistant to radiation, shock, and heat. When coupled with low refractive indices and D.C. dielectric constants, the collective properties of these extraordinary materials show great promise towards improving the performance of existing electro-optic and nonlinear optical devices, as well as allowing new kinds of device architectures to be envisioned.

However, as with any new class of materials, the existance of certain promising samples does not imply that real applications will necessarily be possible. Real optical materials must exhibit some basic properties, including optical clarity (very-low scattering and absorption losses), fabricability, and the potential for mass production. These secondary properties are generally not addressed by fundamental research, but it is the secondary properties which will determine whether the materials can have any practical use. For organics as a class, including pi-electron systems, conducting polymers, and other nonlinear optical polymers, a great deal of research and development remains to achieve usable, exciting new materials.

This paper provides a survey of the current research and development underway at Lockheed in organic and polymeric devices. In particular, we examine organics as a new class of nonlinear and electro-optic materials, delineate their good and bad features, and present a discussion of certain existing organic materials, their use in devices, and their impact on device performance. Along the way, we try to provide the motivation for using an organic material rather than some other material, and point toward the specific new fabrication requirements for devices, and toward the further materials development required for optimum device performance. Our approach is a positive one, as we are firm believers that organics will have a major impact on the optoelectronic device technology of the next decade. However, we moderate our optimism with the realization that much materials research and development remains to be done before organics can even be established as an important new class of practical nonlinear optical materials.

Section II reviews some of the important structural and optical properties of organic materials for nonlinear optical devices, and points toward specific properties that have been demonstrated, and those that remain to be proved. Section III details the application of certain organic and polymeric materials to practical optical devices, including photo-addressed spatial light modulators and integrated optical devices. We review our own work in these areas, and point toward specific performance and manufacturing advantages of organics over inogranic materials. Section IV summarizes our conclusions regarding organic and polymeric materials for device applications, and provides some directions for future research in this exciting new area of nonlinear optical devices and applications.

Our research effort is a cooperative effort with the Hoechst-Celanese Research Company, and we wish to acknowledge their excellent cooperation, support, and teamwork. In particular, we thank Dr. James Stamatoff and his group for providing samples, working with us to develop new device fabrication techniques, and for his equal enthusiasm for the research underway in this new field.

II. REVIEW OF MATERIAL PROPERTIES

For the purposes of this paper, we shall refer by organics to nonresonant pi-electron organic and polymeric materials, including single crystals, thin-films, and other composite materials. As a class, organics offer a number of exciting optical structural properties for devices, summarized in table 1. Some of these properties, such as the capability to alter linear and nonlinear optical properties in dimensions approaching visible optical wavelengths, are the result of the ability to molecularly engineer the active NLO units within a given structure[2]. This feature appears to be unique to organics, although molecularly tailored semiconductor multiple quantum wells offer some tunability in wavelength and response. However, the capability to engineer materials with desired linear and nonlinear properties is unique to organics, and is the major reason they are, as a class, important for device applications.

Table 1. Some Structural and Optical Properties of Organics

STRUCTURAL	OPTICAL
* MOLECULAR ENGINEERING	* LARGE NONRESONANT NONLINEARITIES
* THIN FILMS/BULK XTAL	* LOW DC DIELECTRIC CONST.
* ROOM-TEMPERATURE	* FAST NLO RESPONSE
* CHEMICAL/STRUCTURAL STABILITY	* HIGH OPTICAL DAMAGE THRESHOLDS
* INTERNAL GRATINGS	* BROADBAND
* INTEGRATED OPTICS	* LOW ABSORPTION

Many of the features in table 1 have been demonstrated in a number of different organic systems. Damage thresholds as high as a GW/cm^2 have been measured in single crystal MNA[3] (2-methyl-4-nitroaniline) and in Urea[4]. Large nonresonant susceptibilities have been reported in certain diacetylene polymer systems[5] ($\chi^{(3)}$ of order 10^{-10}-10^{-9} esu) and single crystal MNA[6] ($\chi^{(2)}$=500+/-100 pm/V). Other recent measurements of large electro-optic coefficients have been reported in poled organic films[7], and show great promise for optical waveguides. It is noteworthy that current synthesis efforts underway are moving away from crystal growth and toward thin-film fabrication of nonlinear polymers, and that none of these systems has a $\chi^{(3)}$ or $\chi^{(2)}$ as large as the original reported susceptibilities of the diacetylenes and MNA crystals. For all of the problems of growing crystals, they yield the greatest degree of orientation, and, therefore, the largest macroscopic nonlinearities for a given NLO moiety.

It is also true that most organics fabricated to date do not yet exhibit most or all of the properties in table 1. In particular, absorption and scattering losses are usually higher than desired, and must be minimized by building materials in clean-room facilities or by filtering polymer solutions before using them. Many polymer films exhibit poor optical quality, and techniques for fabrication of optically flat surfaces need to be refined. Some polymer structures are inherently grainy or contain fibrous components which scatter light excessively. Finally, a few researchers are reporting new materials, such as conducting polymers, with large $\chi^{(3)}$ values, without recognizing the requirement that the material must inherently be able to transmit or guide light! These reports indicate a lack of understanding of what constitues an optical material, as opposed to an electrical one. Nonetheless, we remain confident that the field of organics will develop into an important new area of device applications, once the materials and device scientists jointly address the problem.

In this light, it is straightforward to examine the current status of $\chi^{(2)}$ and $\chi^{(3)}$ materials[8], and determine what materials parameters ought to be improved for real applications.

A. SHG and Electro-optic Materials

The largest reported SHG[9] and E-O[6] coefficients in an organic material exist in the organic solid MNA, in single crystal form. This result is interesting for several reasons. First, while the NLO moiety, the MNA molecule, has a large second-order molecular hyperpolarizability β, many other molecules with larger β have been reported. However, single crystals of MNA have been grown and characterized, illustrating the effect of macroscopic orientation on the macroscopic polarizability $\chi^{(2)}$. Thus, we learn that it is critical that proper orientation be achieved to take advantage of the large β in certain molecular systems. This just illustrates the difference between a molecule and a material. Recent results in poled polymer films suggest that we are well on our way toward achieving large, usable $\chi^{(2)}$s from partially oriented films of different polymer systems. Both AT&T and Celanese[7] have reported new polymers at this meeting exhibiting interesting $\chi^{(2)}$s, and it is likely that materials currently unreported upon exist and have even greater $\chi^{(2)}$s. In our view, it is only a matter of time before poled polymer films will appear regularly in real optical devices. As reported below, we have already used certain poled films to produce "proof of concept" prototypes of optical waveguide modulators with great success.

A major drawback to the current poled polymer films is that the $\chi^{(2)}$ is perpendicular to the film surface. For waveguides, this is no problem. However, for bulk E-O device applications, it is desirable to have a film with $\chi^{(2)}$ parallel to the film surface to achieve E-O modulation for two-dimensional device applications. It remains to be seen whether new orientation techniquies, equivalent to surface alignment techniques for nematic liquid crystals, can be developed.

There is one major observation worthy of note when comparing organic pi-electron materials with inorganic crystals in terms of their $\chi^{(2)}$. In both materials, the contributions from $\chi^{(2)}$ to SHG arise solely from the electronic contributions, since these are the only ones capable of responding on femtosecond time scales. In organics, the same contributions also contribute to the entire E-O coefficient, while inorganics gain most of their contribution from lattice phonons. Thus, an organic material with the same $\chi^{(2)}$ as an inorganic material, as measured by SHG, will probably have a much smaller E-O coefficient than the inorganic material. This is illustrated in table 2, where it is clear that for KDP and LiNbO$_3$, most of the contribution to the electro-optic coefficient r arises from phonons. This result must be borne in mind when comparing organics, as characterized by SHG, with E-O inorganic materials.

B. Third-Order Materials

The largest nonresonant $\chi^{(3)}$ yet reported[5] for an organic material is in the diacetylene systems, and has a $\chi^{(3)}$ of order 10^{-9} esu. This value is large by comparison with CS$_2$, but is still too small for most applications. In devices based on $\chi^{(3)}$, it is generally the product of $\chi^{(3)}$ with the light-medium interaction length and the optical intensity that determines the net nonlinear phase shift that can be obtained and exploited for device operation. Small $\chi^{(3)}$ necessarily implies large optical intensities and/or long interaction lengths. For nonresonant organics like PTS diacetylene polymers, intensities of order MW/cm^2 would be required over a length L=1 cm. Thus, it is likely that third-order organics will find applications in optical waveguide devices, such as bistable optical switches and optically-controlled modulators and couplers. It is unlikely that thin-film applications, such as etalons, will be likely with organics unless significant advances toward achieving larger nonlinearities can be made.

Organic thin films can, however, offer some really new features for third-order devices, if the coefficients can be made larger. Through molecular engineering, it should be possible to produce anisotropic $\chi^{(3)}$ materials for optically-activated birefringent film applications. It should even be possible to build layered structures with a $\chi^{(3)}$ that varies along the thickness of the layers, thus producing nonlinearly-activated optical gratings for fixed and tunable filter applications.

Recent synthesis work reported at this meeting[10] includes off-resonance side and main chain polymers with reasonable nonlinearities (10^{-11}-10^{-10} esu). Such materials could be useful in degenerate four-wave mixing[11] for optical phase conjugation, or as self-focusing media for optical shutters. However, it is still necessary to achieve optically good materials (clarity and optical flatness) and larger susceptibilities for most applications.

In summary, second-order poled polymer films appear to be close to achieving interesting levels of response for waveguide electro-optic device applications. Third-order NLO polymers still have a ways to go, however. In both cases, the materials must necessarliy qualify as optical materials before real applications can be envisioned. In light of the recent success in producing poled polymer films, we discuss next two important device application areas for these materials.

Table 2. Values of $\chi^{(2)}$ and r (as determined by SHG) and r (as determined by the linear electro-optic effect) illustrating the large phonon contribution to r in inorganic materials.

Material	ij	n	ε	r_{ij} $(-\omega;\omega,0)$ x 10^{-12} m/V Measured From LEO	d_{ij} $(-2\omega;\omega,\omega)$ x 10^{-12} m/V Measured From SHG	$\chi^{(2)}_{ij}(-2\omega;\omega,\omega)$ x 10^{-12} m/V Calculated From d	r_{ij} $(-\omega;\omega,0)$ x 10^{-12} m/V Calculated From d	Percent Electronic Contribution to Electro-Optic Coeff. r
KDP	63	1.47	42	10.6	0.45	0.9	0.02	0.2%
LiNbO$_3$	22	2.3	78	3.4	3.0	6.0	0.03	0.9%
	33	2.3	32	32.	40.	80.	0.9	3.0%
MNA	11	2.0	4	~67.	250.	~500.	60.	~100.%
PC6S (Preliminary)		1.7	3.3	~2.8	?			

III. DEVICE APPLICATIONS

In this section, we report recent work toward achieving useful devices based on second-order, poled polymer films. The poling process produces a material with a $\chi^{(2)}$ perpendicular to the film surface. Our major experimental results to date have been obtained in E-O waveguide devices using a polymer known to us as PC6S, and provided by the Hoechst Celanese Research Company. This polymer is optically transparent above 0.6 μm, and has a measured value of $r_{33} = 2.8$ pm/V (table 2). We report next on our investigations into specific devices for exploiting poled polymer films.

A. SPATIAL LIGHT MODULATORS

Spatial light modulators (SLMs) are two-dimensional optical modulation devices[12]. Such devices are of great interest in optical processing, computing, and beam control, and are usually photoaddressed devices. A typical electro-optic SLM, operating in reflection, is illustrated in figure 1. This device is an electro-optic modulator, composed of a photoreceptor, dielectric mirror, and an electro-optic material. The SLM operates as follows: An incident optical field addresses the photoreceptor, creating a two-dimensional charge distribution which is proportional to the intensity of incident light. Under the influence of a bias voltage, the charge distribution $\sigma(x,y)$ migrates to the dielectric mirror-photoreceptor interface, and modulation of the E-O material due to the charge distribution can occur. If the modulation is due to the difference between σ (and the bias voltage) and the ground plane, the device is a longitudinal modulator. If the modulation is due to local differences in σ at the surface of the E-O material, the device is a transverse modulator. In the longitudinal modulator, a readout field incident from the right acquires a phase shift proportional to the induced E-O modulation due to $\sigma(x,y)$, and the device operates as an intensity to phase converter. In the transverse device, it is the local field gradient along the dielectric mirror and E-O material interface which creates the modulation field, and the device produces a phase shift upon readout proportional to the gradient of the intensity. Thus, the fundamental operation of the E-O modulator depends on the tensorial nature of the E-O medium in a fundamental way.

Electro-optic SLMs can be seen to be important optical devices because they allow the transfer of optical information from one beam to another. In particular, a longitudinal modulator can be used as an optical correlator, convolutor, or optical phase conjugator, while a transverse modulator can be used as an edge detector or intensity-to-position encoder. In either case, SLM performance parameters are determined primarily by the optical properties of the photoreceptor and E-O material, and the electrical properties of the entire unit.

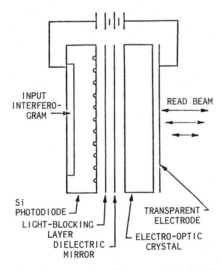

Fig. 1. Schematic Diagram of an Electro-optic Spatial Light Modulator

Photoaddressed E-O SLM technology has been under development at Lockheed for the past several years. The Lockheed photoreceptor is a high-resistivity silicon device, with very fast response times (of order a nsec) and good lateral charge confinement. Lockheed has mated this technology to thin (100 μm) KD*P crystals to produce fast E-O modulation devices with spatial resolution approaching 10 line pairs/mm, frame grabbing times under 10 μsec, and frame rates approaching a kHz[13]. This architecture has formed the basis for all of our subsequent SLM prototypes, including a longitudinal nematic liquid crystal device[14], a transverse nematic liquid crystal edge detection device[15], and the current organic transverse SLM under development.

The impact on using an organic material in the Lockheed SLM architecture is currently under experimental evaluation, and no results are available at the time of this paper. However, we have evaluated theoretical performance for an organic SLM, and it is worthwhile to explore this briefly to understand the importance of organics in SLM technology.

In the Lockheed device, response time for a single frame is determined by the integration time of the input light required to generate sufficient charge to get a desired modulation (say half-wave) on the E-O material, the RC time constant of the device, and the intrinsic response time of the E-O material. The response of the KD*P is sub-nanosecond, so the device frame time is determined solely by the write light intensity and the device capaticance. The KD*P is operated near its Curie temperature of -57C, so that the enhancement in the E-O coefficient can be obtained, allowing the bias voltage to be kept below the breakdown voltage of the photoreceptor. The dielectric constant of KD*P, however, also rises dramatically near the Curie temperature and no benefit in device efficiency is achieved. Since the polarization optic coefficient is not strongly temperature dependent, the same amount of photo-generated charge (σ) is required to produce an equivalent phase modulation, and the required integration time is not reduced. Large dielectric constants are an inescapable feature of inorganic ferroelectric E-O materials, arising from the electron-phonon coupling origin of the electro-optic effect. The resultant large device capacitance adversely effects the speed and sensitivity that can be achieved in a photoaddressed device. Finally, device resolution is limited, by capacitative effects, to about 10 lp/mm. Organic thin films exhibiting a large E-O coefficient and a small dielectric constant should offer significant improvements.

A nematic liquid crystal version of the same device has been developed at Lockheed, in the hopes of achieving much more sensitiviy and greater spatial resolution. This device, operated at room temperature, exploits the birefringence induced by an applied electric field. The nature of the device depends on the surface alignment of the liquid crystal. Parallel alignment produces a longitudinal modulator, while perpendicular alignment produces a transverse modulator. In both cases, a significant gain in device sensitivity is achieved, due to the much lower dielectric constant of the nematic liquid crystal relative to cooled KD*P and the larger birefringence of the nematic liquid crystal. Higher resolution is achievable because thinner films can be utilized. However, the nematic alignment must be driven off to achieve msec response times, and this device is never as fast as the KD*P device.

Organic E-O films can probably achieve higher resolution, higher speed and greater sensitivity in an E-O SLM than either inorganic E-O crystals or liquid crystals. A poled organic film of order 10 μm thick, with an E-O coefficient comparable to KD*P, and a dielectric constant ε=4, should produce a longitudinal SLM with sensitivity comparable to a nematic liquid crystal device, speed of a KD*P device, and resolution much better than both KD*P and nematic liquid crystals. This increased performance results from the lower film capacitance at room temperature, for a given E-O coefficient. Such films, available now as poled polymer films oriented perpendicular to the surface of the film, cannot be used in normal incidence and must be read out at an angle. We expect to have the first prototype SLM of this type operating my mid-1987.

A major drawback to the use of organics in E-O devices will arise from the poor thermal conductivity of polymers. Thus, repetition rates and optical loads will be limited by the device architecture, and careful design of heat-sinks and thermal conditions will be necessary. If properly designed, the ultimate optical load of the device should be much greater than inorganic devices, due to the potentially larger optical damage threshold of organic materials.

B. INTEGRATED OPTICAL DEVICES

Organic electro-optic materials offer a variety of potential advantages over conventional materials for integrated optical device applications. Table 3 provides a comparison of the potential of organic materials with the current frontrunner technology, Ti-indiffused $LiNbO_3$ in three major areas of importance: materials parameters, processing technology, and fabrication technology. Some of these advantages have already been realized in our work on E-O modulators using poled polymer films, described below.

Table 3. Comparison of Integrated Optics Technologies: Current Ti-LiNbO$_3$ and Projected Organics Technologies

CURRENT TECHNOLOGY, Ti:LiNbO$_3$	POTENTIAL ADVANTAGES OF ORGANIC E-O MATERIALS
• $r = 32 \times 10^{-12}$ m/V	• $r = 67 \pm 25 \times 10^{-12}$ m/V (MNA)
– LARGER MODULATING VOLTAGE	– LOWER MODULATING VOLTAGE
– LITTLE IMPROVEMENT EXPECTED	– POTENTIALLY MUCH LARGER r
• LIMITED FABRICABILITY	• FLEXIBLE FABRICATION
– 1000°C PROCESSING	– LOW TEMPERATURE PROCESSING
– DEPTH LIMITED TO ~5 μm	– FLEXIBLE DIMENSIONS
– LOW Δn	– CONTROLLABLE Δn
– LOSS ~ 0.1 dB/cm	– POTENTIAL LOW LOSS
– OPTICAL DAMAGE (PHOTOREFRACTOR)	– HIGH OPTICAL DAMAGE
• LARGE DIELECTRIC CONSTANT ($\epsilon \sim 28$)	• LOW DIELECTRIC CONSTANT ($\epsilon \sim 4$)
– LONGER TIME CONSTANTS τ = RC	– SHORTER TIME CONSTANTS τ = RC
– LARGE VELOCITY MISMATCH IN TRAVELING WAVE MODULATOR	– SMALLER VELOCITY MISMATCH
• MASS PRODUCTION DIFFICULT	• POTENTIAL FOR MASS PRODUCTION

The most obvious potential advantages are due to the intrinsic differences in E-O mechanisms in inorganic and organic materials. Organics should provide flat E-O response well beyond a GHz, and, indeed, measurements of the E-O coefficients and SHG coefficients of cetain poled polymer films show little or no dispersion. Second, it is likely that the E-O coefficients of poled polymer films can be made nearly as large as LiNbO$_3$, and it is obvious that many materials efforts underway in the U.S. and abroad are attempting to achieve exactly that. Pure MNA crystals already exhibit larger E-O coefficients than LiNbO$_3$, as indicated in table 3, but it is our sense that real device advances will be made with films, not crystals. It is also true that the dielectric constant of poled polymer films is substantially lower than that of LiNbO$_3$, implying smaller RC time constants and wider frequency bandwidths. Finally, the processing technology for integrated optical devices based on poled polymer films is relatively straightforward and fast, requiring only moderate temperatures (100C) for poling, and standard semiconductor fabrication equipment for fabrication of layered waveguide structures.

As with all other materials, the secondary issues will be the drivers for organic integrated optical devices. Low absorption and scattering losses need to be designed into the material and fabrication technology, and good quality films must be regularly producable. However, it seems likely that these achievements can be met with further research and development. It is worth noting that the current inorganic crystal growth technology is a very old technology, representing thousands of man-years of research. Much work remains to be done with organics!

The Lockheed organic integrated optical devices effort has recently focused on the fabrication of simple slab guided wave structures made from polymer films. These structures allow the determination of fabrication techniques for organic devices, as well as direct measurement of Kerr and E-O coefficients in the waveguide configuration. By way of example, we discuss next two guided wave structures made from organics: a Kerr effect modulator based on an unoriented MNA film and an electro-optic modulator based on the PC6S poled polymer film described earlier.

A. Kerr-Effect Modulator

The first organic guided wave device fabricated at Lockheed was a Kerr effect modulator based on an MNA/PMMA guest-host film[16]. This structure, illustrated in figure 2, is a slab modulator consisting of a thin film of SiO$_2$ as a lower buffer layer, the MNA/PMMA film as the guide, and another thin buffer layer film composed of polysiloxane. The entire structure is built on an aluminum-coated glass substrate with an aluminum electrode at the top. Layer dimensions and composition are illustrated in figure 3. An HeNe laser was prism-coupled into the device, which guided both a TE and TM mode. Intermode interference was observed on output, and the frequency response of the device is illustrated in figure 4. The low frequency response is exactly what one would expect from a Kerr effect in this material. The output was measured with a lock-in amplifier, set to lock to twice the modulation frequency of the applied voltage, as appropriate for a Kerr effect. A linear effect was also sought, but was not measurable, as expected for an unpoled film.

Although this device is not useful for applications, its fabrication helped define procedures for the fabrication of E-O slab waveguide'devices, described next.

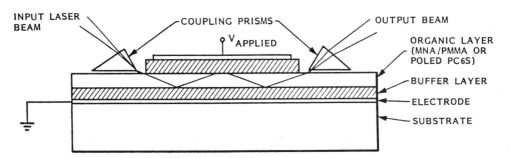

Fig. 2. Schematic of Kerr-Effect Waveguide Modulator

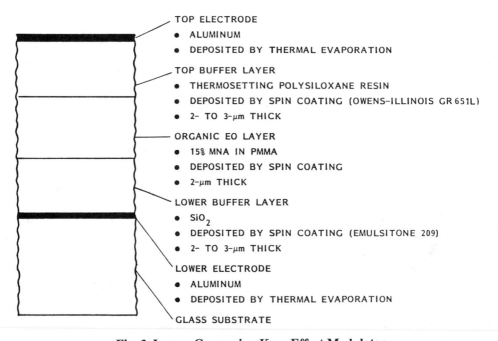

Fig. 3. Layers Composing Kerr-Effect Modulator

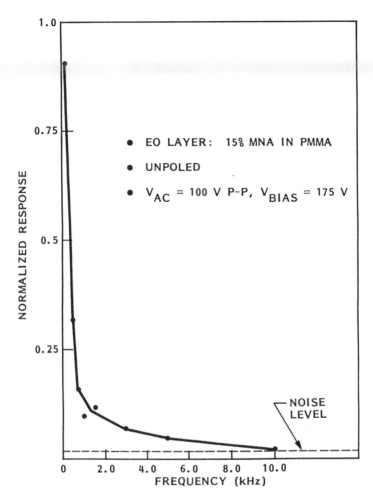

Fig. 4 Measured Frequency Response of Kerr-Effect Modulator

B. Electro-optic Modulator

The next slab modulator constructed at Lockheed consisted of a layered structure similar to that in figure 3, with the PC6S polymer film replacing the MNA/PMMA film[16]. The poling procedure consisted of first spin-coating the electrode-coated glass with the bottom buffer layer and then the PC6S, applying a top electrode, and applying a voltage of order 1 MV/cm to the structure. Poling was observed with a polarizing microscope, and the entire procedure was monitored in real time to optimize the poling. The top electrode was then removed, and a top buffer layer and new electrode were applied to the poled structure. Guiding of 830 nm light from a semiconductor laser over a 1.5 cm dimension with minimal loss was then achieved by locating the prism couplers directly over the ends of the poled region.

The optical response was measured interferometrically, and is illustrated in figure 5. The device half-wave voltage was about 23 volts. Figure 6 illustrates the measured frequency response out to about 100 kHz, and shows it to be flat. The electro-optic coefficient of the poled PC6S film was measured to be 2.8 pm/V, as reported in table 2. Higher frequency measurements are currently underway with a new optical test system, and will be reported later in 1987.

The fabrication of such poled slab waveguide devices from a spin-coated polymer films represent the first major steps toward the development of organic integrated optical devices. Many issues remain to be studied, and include measurement and minimization of scattering losses, measurement of higher frequency response, development of procedures to form two-dimensional guides, and development of simple optoelectronic devices. However, the implications of this work are significant and promising. Simple fabrication techniques, requiring only standard spin-coating tools and chemical etching, can produce organic electro-optic devices from unoriented polymer films.

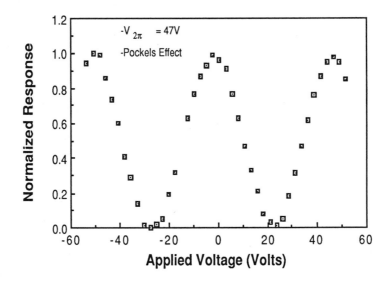

Fig. 5. Interferometric Measurement of E-O Modulator Response

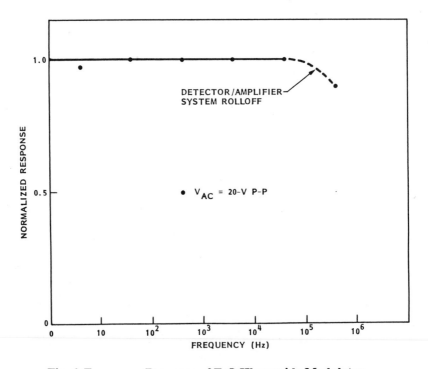

Fig. 6. Frequency Response of E-O Waveguide Modulator

424

IV. CONCLUSIONS

We have described certain features of organic and polymeric materials and their applications to nonlinear and electro-optic devices. As a class, the materials offer an important number of structural and optical properties for device applications, including flexible and straightforward fabricatibility, large E-O coefficients, low D.C. dielectric constants, and high optical damage thresholds. For second-order materials, current advances indicate that materials with performance approaching that of $LiNbO_3$ should soon be available, and a major impact on E-O device technology will be forthcoming. For third-order materials, it is likely that guided wave devices, with their long interaction lengths, will be the most practical for organics. At this time, it is still true that MNA (second-order) and diacetylene polymers (third-order) have the largest optical nonlinearities, as it was when original results were reported. However, it is clear that the fundamental material science is producing a working knowledge of how to engineer materials with larger nonlinearities, and it is only a matter of time before organics emerge as a promising class of practical nonlinear optical materials.

It is our view that future research in $\chi^{(3)}$ materials be carefully examined to ensure that claims of large nonlinearities in optically poor media do not spoil the real progress being made in the development of new materials. Nonlinearities of order 10^{-9} esu are not large at all. Instead, they are large for a nonresonant material, but still imply devices with large operating intensities. Nonresonant nonlinearities offer broadband response, but operating intensities must be reasonable, and should be obtainable from diode laser sources, as the latter can already be used with the large resonant response of III-V semiconductor multiple quantum well structues (10^{-1} esu). Care must be taken not to overpromote the $\chi^{(3)}$ of a new material without first carefully comparing its material parameters to other materials, and then comparing its performance in a device. As we have seen, device performance depends on materials parameters in a very complicated way, and is often determined by the device architecture or electrical properties, rather than the material.

REFERENCES

1. For a recent review, see "Nonlinear Optical Properties of Organic Molecules and Crystals", Vol. 1 and 2, D. Chemla and J. Zyss, ed. (Academic Press, MO) 1986. See also "Origin of the Nonlinear Second-order Optical Susceptibilities of Organic Systems", S.J. Lalama and A.F. Garito, Phys. Rev. A 20, 1179 (1979); and "Nonlinear Optical Properties of Organic and Polymeric Materials", D.J. Williams ed., ACS Symposium Series 233 (American Chemical Society), 1983.

2. "Nonlinear Optics: Organic and Polymeric Systems", A.F. Garito and other papers in this volume.

3. A.F. Garito, private communication.

4. "Nonlinear Optical Properties of Urea", C. Cassidy, J.M. Halbout, W. Donaldson, and C.L. Tang, Opt. Comm. 29, 243 (1979).

5. "Optical Nonlinearities in One-Dimensional Conjugated Polymer Crystals", C. Sauteret, J.-P. Hermann, R. Frey, F. Pradere, and J. Ducuing, Phys. Rev. Lett. 36, 956 (1976); "Intensity-dependent Index of Refraction in Multilayers of Polydiacetylene", G.M. Carter, Y.J. Chen, and S.K. Tripathy, Appl. Phys. Lett. 43, 891 (1983).

6. "An Exceptionally Large Linear Electro-optic Effect in the Organic Solid MNA", G.F. Lipscomb, A.F. Garito, and R.S. Narang, J. Chem. Phys. 75, 1509 (1981).

7. "Electric-Field Poling of Nonlinear Optical Polymers", C.S. Willand, S.E. Feth, M. Scozzafava, D.J. Williams, G.D. Green, J.I. Weinshenk, H.K. Hall, and J.E. Mulvaney; "Development of Polymeric Nonlinear Optical Materials", J.B. Stamatoff, A. Buckley, G. Calundann, E.W. Choe, R. DeMartino, G. Khanarian, T. Leslie, G. Nelson, D. Stuetz, C.C. Teng, and H.N. Yoon; "Orientationally Ordered Electro-optic Materials", K.D. Singer, J.E. Sohn, and M.G. Kuzyk; this volume.

8. It is not as straightforward as we imply to evaluate the status of nonresonant organic materials. We feel certain that the latest industrial developments are unreported, especially for second-order materials, because they should have significant value to the developers.

9. "An Organic Crystal with an Exceptionally Large Optical Second-Harmonic Coefficient: 2-methyl-4-nitroaniline", B.F. Levine, C.G. Bethea, C.D. Thurmond, R.T. Lynch, and J.L. Bernstein, J. Appl. Phys. 50, 2523 (1979).

10. "Side-Chain Liquid Crystalline Copolymers for NLO Response", A.C. Griffin, this volume.

11. "Optical Nonlinearities in Organic Materials: Fundamentals and Device Applications", G.F. Lipscomb, J Thackara, R. Lytel, J. Altman, P Elizondo, E. Okasaki, M. Stiller, and B. Sullivan, Proc. SPIE Vol. 682, 125 (1986).

12. For a comprehensive review, see Optical Engineering, Vol. 25, No. 2 (1986).

13. "High-speed Spatial Light Modulator", D. Armitage, W.W. Anderson, and T.J. Karr, IEEE J. Quant. Elec. QE-21, 1241 (1985).

14. "Fast Nematic Liquid-Crystal Spatial Light Modulator", D. Armitage, J.I. Thackara, W. Eades, M. Stiller, and W.W. Anderson, submitted to SPIE San Diego, August, 1987.

15. "Liquid-crystal Diffrentiating Spatial Light Modulator", D. Armitage and J.I. Thackara, Proc. SPIE 613, 165 (1986); "Ferroelectric Liquid-crystal and Fast Nematic Spatial Light Modulators", D. Armitage, J.I. Thackara, N.A. Clark, and M.A. Handschy, Proc. SPIE 684, 60 (1986).

16. "Optoelectronic Waveguide Devices in Thin-Film Organic Media", J.I. Thackara, G.F. Lipscomb, R. Lytel, M. Stiller, E. Okasaki, R. DeMartino, and H. Yoon, CLEO-87, paper ThK29 (1987).

NON-LINEAR OPTICAL PROCESSES IN OPTICAL FIBRES

B.K. Nayar, K.I. White, G. Holdcroft and *J.N. Sherwood

British Telecom Research Laboratories
Martlesham Heath, Suffolk IP5 7RE, UK
*Department of Pure and Applied Chemistry
Strathclyde University, Glasgow G1 1XL, UK

INTRODUCTION

Optical fibers are now the transmission medium of choice for many applications and there is an increasing interest in optical signal processing using electric, optical, magnetic or acoustic fields. In particular 'all-optical' signal processing, the optical control of optical signals, is a preferred approach as it circumvents the need of transducers for conversion of signals between the optical and electronic domains. This approach results in non-blocking transparent networks which are independent of the modulation format and the data bit rate. Optical devices capable of modulation, amplification, switching and generation of tunable radiation in the near UV to near IR have applications in telecommunications, sensors, reprographics, spectroscopy, optical logic and optical data storage.

These parametric interactions occur in non-centrosymmetric materials by virtue of the material's second order susceptibility. Organic materials[1] are particularly attractive as unlike the presently used inorganics they have large optical non-linearities (100×10^{-12}m/V) and high optical damage thresholds (>100MW/cm^2). Semiconductor laser diode pumped devices can be realised using waveguiding structures as they permit high optical intensities with modest powers, long interaction lengths and phase-matching using modal dispersion. A particularly attractive wave-guiding structure for organic materials is the crystal cored fiber (CCF). CCFs are cylinderical waveguiding structures, like optical fibers, with a single crystal organic core and glass cladding. They have the advantages of providing physical protection to the crystal core, compatibility with optical fibers and uniform core dimensions for sustained phase-matching. In this paper we describe fabrication of CCFs for optical second harmonic generation (SHG) and parametric amplification, measurement of linear optical properties and SHG, and discuss areas requiring further investigation to achieve the predicted performance.

CRYSTAL CORED FIBERS

The growth of CCFs was pioneered by Stevenson and Dyott[2]. This work has been followed up by a number of researchers to study crystal growth mechanism in confined structures[3-7] and to investigate the fabrication of non-linear devices[8-13].

The CCFs are fabricated by growth of a single orientated crystal from the melt in glass capillaries using a modified Bridgman method. For a given material a cladding glass with a lower refractive index is selected from a wide range of commercially available glasses[14-15]. The cladding glass is machined to produce a cylinder with a hole through the middle which can then be used to draw glass capillaries using conventional optical fiber drawing equipment. The capillaries are filled with the crystal melt by the capillary action in the 'hot' zone of a furnace maintained at a temperature few degrees higher than the crystal melting point. The single crystal growth is then achieved by slowly moving (typically 5-20mm/hr) the capillaries, filled with the melt, through a sharp temperature gradient, Fig 1. For single crystal growth it is necessary to optimise both the crystal growth rate and the temperature gradient. Also, during the melt filling process it is necessary to ensure that there are no bubbles present as these can 'freeze-in' on crystal growth as voids. Single crystals of acetamide, benzil[5,8-9], 2-bromo-4-nitroaniline[4], 4-dimethylamino-3-acetamidonitrobenzene[10] (DAN), formyl-nitrophenylhydrazine[4], m-dinitrobenzene[4-7] (mDNB), m-nitroaniline[2-3] (mNA), 2-methyl-4-nitroaniline[11-12] (MNA) and N-(4-nitrophenyl)-(L)-prolinol[13] (NPP) have been grown in glass capillaries of up to 50mm long and with bore diameters in the 2-125µm range.

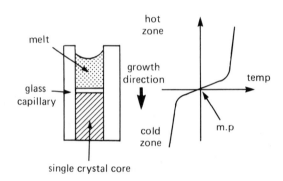

Figure 1: Schematic showing crystal cored fiber fabrication

The crystal orientation in the glass capillary is important for non-linear interactions as it determines both the pertinent coefficient of the second order susceptibility tensor and the required electric field polarisations of the interacting modes. A number of the above organic materials grow in glass capillaries such that the highest non-linear tensor co-efficient is along the crystal growth direction eg mDNB and mNA. Since the electric field distribution of the modes in CCFs is very nearly transverse to the propagation direction the non-linear response is small. This limits the choice of crystal materials.

In this paper we report optical measurements on acetamide, MNA and DAN CCFs in silica capillaries with bore diameters in the 2-25µm range. Acetamide has a modest non-linearity[16] (powder SHG efficiency of 0.35x urea) and was selected primarily for its ease of crystal growth. MNA[17] is particularly attractive as its SH tensor co-efficient, d_{11} (250pm/V), is very large and DAN has been reported[18] to give very high powder SHG efficiency (115x urea cf 22x urea for MNA).

LINEAR OPTICAL PROPERTIES OF MNA CRYSTAL CORED FIBERS

The linear optical properties such as polarisation dependent propagation, number of propagating modes, effective mode refractive indices and insertion loss are important pre-requisites for the design of efficient non-linear devices. Here as an example we consider propagation in MNA CCFs with silica cladding.

MNA crystal orientation in silica glass capillaries was determined using X-ray techniques[12]. The 125μm capillary was etched to within few microns of the crystalline core and then photographed transversely in a powder X-ray camera. The crystallographic b-axis of the MNA is parallel to the fibre direction (same as the growth direction). MNA belongs to the monoclinic system with the space group Cc and point group m. In this case only one of the principal axis of the dielectric tensor is fixed by the symmetry ie it is parallel to the twofold b axis of the crystal. The other two axes lie in the (010) crystallographic plane. Here an arbitrary orthogonal coordinate system is assumed with y along the crystal b axis, Fig 2. The examination of MNA cored fibres between the cross polars in a polarising microscope showed complete extinction indicating single crystal orientation and an optical axis parallel to the fibre axis. This crystal orientation is different from that in reference [11] and is potentially advantageous as it allows access to the large d_{11} coefficient.

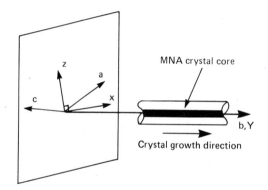

Figure 2: MNA crystal orientation in crystal cored fibers

MNA is a biaxial crystal and to date its refractive indices have not been accurately measured as its bulk crystal growth has been difficult and yielded mostly thin (30-300μm) (010) platelets. The refractive index measurements on these platelets have given n_x=1.8 at 1064nm wavelength and n_z=1.453±0.005 at 589nm wavelength and our measurements are in close agreement. For MNA cored fibres with silica cladding the light polarised along the z-axis cannot be guided as the silica refractive index (1.458 at 589nm wavelength) is higher than the n_z index of MNA. Hence the fibre acts as a polariser allowing the transmission of only the light polarised along the x-axis. The fibres used had core diameters in the 4-10μm range, a large refractive index difference (0.35) and thus were highly multimode at the 1064nm wavelength.

The polarisation dependant transmission characteristics of MNA CCFs were observed using the set-up shown in Fig 3. Light from a low power linearly polarised He-Ne laser at 633nm wavelength was launched, using a microscope objective, into the core of 10-15mm long fibres immersed in refractive index-matching liquid to strip cladding modes. At the output a microscope objective was used to collimate the light and a variable

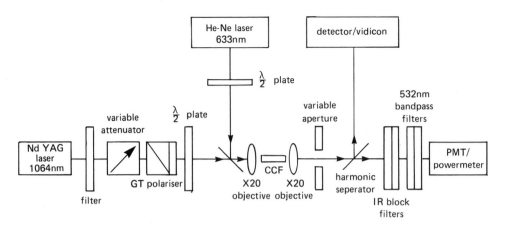

Figure 3: Experimental set up for optical characterization of crystal cored fibers

aperture was used to screen the light from the cladding. This arrangement permits examination of the fibre near field and determines, in a low moded fibre, the number and the type of the modes guided by the fibre. The input polarisation angle was varied by rotating a half-wave plate in front of the laser. In Fig 4 the variation of the core output power as a function of the input polarisation angle is shown. This fibre has an extinction ratio of 12.8dB. Higher values for extinction ratio, up to 25dB, were obtained in fibres in which the light was launched using butt-coupling. In butt-coupling the light from the laser is first launched into a primary fibre which is then butted against the CCF with a thin film of index-matching liquid between the two. Butt-coupling is a more efficient method of launching light as it is possible to match the mode field profiles of the two fibres and it is also the most convenient coupling scheme for systems applications. In this work a polarisation-maintaining primary fibre was used in order to enable the investigation of the polarisation dependant transmission of the MNA CCFs. In Table 1 the summary of the results obtained by using both the end-fibre and butt-coupling schemes is given. Lower extinction ratios obtained with the end-fibre coupling are due to the detector sensitivity limitation as a result of the low core throughputs, $P_{core}(out) / P(in)$, and the difficulty of cladding mode stripping.

The poor core throughputs of MNA CCFs are primarily due to the poor crystal end quality and crystal defects/voids. Input and output fibre ends were made by scribing the cladding with a diamond blade and breaking. Typically, the MNA crystal breaks either inside or outside the cladding with jagged ends. This results in high scatter loss at the input. An explanation for this behaviour could be that the MNA crystal has a plate like growth habit and the easy cleavage plane along the [103] crystal edge[19]. Materials having needle like growth habit and with cleavage plane perpendicular to the fibre axis eg benzil, give good quality ends and low insertion losses. Earlier work on benzil and meta-nitroaniline CCFs resulted in insertion losses of 2dB/cm. The other source of high insertion loss of MNA CCFs, 13-25dB/cm, is scatter due to defects/voids in the crystal core. This was easily verified by using a travelling microscope

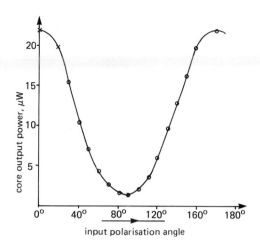

Figure 4: Transmission characteristic of a MNA crystal cored fiber as a function of the input polarisation angle

TABLE 1

Launch	Core throughput	Extinction ratio
end-fire	0.25-1%	10-25dB
butt-coupled	1-5%	12-25dB

Core diameter: 4-8µm
Cladding: Silica
Wavelength: 0.633µm

focussed on the crystal core. At present we are addressing the problems of both the MNA crystal end preparation and the crystal quality. Low insertion losses are pre-requisite for efficient non-linear interactions.

SECOND HARMONIC GENERATION MEASUREMENTS IN CRYSTAL CORED FIBERS

Optical second harmonic generation in MNA, Acetamide and DAN CCFs with silica cladding was studied using the experimental set up illustrated in Fig 3. A Q-switched Nd:YAG laser at 1064nm wavelength with laser pulse width of 200ns and repetition rate of 1KHz was used. A Glan-Taylor polariser is used to improve the input polarisation fidelity and a half-wave plate to rotate the input polarisation. The SH signal was measured after KG3 filters (to block the fundamental) and 532nm band pass filters using a photomultiplier tube or a power meter.

In Fig 5 the normalised SH power, P_ω, from a MNA CCF is plotted as a function of the input polarisation angle. The maximum SH power is

generated with the input polarisation parallel to the crystal x-axis. On
rotation of the input polarisation the launched power along the x-axis
varies as A.sin(x), where A is the input optical power and x is the
change in the input polarisation angle. Since the MNA CCF with silica
cladding behaves as a polariser the light polarised along the z-axis is
not guided by the core and couples out into the cladding. Thus, the SH
power should vary as $[A.sin(x)]^2$ with the rotation of the input
polarisation and hence demonstrating the square law dependance on pump
power. From Fig 5 it can be seen that the experimental results are in
close agreement with the predicted theoretical response. Typical values
of SHG efficiency, $\eta = P_\omega(out) / P_{\omega/2}(in)$, in MNA CCFs having core
diameter in the 6-10μm range, 10-15mm long and with average input power of
10mW (peak power = 50W) were $0.2-4.0 \times 10^{-5}\%$. Core throughput was typically
0.2-0.6%. Our best result was $\eta = 1.6 \times 10^{-4}\%$ for a 12mm long fibre with
19μm core diameter. Recently in the literature[11], an alternative
definition for the SHG efficiency has been used which allows for the
launch and propagation losses ie $\eta = P_\omega(out) / P_{\omega/2}(out)$. Using this
definition our best SHG efficiency value for MNA CCF becomes $1.5 \times 10^{-3}\%$.
However, we prefer the generally accepted definition of the SHG efficiency
as it makes comparisons of SHG efficiencies of CCFs with different crystal
core materials more meaningful and presents a true picture of the state
of art. The fundamental and the SH polarisation fidelity at the output
was measured using a Glan-Taylor polariser and was found to be typically
5-7dB for the fundamental and 0.5dB for the SH. The poor polarisation
fidelity at the output points to the de-polarising effect of the defects/
voids in the crystal core. The output near-field pattern of the SH showed
that the SH was present in both the core and cladding, however,
predominantly in the cladding. The poor crystal end quality and low power
levels made it difficult to make quantitative measurements.

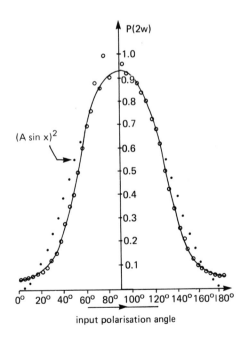

Figure 5: Optical SH output power of a MNA crystal cored fiber as a
function of the input polarisation angle.

Acetamide and DAN CCFs having core diameters in the 6-10μm range and 10-15mm long were similarly assessed for optical SHG. Typical SHG conversion efficiencies for acetamide and DAN CCFs with 10mW average fundamental power were $<1\times10^{-6}\%$ and $2\times10^{-3}\%$ respectively. Typical core throughputs for acetamide and DAN CCFs were 5% and 0.5% respectively.

NON-LINEAR INTERACTIONS IN CRYSTAL CORED FIBERS

The above measurements of optical SHG in CCFs with silica cladding are useful for determining the crystal orientation in the glass capillaries (X-ray techniques are not suitable for routine characterisation) and the overall crystal quality. However, they are not representative of the performance expected of engineered devices as, in three wave non-linear mixing processes, it is necessary to phase-match the interacting modes and to optimise the overlap integral. Device design takes these factors into account and these are discussed in this section.

The optical second harmonic generation in CCFs can be theoretically modelled as a coupled wave process and for undepleted fundamental power the SH conversion efficiency is given by[20]

$$\eta = \{\kappa.L.\sqrt{P}_{(\omega)/2}\}^2.\text{sinc}^2[\Delta\beta L/2]$$

where,

$\kappa = (\omega/4).\varepsilon_o.d_{im}.\int[E_{(\omega/2)}]^2.E_\omega.dA$ is the coupling coefficient.

L is the interaction length.

$P_{(\omega/2)}$ is the input power.

$\Delta\beta = \beta_\omega - 2\beta_{(\omega/2)}$ is the phase-mismatch factor.

The above formula for the SH conversion efficincy is same as for the bulk crystals with the exception of the difference in the detail of the coupling constant. The phase-matching requires $\Delta\beta=0$ ie equal phase velocities and hence equal effective refractive indices of the fundamental and the SH waveguide modes. In multimode CCFs the 'modal dispersion' can be used for phase-matching as the optical signals at different frequencies can have the same value of the effective refractive index when they propagate in different mode orders. However, if the SH mode order is high the overlap integral, in the coupling coefficienct, which determines the spatial overlap of the electric field distributions of the interacting modes has a low value resulting in poor SH conversion efficiency.

Optimisation of the overlap integral[21] has been considered in detail and it has been shown that an optimum device design requires phase-matching between the lowest order orthogonally polarised fundamental and SH modes. This can be achieved, for example in birefringent materials[22] belonging to certain crystal classes and having modal birefringence greater than the modal dispersion. In addition, for some materials it is necessary for the crystal to have a specific orientation relative to mode propagation directions. The overlap integral can be optimised for other mode combinations, however, with comparatively lower conversion efficiencies. For frequency doubling with pump depletion, the SH conversion efficiency is given by[20]

$$\eta = \tanh^2\{\kappa.L.\sqrt{P}_{(\omega/2)}\}$$

In Fig 6 the above expression has been used to compute the SH power as a function of the CCF length for two core materials having the SH tensor coefficient, d_{im}, values of 20pm/V and 250pm/V respectively. In this example the phase-matching between the lowest order orthogonal fundamental and SH modes is assumed and the fibre is considered to be lossless. Fibre core diameter has been taken to be 1.65μm and the core-cladding refractive index difference of 0.1. With the input CW power of 10mW at the fundamental wavelength of 1.5μm the CCF length for nearly complete harmonic conversion is 40mm and 3.4mm for the d_{im} values of 20pm/V and 250pm/V respectively. The d_{im} value of 20pm/V corresponds to a material having modest non-linearity eg $LiNbO_3$, while the 250pm/V value corresponds to the d_{11} of MNA.

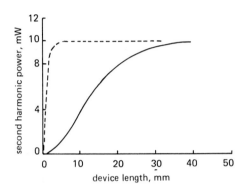

Figure 6: SH output power of crystal cored fiber frequency doublers as a function of fiber length for core materials having SH tensor coefficient values of 1. ——— 20pm/V and 2. - - - 250pm/V.

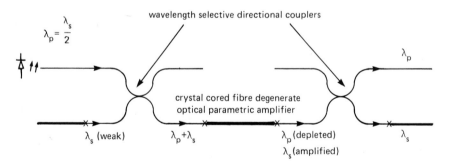

Figure 7: Schematic of an 'all-optical' regenerator.

An important requirement for the optical fibre communication systems in both the local area and trunk networks is an optical amplifier/regenerator. Optical parametric amplifier in a CCF form is an attractive device as it has the advantages of being independent of the modulation format (analogue or digital), independent of the data bit rate and it is a fast passive device with no stored energy. In Fig 7 a schematic of an 'all-optical' regenerator with a CCF degenerate optical parametric amplifier is shown. In this case the pump laser wavelength is the SH of the signal wavelength and for fiber systems in the 1500-1600nm wavelength

range GaAs semiconductor laser diode could be used as the pump laser. The
pump and the signal wavelengths are combined and separated using wavelength
dependent fiber directional couplers (commercially available). In Fig 8
the gain of a degenerate parametric amplifier is plotted as a function of
the CCF length. The signal wavelength is 1500nm and the CW pump power is
10mW. The CCF parameters used are same as in the previous SHG example.
It can be seen that a gain of 35dB, required in most systems applications,
can be achieved for 57mm and 4.5mm long CCFs having non-linearity of
20pm/V and 250pm/V respectively. These calculations demonstrate the
potential of CCFs, even with materials having modest non-linearities, for
three wave non-linear mixing processes.

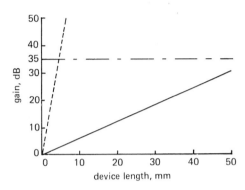

Figure 8: Optical gain of crystal cored fiber degenerate optical
parametric amplifiers as a function of fiber length for core materials
having SH tensor coefficient values of 1. ———— 20pm/V and 2. - - - 250pm/V.

CONCLUSIONS

The linear transmission studies on MNA and DAN CCFs demonstrate the
need for reduction of insertion losses to allow the study of their modal
properties and hence design efficient non-linear devices. The present
high losses are as a result of scatter loss due to the crystal defects
and poor coupling efficiency rather than the absorption loss. The optical
loss due to crystal defects is the more serious and requires better understanding of the crystal growth process for these materials. The earlier
work on both benzil and meta-nitroaniline CCFs has shown that 2dB/cm
insertion losses can be achieved and more recently throughputs as high as
60% have been obtained for some acetamide CCFs with the core diameter in
the 17-20μm range and lengths up to 20mm.

Optical second harmonic measurements have been carried out in multimode fibers to demonstrate three wave mixing and qualitatively determine
the crystal quality and orientation. The measured SH conversion
efficiencies are low due to the poor crystal quality and un-optimised
waveguide structure. A difficulty in device design using many organics
eg DAN, is that their optical properties such as absorption, refractive
indices, dispersion and non-linear co-efficients are not known. Some
materials like mNA are well characterised but have been found to grow in
unsuitable orientations. The control of crystal orientation, during the
growth, by the use of electric fields, transverse temperature gradients
and seeding is under investigation. This approach is considered more
promising for the early realisation of semiconductor laser diode pumped
parametric devices.

ACKNOWLEDGEMENTS

The authors acknowledge H. Morrison for the growth of acetamide
crystal cored fiber, D. O'Brien for help with the linear transmission
measurements, M. Halliwell for the X-ray work and P. Dunn, R. Kashyap and
J. D. Rush for discussions. Acknowledgement is also made to the Director
of Research British Telecom Research Laboratories for permission to
publish this paper.

REFERENCES

1. D. J. Williams (Editor), "Nonlinear optical properties of organic and
    polymetric materials" (Symposium Series 233), ACS, Washington D.C.
    (1983).
2. J. L. Stevenson and R. B. Dyott, Optical-fibre waveguide with a
    single-crystal core, Electron. Letts., 10:449 (1974).
3. J. L. Stevenson, Growth and characterisation of single crystal
    optical fibre waveguides- meta-nitroaniline, J. Crystal Growth,
    37:116 (1977).
4. F. H. Babai, R. B. Dyott and E. A. D. White, Crystal growth of
    organic materials in glass capillaries, J. Mats. Sci., 12:869
    (1977).
5. F. H. Babai and E. A. D. White, The growth of void-free crystal
    cored fibres of organic materials, J. Crystal Growth, 49:245
    (1980).
6. D. W. G. Ballentyne and S. M. Al-Shukri, The growth of electrooptic
    crystals in monomode optical fibres, J. Crystal Growth, 48:491
    (1980).
7. D. W. G. Ballentyne and S. M. Al-Shukri, The growth of single crystals
    of electro-optic organic compounds in monomode optical fibres, J.
    Crystal Growth, 68:651 (1984).
8. B. K. Nayar, Optical second harmonic generation in crystal cored
    fibers, paper ThA2 in : "Tech. Digest 6th Topical Mtg. on
    Integrated and Guided Wave Optics", OSA, Washington D.C. (1982).
9. B. K. Nayar, Organic crystal cored fibers, chap. 7 in : Ref 1.
10. S. Tomaru and S. Zembutsu, Organic crystal growth in glass
    capillaries, in : "Preprints 2nd SPSJ Int. Polymer Conf.", The
    Soc. of Polymer Science, Japan (1986).
11. S. Umegaki, A. Hiramatsu, Y. Tsukikawa and S. Tanaka, Crystal growth
    of organic material and optical second harmonic generation in
    optical fiber, in : "Proc. vol.682- Molecular and Polymeric
    Optoelectronic Materials: Fundamentals and Applications",
    G. Khanarian Ed., SPIE, Bellingham (1986).
12. G. E. Holdcroft, B. K. Nayar, D. O'Brien, J. D. Rush, M. Halliwell,
    R. Kashyap and K. I. White, Preparation and characterization of
    organic crystal cored fiber devices, paper THE4 in : "Proc. of
    Conference on Lasers and Electro-optics", OSA, Washington D.C.
    (1987).
13. P. Vidakovic, J. Badan, R. Hierle and J. Zyss, Highly efficient
    structures for wave-guided nonlinear optics, paper PD-C5-1 in :
    "Post-deadline Papers XIII Int. Quant. Electron. Conf.", OSA,
    Washington D.C. (1984).
14. Catalogue of optical Glasses, Jenar Glaswerk, Schott and Gen.,
    Mainz, W. Germany.
15. Hoya Optical Glass Catalogue, Hoya Corp., Akishima-shi, Tokyo, Japan.
16. J. M. Halbout, S. Bilt and C. L. Tang, Evaluation of the phase-
    matching properties of nonlinear optical materials in powder form,
    IEEE J. Quantum Electron., QE-17:513 (1981).
17. B. F. Levine, C. G. Bethea, C. D. Thurmond, R. T. Lynch and
    J. L. Bernstein, An organic crystal with an exceptionally large
    optical second harmonic coefficient: 2-methyl-4-nitroaniline,
    J. Appl. Phys., 50:2523 (1979).

436

18. R. J. Tweig and K. Jain, Organic materials for optical second harmonic generation, chap.3 : in reference 1.

19. G. F. Lipscomb, A. F. Garito and R. S. Narang, An exceptionally large linear electro-optic effect in the organic solid MNA, J. Chem. Phys., 75:1509 (1981).

20. B. K. Nayar, Optical fibres with organic crystalline cores, in : "Nonlinear Optics : Materials and Devices", C. Flytzanis and J. L. Oudar Eds., Springer-Verlag, Heidelberg (1986).

21. K. I. White and B. K. Nayar, Nonlinear fiber waveguides: the field overlap, in : "Proc. Conf. on Lasers and Electro-optics", paper WK46, OSA, Washington D.C. (1986).

22. B. K. Nayar, R. Kashyap and K.I. White, Design of efficient organic crystal cored fibres for parametric interactions: phase matching requirements, in : "Proc. vol. 651- Integrated Optical Circuit Engineering III", R.Th. Kersten Ed., SPIE, Bellingham (1986).

NONLINEAR OPTICAL PROCESSES AND APPLICATIONS IN THE
INFRARED WITH NEMATIC LIQUID CRYSTALS

I. C. Khoo

Department of Electrical Engineering
The Pennsylvania State University
University Park, PA 16802

ABSTRACT

Nematic liquid crystal films possess several unique characteristics
for applications in optical wave mixings and beam amplifications in the
infrared spectral region. We present new theoretical understandings of
optical multiwave mixing and beam amplifications in nematic liquid
crystal films. Low power laser beams, with intensities of the order of
a few Watts/cm^2, are found to be sufficient to generate large useful
effects, in conjunction with the director axis reorientational and
thermal nonlinearities.

INTRODUCTION

The unique characteristics of nematic liquid crystals[1] and nematic-
cholesteric[2] mixtures for infrared applications have been recognized and
applied in several electro-optical switching and modulation devices.
They are relatively transparent in the infrared; they possess large
birefringence ($\Delta n \approx 0.2$), and large operating temperature range ($-20°C$
to $110°C$) nematics are readily available commercially, and larger
temperature ranges can be achieved with suitable mixtures. This is in
addition to their well established fabrication technique, stability and
low cost.

Nematic liquid crystals are also potentially excellent candidates
for nonlinear optical processes. In addition to the characteristics
mentioned above, they naturally possess two mechanisms for strong
optical nonlinearities, namely, the director axis reorientational
nonlinearity[3] and their large thermal index gradients near the phase
transition temperature T_c[4]. In conjunction with visible lasers, these
nonlinearities have been successfully utilized in a myriad of optical
processes including bistability, switching, phase conjugation,
stimulated scatterings, etc.[5] In some of these processes, for example,
self-diffraction and beam amplifications[7], it was theoretically and
experimentally shown that the efficiency of the process (via
orientational nonlinearity) is critically dependent on the grating
spacing; the efficiency increased tremendously with the increase in
grating constant. With visible lasers, larger grating constants require
correspondingly smaller crossing angles between the interference beams;

large grating constants ($\geq 200\mu m$) require that the visible laser be crossed at very small angles ($\leq 2 \times 10^{-3}$ rad), making any practical applications very cumbersome. As we will presently see, thermal grating mediated nonlinear effects are also critically dependent on the grating constant in a similar fashion. This special requirement on the grating constant is easily fulfilled with long wavelength (e.g., CO_2) lasers.

The underlying mechanism for phase conjugation, bistability, and other processes mentioned above are quite well known. In this paper, therefore, we will discuss in detail a new wave mixing processes that have recently been demonstrated using nematic liquid crystal films, namely, thermal grating mediated beam amplification via four wave mixing. As we will presently see, the process is particularly relevant to infrared application because of its requirement for a large grating constant for efficient wave mixing.

THERMAL GRATING MEDIATED MULTIWAVE MIXING

Consider the geometry of the laser-nematic interaction as depicted schematically in Fig. 1. The two lasers (pump and probe) produce an intensity grating in the x-direction (on the plane of the paper). We shall consider the case where the pump beam is much stronger than the probe beam in intensity. The physics of the multibeam couplings and heat diffusion in this system are described by the heat diffusion equation and the Maxwell wave equation, respectively.

$$-D_\perp \frac{\partial^2 T}{\partial x^2} - D_{//} \frac{\partial^2 T}{\partial z^2} = \frac{c\alpha |E|^2}{4\pi} \tag{1}$$

and

$$\nabla^2 E - \frac{n^2}{c^2} \frac{\partial^2 E}{\partial t^2} = \frac{4\pi w^2}{c^2} P_{NL} \tag{2}$$

where $\vec{E}$ is the optical electric field, $\vec{P}_{NL}$ the nonlinear polarization, T the temperature, α the absorption loss coefficent, and $D_\perp$ and $D_{//}$ are the thermal diffusion constant for direction perpendicular and parallel to the director axis, respectively.

In the plane wave approximation, the pump (E_0), probe (E_1) and diffracted beams (E_2) may be expressed as follows:

$$E_0(z) = \varepsilon_0(z) \, e^{i\vec{k}_0 \cdot \vec{r}} = \varepsilon_0 \, e^{ikz}$$

$$E_1(z) = \varepsilon_1(z) \, e^{i\vec{k}_1 \cdot \vec{r}} = \varepsilon_1 \, e^{ikz \cos\theta + ikx \sin\theta} \tag{3}$$

$$E_3(z) = \varepsilon_3(z) \, e^{i\vec{k}_2 \cdot \vec{r}} = \varepsilon_2 \, e^{ikz \cos\theta + ikx \sin\theta}$$

where $k = n\omega/c$. Note that k_0 is along the $\hat{z}$-direction.
The generated polarization P^{NL} is given by

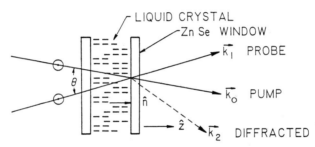

Fig. 1   Schematic of the pump-probe beam interaction in a nematic
          liquid crystal film. Both beams are linearly polarized in
          the direction of the plane of the paper.

$$P^{NL} = \frac{2n}{4\pi} \left(\frac{\partial n}{\partial T}\right)_{T_0} (E_1 + E_2 + E_3) \Delta T \tag{4}$$

where T_0 is the initial equilibrium temperature, and ΔT is the temperature rise.

Consider the $|E|^2$ term on the right hand side of (2). Using (3), we have

$$|E|^2 = \{|E_0|^2 + |E_1|^2 + |E_2|^2 + E_0 E_1^* + E_1 E_2^* + E_0 E_2^*\} + [c.c]$$

$$= \{|\epsilon_0|^2 + |\epsilon_1|^2 + |\epsilon_2|^2 + \epsilon_0 \epsilon_1^* e^{ikx \sin\theta} + \epsilon_0 \epsilon_2^* e^{ikx \sin\theta}$$

$$+ \epsilon_1 \epsilon_2^* e^{2ikx \sin\theta}\} + c.c. \tag{5}$$

where for small θ, we have approximated $1-\cos\theta \approx 0$, i.e., we shall include effects coming from the first order in θ. Furthermore, in view of the fact that ϵ_1 and ϵ_2 are much smaller than ϵ_0 (usually ϵ_0 is about 10^2 times ϵ_1 or ϵ_2), we have

$$|E|^2 = |E_0|^2 + |E_1|^2 + |E_2|^2 + (E_0 E_1^* + E_2 E_0^*) e^{-i\beta_0 x}$$

$$+ (\epsilon_1 \epsilon_0^* + \epsilon_0 \epsilon_2^*) e^{i\beta_0 x} + \text{negligibly small terms} \tag{6}$$

where $\beta_0 \equiv k \sin\theta$.

If the laser beam sizes are much larger than the grating constant and the film thickness, then the solution for the temperature rise ΔT may be expressed in the form

$$\Delta T(x,z) = f_0(z) + f_1(z) e^{-i\beta_0 x} + f_2(z) e^{i\beta_0 x} \tag{7}$$

Substituting (6) and (7) into (2) gives

$$-D_{//} f_0'' = \bar{\alpha} (|\epsilon_0|^2 + |\epsilon_1|^2 + |\epsilon_2|^2) \tag{8}$$

$$D_\perp \beta_0^2 f_1 - D_{//} f_1'' = \bar{\alpha} (\epsilon_0 \epsilon_1^* + \epsilon_0 \epsilon_2^*) \tag{9}$$

and

$$D_\perp \beta_0^2 f_2 - D_{//} f_2'' = \bar{\alpha} (\epsilon_1 \epsilon_0^* + \epsilon_0 \epsilon_2^*) \tag{10}$$

In standard differential equation forms, (8) - (10) become

$$f_0'' = -\frac{\bar{\alpha}}{D_{//}} (|\epsilon_0|^2 + |\epsilon_1|^2 + |\epsilon_2|^2)$$

$$f_1'' = -\frac{D_\perp}{D_{//}} \beta_0^2 f_1 - \frac{\bar{\alpha}}{D_{//}} (\epsilon_0 \epsilon_1^* + \epsilon_2 \epsilon_0^*) \tag{12}$$

and

$$f_2'' = -\frac{D_{\perp}\beta_0^2}{D_{//}} f_2 - \frac{\bar{\alpha}}{D_{//}} (\varepsilon_1\varepsilon_0^* + \varepsilon_0\varepsilon_2^*) \tag{13}$$

From (4) and (7), we get

$$P_0^{NL} = \frac{2n}{4\pi} \left(\frac{\partial n}{\partial T}\right)_{T_0} (\varepsilon_0 f_0 + \varepsilon_1 f_1 + \varepsilon_2 f_2) e^{ikz} \tag{14}$$

$$P_1^{NL} = \frac{2n}{4\pi} \left(\frac{\partial n}{\partial T}\right)_{T_0} (\varepsilon_1 f_0 + \varepsilon_0 f_1) e^{ikz + ik\beta_0} \tag{15}$$

$$P_2^{NL} = \frac{2n}{4\pi} \left(\frac{\partial n}{\partial T}\right)_{T_0} (\varepsilon_2 f_0 + \varepsilon_0 f_1) e^{ikz - ik\beta_0} \tag{16}$$

Substituting these polarization terms into the corresponding Maxwell equations for ε_0, ε_1 and ε_2 and using the usual slowly varying envelope approximation[8] yields

$$\frac{\partial \varepsilon_0}{\partial z} = -i \frac{2n\omega^2}{c^2 k} \left(\frac{\partial n}{\partial T}\right)_{T_0} (\varepsilon_0 f_0 + \varepsilon_1 f_1 + \varepsilon_2 f_2) - \frac{\alpha}{2} \varepsilon_0 \tag{17}$$

$$\frac{\partial \varepsilon_1}{\partial z} = -i \frac{2n\omega^2}{c^2 k} \left(\frac{\partial n}{\partial T}\right)_{T_0} (\varepsilon_1 f_0 + \varepsilon_0 f_2) - \frac{\alpha}{2} \varepsilon_1 \tag{18}$$

$$\frac{\partial \varepsilon_2}{\partial z} = -i \frac{2n\omega^2}{c^2 k} \left(\frac{\partial n}{\partial T}\right)_{T_0} (\varepsilon_2 f_0 + \varepsilon_0 f_1) - \frac{\alpha}{2} \varepsilon_2 \tag{19}$$

where we have included the effect of the absorption loss in the last term on the R.H.S. of (17) - (19).

The coupled equations for the temperature distribution (11) - (13) and the Maxwell equations for the pump, probe and diffracted beams (17) - (19) form the basis for our theoretical and experimental studies of the various wave mixing processes (such as diffraction, beam amplification) that are mediated by the thermal grating. In this paper, we shall limit our discussion to the possiblity of probe beam amplification via the pump-diffracted beam coupling, and some experimental confirmation.

In analogy to similar wave mixing effects in a Kerr medium[7], the optical intensity grating terms on the R.H.S. of (11) - (13) may be classified into two distinct types. One arises from the interference between the pump and the probe beam (e.g., $\varepsilon_0\varepsilon_1^*$ and $\varepsilon_1\varepsilon_0^*$), which can give rise to probe beam gain only if the resulting refractive index is appropriately phase shifted to the intensity grating. The other arise from the interference between the pump and the diffracted beam ($\varepsilon_0\varepsilon_2^*$ and $\varepsilon_2\varepsilon_0^*$), which can give rise to probe beam gain without the phase shift requirement. This observation is borne out in the numerical solutions of equations (17) - (19).

Figure 2 shows a plot of the probe beam intensity as a function of the reduced distance z/d, for the parameters I_0 = 80 watt/cm^2, I_1 = 0.8

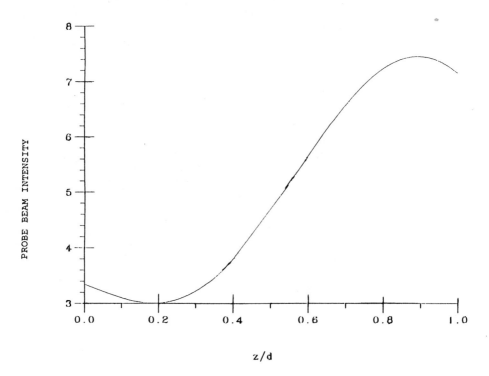

Fig. 2   Plot of the probe beam intensity as a function of the
reduced distance z/d, showing amplification effect.

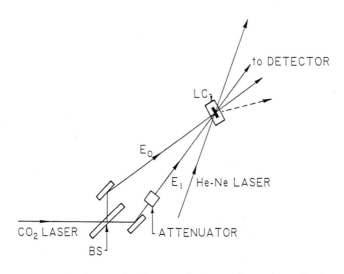

Fig. 3   Plot of the probe beam gain as a function of the grating
constant.

watt/cm^2 (i.e, a pump/probe ratio of 100 to 1), a grating constant of

$$\Lambda = \frac{\lambda}{2 \sin(\theta/2)} = 300\mu m,$$

$d = 100\mu m$, $\alpha = 80$ cm^{-1} and $dn/dT = dn_0/dT = 10^{-3}K^{-1}$.

It shows a rather interesting result, i.e., that the probe beam intensity can be amplified by more than 200% in traversing this high-loss, but also highly nonlinear medium. For a different value for $\alpha[\alpha = 20$ cm$^{-1}]$, we have also studied the dependence of the probe gain as a function of the grating constant (c.f., Fig. 3, which shows that larger gain can be achieved at larger grating constant). The physical reason is that larger grating constant can sustain larger temperature modulation, and therefore, a higher refractive index grating modulation.

Figure 4 is a schematic of the experimental setup. A TEM$_{00}$ CO$_2$ laser beam (Advanced Kinetic Power-Lase 50) operating at very modest powers is split into a strong pump and a weak probe beam (Pump/probe intensity ratio is 60) and is overlapped on the sample at an (external) crossing angle of 2.3° (corresponding to a grating constant of about 500μm). The beam diameter on the sample is 4mm. The pump power is varied from 1.7 to 3.3 watt, and the power of the transmitted probe beam is monitored. The film is fabricated using ZnSe plates coated with surfactant for homeotropic alignment. The liquid crystal used is PCB (Pentyl-cyano-biphenyl) and the film is 120μm thick, maintained at a room temperature of 22°C. PCB has been shown in a previous study[2] to have good transmission characteristics at 10.6μm. However, it does absorb appreciably at 10.6μm (for $d=100\mu m$, $\alpha \approx 80$ cm^{-1}) for thermal grating effects to occur readily.

At an input pump power of 1.7 watt (which amounts to 0.8 watt on the liquid crystal because of the 50% reflection loss at the air-(uncoated) ZnSe window interface), an increase of the transmitted probe beam of about 10% is observed. The probe beam gain is observed to be rather nonlinear with respect to the pump power (Fig. 5). At a pump power of 3 watt (intensity on the order of 25 watts/cm^2), a gain of 40%, is observed. Higher probe gain (>100%, not shown in Fig. 5) can be easily obtained by simply increasing the pump intensity. The magnitude of these probe gains for the parameters used in the experiment are in good agreement with the numerical solutions.

The beam amplification effect as described and demonstrated above can be applied in the construction of a ring oscillator[9] (c.f. Fig. 6). The liquid crystal is placed within a ring cavity. Noise originating from scattering from the pump laser (by the liquid crystal film) traveling along the direction of the axis of the ring cavity will form a grating with the pump beam and get amplified. When the amplification is greater than the loss (e.g., by increasing the pump laser power), a ring laser will be created, reaching a steady state value. The steady state corresponds to a probe beam intensity such that at that pump/probe ratio, the gain is just equal to the loss. In some preliminary study, we have observed such self-oscillation effects. Currently, efforts are underway to study this effect in greater details.

One other interesting application of these thermal grating (formed by the pump and the probe beam) is that it can be easily probed with a

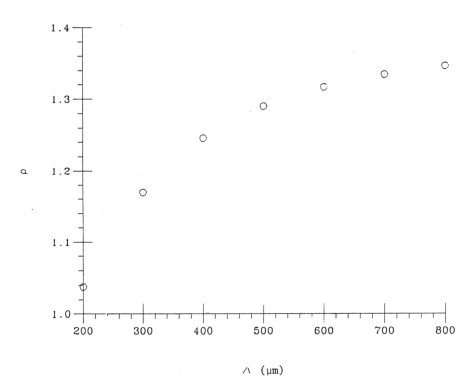

Fig. 4   Schematic of the experimental setup for CO_2 beam
         amplification. The He-Ne laser is for probing the grating
         formed by the CO_2 pump and probe beams.

PROBE BEAM VS. PUMP BEAM POWER

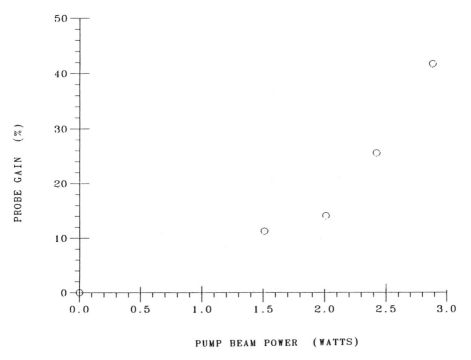

Fig. 5   Observed probe beam amplification dependence on pump beam power.

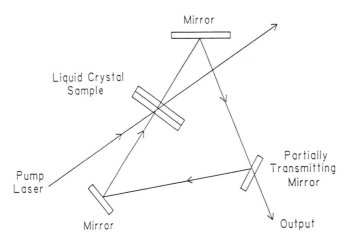

Fig. 6   One of several possible ring oscillators that can be
constructed using the beam amplification effect.

visible laser. If the infrared probe beam is a image-bearing beam, obviously one can perform infrared to visible image conversion in a manner analogous to what has previously been observed with near-infrared (1.06μm) laser.[10] A preliminary experiment using this wavelength conversion process has been successfully tested with simple objects (lines, square aperture) and reconstruction with He-Ne. Currently, we are investigating imaging processes involving higher resolution objects.

The response time for all these processes depend inversely on the square of the characteristic thermal diffusion length (either the film thickness, or the grating constant, whichever is smaller). Our previous study has shown that typically for a 17μm grating constant, the decay time constant is 50μs. For a grating constant of 170μm, for example, the (decay) time is about 5ms, which is fairly fast for image processing or laser self-oscillations (compared to photorefractive crystals, for example). The build-up time, of course, depends on the type of laser used. With pulsed laser, it is possible to have a build-up time on the order of a few nanoseconds.[4]

This research is supported by the Air Force Office of Scientific Research under grant no. AFOSR 840375.

REFERENCES

1. See, for example, U. Efron, S. T. Wu and T. D. Bates, J. Opt. Soc. Am. B3, 247 (1986) and references therein.

2. See, for example, J. G. Pasko, J. Tracy and W. Elser, SPIE Proceedings on Active Optical Devices, Vol. 202, 82 (1979).

3. I. C. Khoo, Phys. Rev. A 25, 1637 (1982).

4. I. C. Khoo and R. Normandin, IEEE J. Quant. Electron. QE-21, 329 (1985).

5. See, for example, I. C. Khoo, IEEE J. Quant. Electron. JQE22, 1268 (1986).

6. I. C. Khoo, Phys. Rev. A 27, 2747, 1983.

7. I. C. Khoo and T. H. Liu, IEEE J. Quant. Electron. JQE23, 171 (1987).

8. J. F. Reintjes, "Nonlinear Optical Parametric Processes in Liquids and Gases," Academic Press, NY, 1984.

9. J. P. Huignard, H. Rajbenbach, Ph. Refregier and L. Solymar, Opt. Eng. 24, 586 (1985).

10. I. C. Khoo and R. Normandin, Appl. Phys. Letts. 47, 350 (1985).

INDEX